EDWARDIAN INVENTIONS

1901~1905

An extraordinary extravaganza of eccentric ingenuity

In the years 1901-1905 over 140,000 British patents were granted; many to private inventors - not only British, but also American, European and Antipodean.

From this rich field, Rodney Dale and Joan Gray have culled a profusely illustrated selection of the most entertaining fruits of the ingenuity of early Edwardian inventors.

In drawing attention to many of the patented preoccupations of the period, the authors provide as significant and absorbing a social commentary as may be found in any of the more familiar sources.

For Judith, Timothy, Henry & Malcolm
— RAMD

For Harry & Naja; Chris, Phillida & Anna
— JKG

A Star Book
Published in 1979 by the Paperback Division of
W H Allen &Co Ltd
A Howard and Wyndham Company
44 Hill Street, London W1X 8LB

Cover drawing: John Holder
Cover photograph: Anne Zepherelli
Design: Ken Vail Graphic Design
Printed in Great Britain by
Hunt Barnard Web Offset Ltd

ISBN 0 352 30345 X

By Rodney Dale
Louis Wain: The Man who drew Cats
Catland
The Tumour in the Whale
The World of Jazz

with George Sassoon
The Manna-Machine
The Kabbalah Decoded

with Ian Williamson
BASIC Programming
Understanding Microprocessors

EDWARDIAN INVENTIONS
1901–1905

RODNEY DALE & JOAN GRAY

Acknowledgements

We would like to thank the many indefatigable librarians and others — notably at the Cambridge University Library and the Patent Office; Joyce Wrightson, who turned our handwriting into amazingly accurate typing; and Judith, Timothy, Henry & Malcolm Dale, Chris Lakin, David Paton, Steve Puttick, David Tassell, Mike Waggettt, and especially Ken Vail, with Jackie Barton, Victoria Squires and Andrew Westley of his eponymous studios, for their help in the preparation of this book.

We are also most grateful to Cambridge Consultants Ltd for making available to us facilities for manufacturing some of the inventions.

Note
References to our bibliography are in square brackets; patent numbers and years are in parentheses.

Foreword

This is what the Edwardians themselves might have termed a 'surprise' book. Our original idea was to classify our inventions, but in making our selection, as we did, numerically, we discovered a certain excitement in the random presentation, which we felt should be passed on to our readers.

The words and phrases we have used are, as far as possible, those used by our patentees or their agents. It is clear from the patents filed without the benefit of an agent that, although some patentese may have rubbed off on our patentees [?22!], there is a distinct period flavour in the phraseology, be it amateur or professional, and we have done our best to convey this.

It is fair to ask how we made our selection, and we return to this question in our introduction. Here, we should say that it was made by our feelings and reactions as we read the vast numbers of patents with which we were faced, rather than by the application of pre-ordained criteria. The finished effect gives, we hope, a feeling for the Edwardian age seen through the eyes of the amateur inventor — not the mechanical engineer developing a new Jacquard system for a loom manufacturer, but the village parson preoccupied with his eggs being boiled to his liking.

At the same time, we have placed some importance on the choice turn of phrase ('The said person may remain in the chair, reading and smoking while washes and douching are proceeding' (12475/1904)), and on happy coincidence (Boddy and Bottomley's Medicinal Compound for the Treatment of Piles (16317/1904)).

Because of the enormous amount of material, we have limited ourselves to the five-year period 1901 to 1905. During these years, the safety-valve of Victorianism blew, so to speak, and the Edwardian Era began with a bang.

Our introduction examines especially the selection of inventions as they reflect the climate of the period. There are, of course, some gaps in this examination — we do not, for example, discuss the attitude of the man in the street towards those who sought to fly through the air, and there are other topics we have left untouched.

In such fields, we allow our inventors to speak for themselves with the elegance — and eloquence — to which you will become accustomed.

Rodney Dale + Joan Gray
January 1979

Introduction

The Edwardian idyll

The 'Edwardian Age' conjures up two contrasting images.

The first is one of the upstairs world, of opulence and extravagance, of the fast, fashionable set surrounding the newly-crowned king; of country-house weekends with their judicious juxtaposition of bedrooms; of shooting parties, of balls and of Lucullan banquets; of a world where 'a race of Gods and Goddesses descended from Olympus lived upon a golden cloud and spent their riches as indolently and naturally as the leaves grow green.'

It was a time when Society was Society, when Britain was the rich centre of an Empire on which the sun never set, and London the heart of that Empire: the heart of the civilised world. It was an age when there was plenty of money — for the top ten thousand, that is — and when the rich could devote themselves to the systematic pursuit of pleasure and competition in ostentatious display in search of the patronage of a King who had transformed the Court into a stage setting of perfect comfort.

Royal patronage made Jews, bankers, even tradesmen, respectable and accepted; they could provide the heights of luxury in which the King delighted. At Alfred de Rothschild's Halton, the tea was served in Sèvres china with gold teaspoons; Reuben Sassoon kept his horses at the *top* of one of his houses in Belgrave Square, whence they were conveyed by a special lift; William Waldorf Astor employed some 800 workmen for five years modernising Hever Castle, and constructing a complete Tudor village outside its walls for the less élite guests; 800 tons of marble were imported from Michelangelo's quarries to line even the kitchens of Sir Ernest Cassell's mansion — such public rooms as the lobby being panelled with lapis lazuli; at Lord Iveagh's Elvedon the coverts were interconnected by telephone so that instant information could be exchanged on where the sport was best.

The second image is that of the downstairs world, the world of the masses as opposed to the classes. One eighth of the working population, and one quarter of the women, were employed in personal domestic service, in situations ranging from the great houses such as Longleat (where there was a total of 42 resident servants with their own rigid hierarchy, known by their masters' or mistresses' titles, and with lower servants to wait on the upper ones) to the vast army of general servants employed in ones and twos by more modest households.

This was a world where a third of the people lived in penury. To travel from West to East London was to encounter a different race of people, 'shorter of stature and of wretched and beer-sodden appearance'. Rowntree's survey of York showed that 28 per cent of the adults and 40 per cent of the children were grossly underfed. By contrast, the rich with their 12-course banquets were so grossly overfed that they needed to take frequent 'cures' at continental spas; nevertheless, their average life expectancy was twice that of the unskilled labourer.

The middle classes

But there was another, and less well known, world in between. The Edwardian Era saw the rise of the middle classes, and the suburbs in which they lived. The demand for office workers rose by 50 per cent in the public service and 33 per cent in commerce; the professions increased their numbers by 17 per cent. Much of this increase was accounted for by the expanded bureaucracies needed to administer the first beginnings of the welfare state — old age pensions, unemployment benefits — and the concomitant rising income tax. And the growing use of the typewriter and shorthand saw the emergence of a whole new profession for respectable young ladies.

With the development of cheap, motorised public transport, the commuter and his suburbs emerged. The new, clean, modern developments might be scornfully dismissed by Lord Abbotsbury as 'places to which one does not go', or derided by G K Chesterton in terms such as 'the rich suggestive life of Wimbledon', but for their inhabitants they offered a thriving round of entertainment — of operatic and dramatic societies; of golf, cricket, tennis and cycling clubs; of evenings at home with the piano, phonograph and games-table.

Our inventors

One of our purposes is to give some insight into the preoccupations of the more eccentric and inventive members of this thriving middle class, and into some of the ways they devised for overcoming their daily difficulties — of art-dealers, indigo-planters, ostrich-feather merchants, wireworkers and the like who found it tedious to stoop and pick up ping-pong balls (23274/1901 &c); of Indian railway contractors who devised automatic means for turning toast (7127/1902); of clergymen who wished to don and doff their boots without soiling their hands (20271 & 20728/1901); of lady teachers of music and drawing who devised caps to improve the set of the ears (944/1903), and perpetual motion machines (11318/1901).

But the labouring classes were not without their inventive streak. Take, for example, the brothers who patented an improved knife for cutting several slices of bread or cake simultaneously (563/1904); the miner who

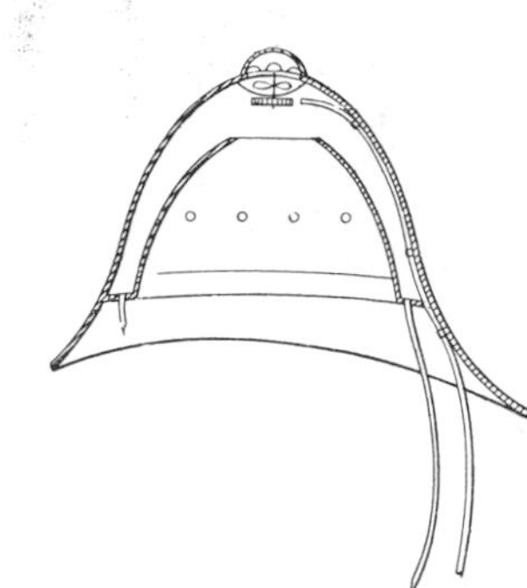

invented an improved fruit picker to obviate the need for ladders (1681/1904); or — a speculation-provoking one, this — the ship's mate who patented an improved device for holding down ladies' skirts (14788/1904).

At the other end of the spectrum are the upper-class inventors. One of our most aristocratic was Prince Hozoor Meerza, whom we first discovered as a Gentleman of No Occupation residing in Lavender Hill (having removed from Colville Square Mansions, Bayswater), where he patented his improved ventilated hat (19015/1902) and, the following year, an entertaining parlour game (538/1903). Eventually he returned to his native kingdom, but kept inventing, filing a patent for his improved suspended rope railway (18064/1905) from The Palace, Murihidabad, Bengal.

Count Vladimir Skorewski must be given credit for one of our most bizarre inventions — an apparatus for facilitating walking or running by storing up and re-using energy resulting from alterations in the height of the body's centre of gravity (14477/1904). Other aristocrats had the more expected preoccupations with huntin', shootin' and fishin' — Baron and Baroness von Heyden, for example, who patented a guard to prevent bitches from copulating (19094/1903); and Sir William Pearce, the proud inventor of an improved shelter or hiding place for fish (1762/1901).

Some of our titled ladies had a surprisingly domestic turn of mind, or perhaps their inventions — such as Countess Borcke-Stargordt's cupboard for containing chamber-pots and the noxious odours therefrom (12280/1904), or Princess Marie at Ysenburg's improved device for holding up sleeves to prevent their coming into contact with dishes at table (25773/1903) — merely give an insight into the more graceless aspects of country house life.

The Edwardian home

The home was the centre of Edwardian family life, and the Age saw the rise of the Garden City and the 'artistic' (or otherwise) suburb. Letchworth,. Port Sunlight, Ilford, Tottenham, Catford, Golders Green . . . all were brought into being as a result of improved public transport, and the rows of Edwardian commuters' villas still dominate vast tracts of outer London. There were two basic types: the solid, 'old-fashioned' terraced house, constructed in dark red, yellow or grey brick under a slate roof, with ponderous bay windows, tiny front lawn neatly bordered with a green or gold privet hedge and small, cast-iron railing, tiled front path leading up to a heavily-varnished front door with stained glass lights and — a final selling point — anaglypta dados to hall and staircase.

Then there was the more modern type, usually detached or semi-detached, of which such suburbs as Purley, Norbury, Southfields and Palmers Green are almost entirely composed. These were 'inspired' by the smaller country houses of the architects C F A Voysey and Norman Shaw, and were altogether lighter in construction and designed to be more cheerful in appearance. They had red-tiled roofs, brick walls covered with cement or pebble-dash, and much outside woodwork, decorative little balconies with white-painted palings being particularly popular.

The Edwardian household

'With the sole exception of the oral contraceptive, there is no birth control method in use today which was not available, and available in considerable variety, in the 1890s.' [40]

How widely contraceptive methods — apart from abstinence and *coitus interruptus* — were used is a matter of some speculation. What is certain is that the Edwardian families were smaller than their Victorian counterparts, and that this was due, in part at least, to improved medical practice reducing the infant mortality rate and removing the fear that, unless one kept producing children, one might be left with none.

The home, however, was not yet designed to be labour saving, and the average middle-class family would expect to have at least one servant. The 'servant problem' was certainly a worry for the middle-class housewife, but it centred more on quality than on quantity. It was still a regular practice for working-class girls to 'go into service', and quite customary for one family to provide a continuity of servants for another for many years. It was not until after the First World War that the shortage of housemaids began to present a problem and the more superior members of the community had to resort to the desperate expedients adopted by the residents of Golders Green in the 1920s, who would have to 'hire a maid, making sure she was in attendance at least for the evening meal, which would be taken at the front of the house with the curtains drawn well back and the lights blazing'. [30]

Young men were, indeed, advised not to marry until their income reached at least £90 per annum if they wished to maintain a securely white-collar standard of living — a fatherly piece of advice given to clerks at the Railway Clearing House. £150-500 per annum, the salary to which most of the lower middle classes — shopkeepers, superior factory foremen, teachers, commercial travellers, small businessmen, office workers and less successful professionals — would aspire, was defined as being 'comfortable', that is to say, able to provide for a household including a servant or two. Mrs Beeton, for example, recommended appropriate domestic establishments for various levels of income: at £150-200 per annum there should be a general servant or maid-of-all-work; at £300 a cook and a housemaid could be added; at £500 a houseboy and a nursemaid. It is a reflection of the low rate of inflation at the end of the nineteenth century that Mrs Beeton's standards held for so long. [5]

It is not surprising, then, that few of our inventors were concerned with labour-saving devices, and that those who were took it for granted that such devices would assist, rather than replace, the servants. Take, for example, the improved vacuum cleaner, designed to be fixed to the operator who could 'easily work the bellows with one hand' (1411/1903); or another improved vacuum cleaner utilising the power readily obtained from 'the ordinary domestic servant', i.e., from a 'pair of exhausting bellows operated by the feet of the maid's assistant' (21304/1905).

Our inventors' domestic preoccupations were more with maintaining the niceties of life; with silent toilet-paper dispensers (339/1903); with improved means of preserving order amongst, and displaying, their holiday souvenirs (1333/1901, 1942/1903); with apparatus for warming and drying gloves and other articles of wearing apparel (534/1903); and with the creation of such entertaining novelties as sewing-machines disguised as statuesque lions (21933/1902); nutcrackers masquerading as squirrels (1213/1903); dragon-shaped fire-blowers which belched forth steam to excite the sluggish embers (14696/1902); and improved Christmas crackers adorned with representations of 'mushrooms, toadstools, . . . elves, fairies, "brownies", imps or similar forest or emblematical representations' (10186/1902).

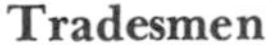

Tradesmen

Our inventors were living in an age long before prepackaged food made a general appearance. Readers will therefore be interested to learn of the practice of packing butter in cabbage leaves 'which are, however, not always clean and at certain times of the year not easily obtainable or in good condition'. To overcome this, there was the collapsible paper wrapper printed and embossed to resemble a real cabbage pack (5001/1905).

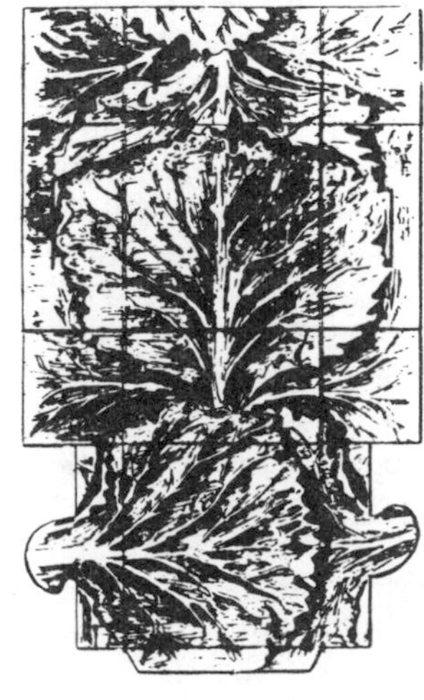

It is not so long ago that grocers packed dry goods, such as sugar, themselves, scooping the material from the sack into a bag of thick, dark blue paper, and sealing it by deftly folding the opening with practised fingers. As one writer wistfully recalls, tradesmen *were* different in those days: 'the customer was always right, and the idea of the tradesman was that they were there to please. Tradesmen of all kinds came to the door, deliveries were daily, and the new customers were keenly competed for; they were waylaid in the street so that their favour could be solicited.' [35]

In view of such over-enthusiasm, it is perhaps not surprising that our inventors were more concerned with patenting devices to save inconvenience by informing tradesmen that their services were *not* required (5230 & 11894/1902).

Any reader puzzled by the curious devices invented to prevent the misappropriation or removal of milk by unauthorised persons — locks to secure cans to the door (1899/1903), or receptacles on the inside of the door filled through a specially-designed delivery funnel carried by the milkman (7014/1905) — should know that this was because 'milk was delivered to the house several times a day, drawn straight from a churn on the milk cart into little pewter cans which were left on the doorstep or hung on the railings until you took it in.' [35] In England, it was not until 1906 that bottled milk was first sold, and it was several decades before the horse-drawn cart with its churns disappeared altogether.

Recreation

Musical evenings, card parties, dinner parties, sessions with the phonograph . . . the home was the centre of the Edwardian family's entertainment. Every family with pretensions to respectability had its piano, and any young lady with pretensions to social accomplishment had to know how to play it, and how to accompany a singer. The 'three-years' system' — which we now call hire-purchase — ensured that even the most lowly paid could afford a piano, and the wealth of singable songs supplied by the Music Hall provided a rich, comic repertoire — not to mention the constant stream of ballads so sentimental as now to appear comic themselves.

Musical evenings were thus a feature of social life; mountains of refreshments would be served — sandwiches, patties, pies, tarts, cakes, custards, jellies and, for the gentlemen, 'whisky and soda, partaken of in stealth and silence in a dimly-lit room elsewhere in the house, it not being considered quite good form to drink when ladies were present.' [35]

We would like to think that it was a disastrous combination of over-enthusiastic amateur singing and piano-playing, coupled with stealthy over-indulgence in clandestine whisky and soda, that led one gentleman to patent an ingenious means for improving such evenings — a trampoline disguised as a grand piano (1840/1901).

Other inventors devised equally lively

attractions: an improved trick device which caused a spray of pulverulent material, such as lamp black, to be blown in the face of the uninitiated person who attempted to turn an attractive windmill (24820/1905); an improved pop-gun designed to be used in the parlour, and invented in an effort to extend the appeal of this 'delightful and harmless toy' to those of 'more advanced years' by improving the accuracy of its aim and the loudness of its explosion (23309/1903); a mask to 'produce curious effects in the wearer and render him or her the cause of much amusement and delight to the beholders' by combining his — or her — face with the horns of a devil or the feathers of a Red Indian in war-paint (20054/1905); and, for a really rousing finale to a warm summer evening, the 'interesting and exciting ball-catch game' (16268/1905).

The phonograph

Being machines, not musical instruments, phonographs were sold mostly in bicycle shops. Their trouble was that they required a great deal of expert attention — and much winding — to produce a very inferior sound. 'You could hear the words "Edison Bell Record" and after a time you got to know the tune by constant repetition, but speech was always difficult to detect, although certain instruments, like the guitar, mandoline or banjo, came over quite clearly.' [35]

By the turn of the century, the novelty of the phonograph had worn off, being replaced by the frustration of its capabilities being as limited as they were. Perhaps it was his frustration with its limitations that led one inventor to patent a phonograph which used plates of edible material — 'the squeaking sweetmeat' (1992/1903) — and caused an engineer and a violin maker to combine their talents to produce an improved phonograph sporting a violin as its sounding box (3723/1903).

In fact, the reproducing equipment was not entirely to blame; the problem lay more in the difficulties of making the recording to start with. It was necessary for the performers to sing or play into a horn, which concentrated the sound on to a diaphragm, causing the recording stylus to vibrate. This is why loud, plangent instruments, such as the banjo, came over comparatively clearly. It was not until the mid-1920s that electrical recording was introduced, whereupon the industry took on a new lease of life.

Crime in fiction and fact

In an era with a comparatively low crime rate, and in which violent offences and murder were sufficiently uncommon for there to be no Homicide Squad at Scotland Yard, the public could relax and revel in the exploits of a squad of fictitious amateur detectives: Conan Doyle's Sherlock Holmes, G K Chesterton's Father Brown, Baroness Orczy's Old Man in the Corner and Lady Molly Robertson-Kirk, and many, many others.

It is recognised that the scientific school of detection, without which no modern police force could exist, owes its foundation — at least in part — to techniques first devised by writers of fiction.

Imaginations aflame with such heady stuff, our amateur inventors were inspired to make their contributions to maintaining law and order: safes to trap and secure burglars by the simple expedient of catching their fingers in toothed blades (5161/1902); improved gloves fitted with talons to deter assailants — and, incidentally, to serve as an aid to falling mountaineers (1567/1904); cunningly-devised letter-boxes fitted with fiendishly sharp prongs to prevent unauthorised removal of their contents (5641/1905); a device for preventing train robberies whereby an arrangement in the engineer's cabin could cause steam to be squirted in the face of an attacking party, so blinding him and compelling him to retire (14039/1905). (And, as a digression, any reader interested in railways cannot fail to enjoy the improved protector for trains: a small pilot motor which precedes the main engine, to which it is connected by a cable, 'several hundred metres long', to give warning to the driver of obstacles on the line ahead (25865/1905); not to mention the improved means for allowing the passing of one vehicle by another (13095/1904).)

The popular press

The Edwardian years saw the rise of a new type of journalism: vulgar, trashy and lacking in the dignity of the old-style press, but established to cater for the tastes of a new class of readers born from the provision of compulsory education in 1876.

The new readers wanted a bright, snappy paper, with short words, short sentences, short paragraphs, and short articles that they could read and enjoy on the train or tram as they journied to and from work. The Prime Minister might scoff that the *Daily Mail* was written 'by office boys for office boys' but Alfred Harmsworth (later Lord Northcliffe — and owner of *The Times* as well as the *Daily Mail*), its creator, made a fortune out of establishing a paper which wrote about matters not previously considered part of a newspaper's chronicles — food, fashion, the human drama behind the news, and sport of all kinds.

Harmsworth's favourite maxim was that his

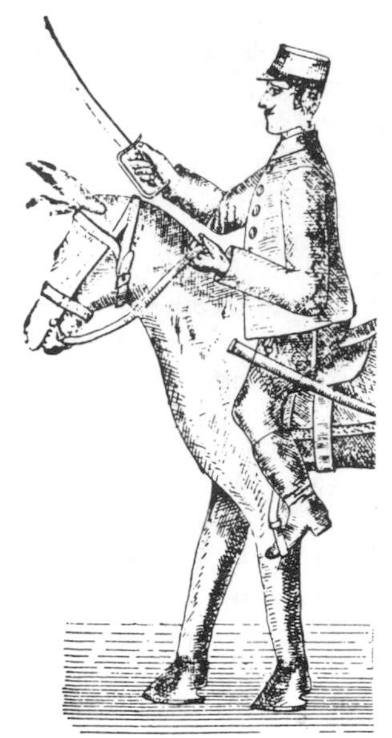

readers liked a good hate; in the interests of mass circulation he played on the nascent anxieties of the aspiring new middle classes and achieved newspaper sales and profits previously undreamed of — for the Edwardian years also saw the rise of the consumer.

Advertising

Industrialists such as Lever and Lipton made their millions by exploiting the vast new market which had opened up as a result of rising wages and improved education; advertisements for soap, tea, cheap groceries and cigarettes proliferated on hoardings and in the press — even Mrs Lillie Langtry herself supplied illustrated testimonials to the efficacy of Edwards's Harlene for the Hair.

This feverish commercialism affected our inventors too. They patented improved advertising table lighters (2858/1902); toy 'cameras', where an unsuspecting child actually *paid* to be presented with an embossed card bearing an advertisement which had supposedly been produced in the magic box (11289/1902); they put advertisements on flycatchers (15258/1902), glove protectors (22696/1902), telephone protectors (paper covers to insure the healthy user against the risk of contagion arising from the instrument's promiscuous use by diseased people) (25948/1902) — and even printed them on toilet rolls (12167/1901) and embossed them on toilet paper, the rough portions thoughtfully placed so as 'not to come into use for toilet purposes' (17066/1903).

One shrewd gentleman attempted to overcome the nuisance of advertisers who preferred their material to appear on the right-hand, rather than on the left-hand, pages of periodicals: he suggested that the left-hand pages should be printed upside-down (10529/1905).

But best of all media was Squires and Morehen's improved advertising hat. This was fitted with a crown which could be caused to move up and down to attract attention and display an advertisement, an electric lamp being arranged so as to light this up when the crown was lifted (7607/1902).

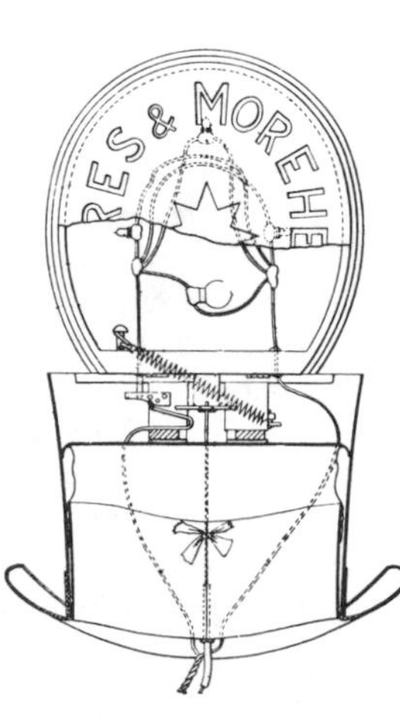

Stage illusions

Squires was manager of the Egyptian Hall in Piccadilly. This was 'a place to which children were taken with delightful anticipation. Incredible feats of magic were performed. Music played without any visible agency. The supreme moment was the Cabinet Trick. Children not cursed with nerves revelled in it. They demanded of their elders how it was done and their elders were shamed and humbled by not being able to tell them. It was no use trying to fob off the children with such statements as "Oh, it's done by mirrors." Many a reputation for infallibility was sacrificed in the Egyptian Hall.' [35]

Many a reputation for infallibility could have been saved if the perplexed parents had perceptively perused the patents, for here they would have found explained the improved stage illusion to give the appearance of a lady being fired from a cannon into a series of locked and bound nested boxes, and of a lady disappearing from a scale pan after she had been enveloped in a cloak and a pistol fired, both in full view of the audience (26307 & 26308/1905). They could have learnt the secrets of an illusion apparatus for effecting the enigmatic appearance of a person in a glass tank under water (19063/1904) and later, a person perhaps having had a narrow escape from drowning, the improved apparatus (20629/1904)

They could have discovered how to enable a performer to appear to stand on one finger (2021/1901); how to construct an improved property quadruped (11705/1901); and the secret of throwing horses and elephants into the air to perform the salto-mortale (8713/1904) and of the improved apparatus using power devices instead of springs (8894/1905).

The music hall and theatre

Once they had been cleaned up after the scandals of the 1880s and 1890s, when the Empire Music-hall and Promenade, particularly, was regarded as a haunt of iniquity and an open market for expensive prostitutes — who paraded 'noiselessly on the rich carpets, wafting in a heady and frightening atmosphere of rich blue cigar smoke, frangipan, patchouli and champagne' — they lost their reputation for being shady, and made more money than ever before.

They were unrivalled as places for working-class entertainment, offering a lively mixture of acrobats, conjurers, singers, dancers, spectacular scenes, sketches, comic turns and songs of all sorts. The most popular performers earned huge sums of money — Pavlova, for example, was paid £750 a week.

The theatre was also immensely popular, as a place to see and be seen in as part of an elegant evening out. Many new theatres were built in our early Edwardian years including, for example, the Globe, the Playhouse, the Imperial and the Coliseum in the West End: masterpieces of fantasy whose architects concentrated on providing magnificent gilt and crimson plush palaces for the suspension of disbelief. Our inventors, however, were preoccupied with rather different matters: with designing buildings provided with ramps or swinging galleries for easy escape in case of fire (16357 &

21280/1904); with developing improved hearing aids particularly for use in auditoria (18944/1901); even with overcoming the problem — itself a music hall joke — of the obstruction offered to one's vision by the large hat of the lady in front (18568/1905).

The funfair

Open-air amusement parks offered scope for more robust entertainment altogether — scenic railways with their attractions heightened by cunning arrangements of mirrors in their tunnels (3288/1903); improved aërial machines designed to shoot off their tracks and land in water or on springs (26821 & 16383/1903); roundabouts provided with cars which became successively submerged in water (1098/1904); and recreational apparatuses designed to give the occupants the authentic sensation of being sucked into a natural whirlpool — 'It might appear that this would be anything but pleasing, but the passengers are conscious that all danger is eliminated and therefore greatly enjoy the trip' — (25623/1904).

Fashion

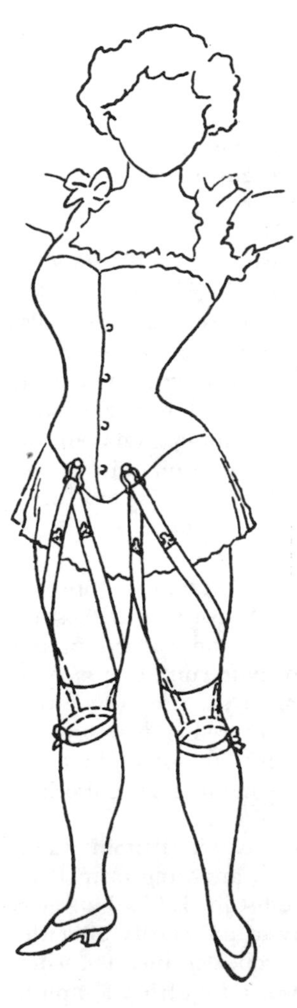

Edward VII was a tyrant in matters sartorial. He took a keen interest in the exact arrangement of orders and decorations, and in the exact degree of mourning correct for a personage of given rank — black cuff-links to be worn to the theatre as a token of respect for a departed Grand Duke for example — and was liable to administer a stern rebuke to anyone who transgressed his strict notions concerning appropriate dress. Lord Harris, resplendent in tweed jacket, was politely asked: 'Goin' rattin'?' Lord Rosebery, having committed the unforgivable *faux pas* of appearing in trousers rather than knee-breeches, was subjected to questioning as to whether he had transferred his allegiance to the American ambassador. An Austrian Count was reproved for wearing a tie in the colours that had belonged to the Brigade of Guards for over 300 years — in this case, however, candidly recollected the Duke of Manchester, the King had made the error: the colours had belonged to the nobleman's family for more than 700 years. [36]

As Prince of Wales, Edward had had four valets to look after his wardrobe — two to travel with him, and two to look after the garments left at home — had popularised the soft Homburg hat, and had introduced the dress lounge, later known as the dinner jacket. The dinner jacket was, incidentally, put on *after* dinner as a smoking jacket when the ladies had retired, so that they would not be rejoined by men impregnated with the offensive smell of tobacco.

Dressing and undressing occupied much of the time of the rich Edwardian gentleman: tweeds for early-morning sporting activities, a lounge suit for the morning, a frock coat for business or luncheon, an evening suit, a dinner jacket, tails for going out . . . The preoccupations of our inventors were less exalted, but they were still bound by the rules of a society which demanded a smartly-arranged frock coat for business, and even the suburbanites — of the better sort, of course — dressed for dinner. A look at their patents gives some idea of the problems they encountered, particularly those gentlemen who had no gentleman's gentleman of their own. We glimpse a world of slipping neckties (321, 818 & 2404/1903); hats needing ventilation (19015, 21754 & 21860/1902) and protection from inadvertent removal by other than their rightful owners (8293/1903); overcoats in need of improvement (807/1901 & 26290/1905); trousers in need of suspension such that the pants beneath did not unfavourably affect their natural hang (13218/1901).

The ideal Edwardian woman was shaped like a swan and possessed of a voluptuous *Directoire* figure of the type popularised by the illustrator Charles Dana Gibson: a full, rounded bottom matched with a magnificent monoprowed bosom, and separated therefrom by the tiniest of tiny waists.

The foundation of this figure was the corset, an indispensable item even for very young girls. Lillie Langtry, an early mistress of Edward, owed a certain element of her considerable notoriety to the fact that she did *not* lace herself in with stays. The horrors and complications of this essential underpinning, and some idea of the difficulty of maintaining it — and the stockings — correctly in place, can be all too clearly seen from a study of our patents (2433/1902, 3161/1903, 1663 & 12210/1904, for example).

The Edwardian lady's opulent S-shaped figure was crowned with a mountain of hair, a pompadour of luxuriant curls, the whole being topped with a vast picture hat. This was *not* perched at a rakish angle; it was placed carefully on top, like a large tray laden with ostrich-feathers, fur, fruit and flowers, and secured to the elaborate *coiffure* beneath with immensely long hatpins. These pins became important fashion accessories in themselves, ornamented with precious stones and intricate designs; they also became a hazard in crowds, and the cause of many accidents. Naturally then, they were the cause of concern for many of our inventors. Patented hat fasteners are legion (see, for example, 3126/1902, 1118 & 1818/1903, 9019, 22209 & 25873/1905) but to the late twentieth-century eye the suggested solutions can look considerably worse than the original problem.

The reader will therefore be relieved to hear

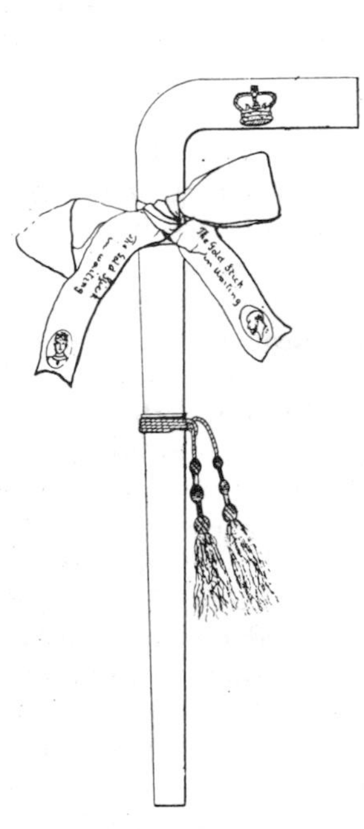

that the mountain of hair was not entirely the property of its owner. The elaborate *coiffure* to which the elaborate hat was so elaborately secured was composed largely of 'rats': pads of horsehair over which the owner's real hair was drawn and supplemented by switches. A vivid description is provided by Vita Sackville-West; the Duchess is dressing for dinner, and her 'rats' have been placed on the dressing table: 'unappetising objects, like last year's birds' nests, hot and stuffy to the head, but indispensable since they provided the foundation on which the coiffure was to be swathed and piled and into which the innumerable hairpins were to be stuck.' [45]

To be a true beauty whose charms were so compelling as to inspire beholders to stand on chairs in Rotten Row the better to catch a glimpse of her presence, the Edwardian lady also had to have the correct deportment: head held high, bust thrust well forward and buttocks thrust well back, a stance maintained with the aid of a Regency beau's walking stick and often, we suspect from our researches, the cause of considerable fatigue; it is, for example, suggested that such sticks could be used to conceal emergency rations of chocolate (11740/1902) or even liquid refreshment (1091/1902).

It is not at all surprising that several of our inventors turned their talents to devising more convenient forms of dress; if any reader should doubt the need for a rational form of clothing (1542/1903), perhaps the following description of a lady of fashion being attired will serve to convince him. The maid 'knelt before the Duchess, deftly drawing on and smoothing the silk stockings, fitting the long heavily boned stays, fastening the busk, clipping the suspenders to the stockings, and then lacing, beginning at the waist and travelling up and down until the desired proportions were achieved; the pads to accentuate the smallness of the waist were added, and then the drawers, and finally the petticoat was spread into a ring on the floor so that she could step into it. The bodice was then held open while the Duchess dived gingerly into the billows of her dress and the maid breathed a sigh of relief as she began doing up the innumerable hooks at the back . . .' [45]

Transport

Our early Edwardian years saw the rise of the petrol engine and the beginning of the end for the horse. Equine numbers had reached a peak at the turn of the century, with one horse for every ten people in Britain; by 1910, 22,000 horses in London alone had been displaced by the newly-introduced petrol-driven cars and omnibuses. *Mayfair* wrote: 'The horse is disowned and forsaken; soon the forlorn animal will betake itself to the Zoological Gardens and beg for the hospitality of a cage.' [17.11.1910]

Nonetheless, this was but the beginning of the end. Horse-drawn vehicles were still the main means of local transport and our inventors — as always — give an unusual insight into some of the lesser-known problems associated therewith: screens to protect the animals' heads against sunstroke (14048 & 15608/1902); traps for catching dung (26710/1902); and — a necessity when the roads were not asphalt but earth, and dung-catchers not in general use — trains and suspenders for holding ladies' skirts clear of the mire (6875, 11106 & 11267/1902).

Light, speedy and kept gleaming by its proud driver, the hansom cab was regarded as a highly romantic form of transport and nicknamed 'the gondola of London'. A young lady who rode in one alone with a gentleman ran a grave risk of ruining her reputation — and more, for if the horse drawing it fell, a not uncommon occurrence, the occupants might be precipitated on to the fallen animal. This possibility was a cause for concern among our inventors: a variety of means for improving the safety of hansom cabs was suggested, perhaps by shaken passengers: an artist and a joiner (9109/1901); a forage contractor (21362/1902); a schoolteacher (17845/1903).

Another ever-present danger was that of the horse bolting, and many and various were the suggested means for preventing this: automatic blindfolds (17294/1902); harness which automatically unharnessed itself (1675/1903); devices to lock the carriage wheels (7973/1903).

But to some extent these were mere details. Desperate measures were needed if the horse were to be competitive in speed with the petrol-driven car or omnibus; thus: a horse velocipede is so constructed that the animal walks on an endless platform so that the friction of its hooves is utilised more fully (8488/1904). Too late, alas.

Motoring

The motor-car was the Edwardians' supreme status symbol. It was expensive to buy (costing more than the average person earned in ten years); it was equally costly to run (a set of tyres alone costing more than a superior domestic's annual wage); and it required a new type of servant, the motor servant, to maintain it. A large, dry and salubrious motor stable also had to be provided to house it.

The first petrol-driven car in Britain was a 4 hp Panhard et Levassor, imported from Paris by the Hon. Evelyn Ellis in 1895. Horseless carriages gained rapidly in popularity after the 'red flag' legislation was repealed in 1896, when the (legal) speed limit was raised from 3 mph to

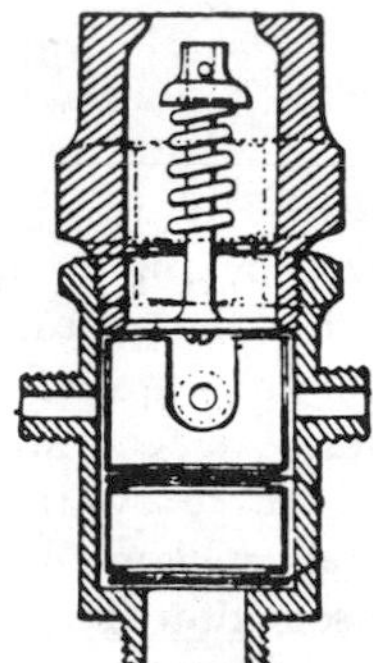

12 mph; it was further raised to 20 mph in 1903.

And, as might be expected, the motor-car was given the Royal seal of approval. After his first ride in one — a Daimler belonging to Lord Montagu of Beaulieu — the King enthusiastically declared: 'The motor-car will become a necessity for every English gentleman.' Equally enthusiastically he became a purchaser of motor-cars, particularly favouring the Daimler, Mercedes and Renault, all of which were painted in the Royal claret-red livery and fitted with a distinctive, four-note cornet horn. It was the special function of the Superintendent of the Royal Cars to play the horn as the King sped along his highway, urging his driver to go faster and faster: he was inordinately fond of speed, and wont to boast of the times he had exceeded 60 mph on the Brighton Road.

Queen Alexandra also loved travelling at speed, and wrote to her son that she 'did enjoy being driven about in the cool of the evening at fifty miles!! an hour! I have the greatest confidence in our driver. I poke him violently in the back at every corner to go gently and whenever a dog, child or anything else crosses our way.'

Her unfortunate driver, nursing his bruises, would no doubt have welcomed the electrical signalling device to enable a person seated in the passenger seat of a motor-car to give instructions to the driver such as 'faster', 'slower', 'stop', 'left', &c (7828/1905).

It took motor-car designers a considerable time to realise that the vehicle was not just an open horseless carriage which had to mirror the latest fashions in drawing rooms; that it was necessary for passengers travelling at speed to be protected from rain, wind and dust. Until this happy realisation, exposure to the elements was a problem, and the need for special protective clothing was quickly seen by our inventors, who patented such weatherproof garments and accessories as face-protectors (85 & 23574/1902); handshields to avoid the danger of the motorist losing control of his vehicle owing to his hands having become numb with cold (1383/1903); aprons (8337/1905); ladies' coverings combining coat, hood, mittens, gaiters and boots (with a whalebone stiffening to ensure that the lower portion hung like a skirt) (26057/1905); and, most ingenious, a footwarmer using the hot exhaust gas to perform its office (2901/1904), and an apparatus using the breath to heat the outer body (7806/1903).

Motor accessories

But civilised standards must not be forgotten. If our inventors had had their way, the Edwardian motorist could also have made use of a patented ash and wind guard for his cigar (3853 & 13887/1905); or an apparatus designed to overcome the problems of holding an umbrella, so troublesome when one is driving oneself (17559/1904).

Thus we see the beginning of the motor-accessory industry. There were devices for inflating tyres using the pressure from the engine (3670/1904, 22000 & 22105/1905); another using the same power source for sounding a warning (26971/1905); there is the driving mirror 'devised specially in view of the recent legislation' (12184/1904).

Early automobiles were not overly efficient, and this problem was tackled also, with shields to diminish the retardation of velocity due to air resistance (9373/1901), and an improved driving apparatus using wind motors to generate electricity for charging batteries (25353/1905). The improved motor-car capable of passing through doors and up stairs (21201/1904) avoided the troublous necessity of providing a motor stable, and, most useful of all, improved bumper-shields (21536/1905) could be fitted to push obstacles — such as horses, dogs or humans — to one side and out of the way.

Velocipedes

Few of our middle-class inventors would have been able to afford a motor, but would undoubtedly have exercised their energies in bicycling, a late Victorian craze which lasted well into this century. Hordes of enthusiasts would pedal out into the countryside every weekend in search of fresh air and inspiration among the flora and fauna, and picturesque hamlets where, in those days, public houses would be open all day for the benefit of *bona fide* travellers — those who had come more than three miles.

Hampered by their voluminous skirts, and their more delicate constitutions, ladies would naturally not be expected to keep up with the gentlemen and, had it been in general use, such a device as the towing arrangement for cycles enabling a gentleman rider to render assistance and relief to ladies when travelling over rough ground or going up hill (13115/1905) would no doubt have been invaluable. And so, no doubt, would have been the various means suggested of improving the efficiency of the two-wheeled velocipede — such as the mechanism for cutting and deflecting the current of air caused by fast riding and utilising it as motive power to assist propulsion (6086/1902); the harness to aid the rider to transmit additional impetus to the free wheel by throwing back his shoulders (25417/1902); the device which employed the pent-up energy of a strung bow to aid forward motion (19344/1901) . . . if only they did not outrage the fundamental principles of physics,

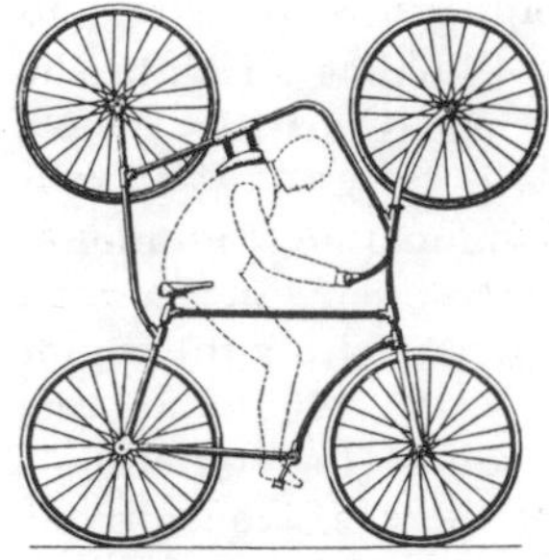

or common sense — or both. In fact, any attempt to employ such aids would probably have been more entertaining than the use of those specifically designed for entertainment: such as the improved illuminated bicycle fitted with electric brushes which contacted lugs on the wheels to produce showers of ornamental sparks (11384/1902); the improved guiding device to enable the action of the hands to be done away with, so leaving them free for any other purpose whatsoever (5508/1904); and the combined upper and lower cycle enabling the rider to complete his journey head downwards (8706/1904).

Golf

In the 1890s, golf became more than an eccentric pastime for Scotsmen: it burgeoned into a popular sport for overweight, middle-aged men — and respectable for women. The turn of the century saw the introduction of the golf tee and, more importantly, the rubber-cored golf ball, an invention which must have done wonders for a game first played with leather balls stuffed with feathers, then with balls of gutta-percha.

Prime Minister Balfour did much to popularise the sport and the King also took a lively interest, ordering courses to be laid out both at Sandringham and at Windsor. Neither his figure nor his disposition, however, were such as to help him become a skilled proponent; unfortunate shots tended to elicit reactions such as: 'What a silly place to build a bunker — see that it is altered tomorrow', keeping his course architect extremely busy moving obstacles to suit the Royal temperament. [37]

Queen Alexandra also adopted her own approach to the pastime, ignoring the customary practice and playing a wild game of her own invention. She and the Princess Victoria would race from tee to hole to see who could get a ball in first. The game usually ended in a hockey-like scrimmage on the green, where the cut and battered balls were once found by the King, who swore that somebody was playing a trick. [25]

Our inventors were also busy trying to overcome the more infuriating features of the game, preoccupied with minimising the amount of energy that had to be expended in its practice, patenting improved golf-bags to avoid the inconvenience of stooping (17450/1905); improved clubs to enable the ball to be removed from the hole without stooping (1045/1904); a whole range of balls and tees marked to betray their position and thus prevent their being lost (for example 826 & 14452/1905); and finally, the club fitted with rollers to prevent the stroke being spoiled should the ground inadvertently intervene (11463/1902).

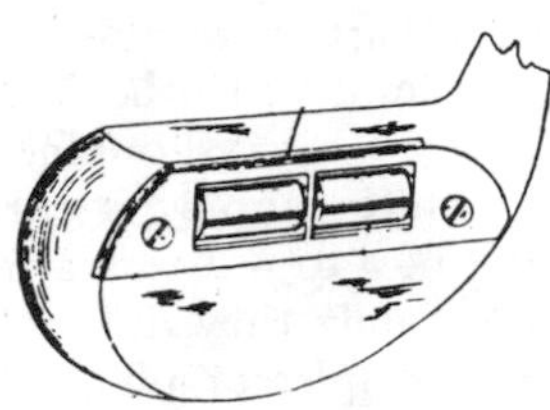

Table-tennis

Table-tennis was marketed as 'Gossima' in 1898, but was slow to take off. In 1901 it was re-christened, onomatopoeically, 'Ping-pong' and became an immediate craze. Picking up balls is a problem today; at the turn of the century when clothes, especially ladies' clothes, were far more restricting, it must have been well-nigh impossible. Accordingly, there were many patents for ping-pong ball picker-uppers, of which we present a wide selection (23274/1901 &c).

In 1902, ping-pong patents were so numerous that the Comptroller-General saw fit to mention the fact in his 'Trend of invention' for 1902, admitting however, that 'the increase had nearly disappeared at the close of the year.' [3]

That year also saw numbers of inventors seeking to 'improve' the game by such means as providing holes in the net through which the ball had to be served or returned for extra points (2516/1902); or by suggesting that it be caught or returned with a scoop or bag instead of a bat (4982/1902). As so often happens, such elaborations were short-lived and the simple original game continues, with increasing popularity.

Shootin'

King Edward was very fond of animals. 'One of the first things to impress itself on the visitor to either Windsor or Sandringham is the universal love of animals which is so deeply implanted in every one of the King's family. The moment one passes through the iron gates and up the long avenue this fact is self-evident, for dogs are waiting to welcome one in that cheerful way of theirs which shows immediately that they are accustomed to be made much of, whilst the entrance-hall is hung with many a superb animal trophy from different parts of the world, collected by the King during his travels.' [9]

Indeed, for most aristocratic Edwardians, a love of animals would most likely concentrate itself on breeding the superior sorts to enable the inferior sorts to be hunted or shot with greater efficiency. Shooting was one of the high-spots of Edwardian country-house living, and to be the possessor of a good shoot was an almost certain way of rising into the highest social circles, since it was clear evidence of the requisite personal wealth.

Cecil Rhodes leased Rannoch Lodge for two months with the expectation of shooting 500 brace of grouse; Baron de Hirsch (a Hungarian Jew who had made his money speculating in Balkan railways) had an estate in Hungary, where the King was thrilled — as only an animal-loving sportsman can be — by the hecatombs of game beaten towards his gun by hundreds of peasants. [31]

But shooting was more than just an entertaining social event and an opportunity for an ostentatious display of wealth; it was also a pastime which presented the hazards of rough country and harsh weather, and our inventors turned their attention to overcoming such hazards, devising becoming cloaks which could be converted into stretchers at need (10244/1903), and alarm calls to be fitted to the ends of the guns so that isolated members of a party could signal to the others (2266/1904) — no doubt an essential safety-precaution in the presence of such sporting enthusiasm.

Patent selection

As we remarked in our Foreword, it is fair to ask how we chose the 600 or so patents from the 140,000-odd available.

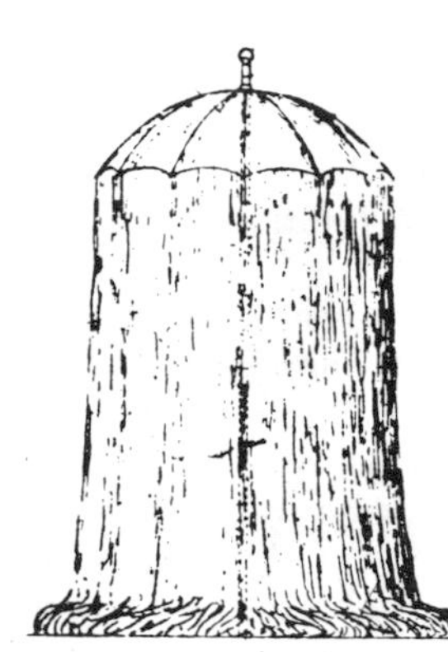

First, there are those which reflect the climate of the times, many of which we have referred to above. Second, there are the bizarre and the gruesome, the Good Ideas which made it all worth while. Third, there are the technologically important patents, to which we will return shortly.

Some of the best unused ideas were surely the umbrella with the insulated tip to prevent the user receiving a shock should its end come into contact with an electrical conductor (18416/1903); the combined umbrella and mosquito net (18749/1903); the device to prevent milk overflowing (19544/1903); the foul breath indicator (16011/1902); the fire-alarm actuated by melting butter (25805/1902); the arrangement to deal with the troublesome tendency of tight knitted garments to become still narrower and shorter after washing (3653/1904); the eraser provided with pneumatic means for clearing away erasings (24547/1905); and the invaluable tongue cleaner (11403/1905).

Any reader interested in the most gruesome or bizarre ideas should perhaps begin by turning to the improved coffin, furnished with a glass window, hammer, matches, candle and bell, for indicating the burial alive of persons in a trance (26418/1903); the husk hat made of maize shucks (11860/1902); the equipment for utilising the electricity in the interplanetary ether (4413/1903); the means for providing pure air for sufferers from consumption (26500/1904); the apparatus for preventing the formation and falling of hail by meteorological utilisation of electromagnetic waves (24688/1905); and the system of wireless telegraphy obviating the need for an aërial by using a tree (25610/1904).

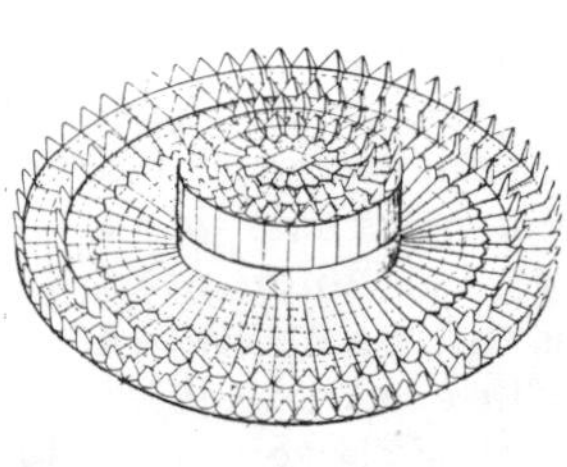

Those in search of amusement, and desirous of laughing at the lengths to which man — or woman — will go in the quest for beauty could turn to the appliance for improving the shape of the finger nails, to be screwed on at night and removed in the morning (19689/1903); the modifying apparatus to impart to the worst-shaped mouth the harmonious curves of the well-modelled embouchure (26012/1904); the head-to-waist backboard to ensure a perfectly erect figure (13550/1905).

Medical practice

From the above, it is but a short step to a range of devices which tell us something of Edwardian medical practice, both legitimate and quack. But which is which? Anyone who takes an interest in medical practice will know that diagnoses of disease, and especially treatments, tend to have a certain periodicity, and what was in favour yesterday may fall from grace today, and is liable to be heralded with a flourish of trumpets tomorrow.

Oliver Wendell Holmes wrote: 'There is nothing men will not do, there is nothing they have not done to recover their health and save their lives. They have submitted to be half-drowned in water, and half-choked with gases, to be buried up to their chins in earth, to be seared with hot irons like galley-slaves, to be crimped with knives like codfish, to have needles thrust into their flesh, and bonfires kindled on their skin, to swallow all sorts of abominations, and to pay for all this, as if to be singed and scalded were a costly privilege, as if blisters were a blessing, and leeches a luxury. What more can be asked to prove their honesty and sincerity?' [27]

Turn, therefore, to any of the gloriously ghastly selection of therapeutic electrical or other ray baths, or apparatuses to cause oxygen or antiseptic vapours to permeate the tissues of the body — lungs, arms, feet, urethra or wherever else desired (7056/1901, 17902 & 23918/1902, for example).

The possibility of using electricity for the treatment of disease dates back to the middle of the eighteenth century. Static electrical machines for electrotherapy were installed at the Middlesex Hospital in 1767, and other London hospitals soon followed suit. In America, Benjamin Franklin (1706-90) treated 'nervous diseases' by electricity — his techniques became known as Franklinism. In Italy, Luigi Galvani (1737-98) propounded his theory that all animals have electricity in their nerves and muscles, and once Galvanic (as opposed to static) electricity became available, a new field was opened for experimental treatment. The inventions of the dynamo, the electric motor, and the sealed electric lamp were all in their turn reflected in medical treatment, and the latter especially made its mark in our period, for it became possible to expose the body or parts thereof to

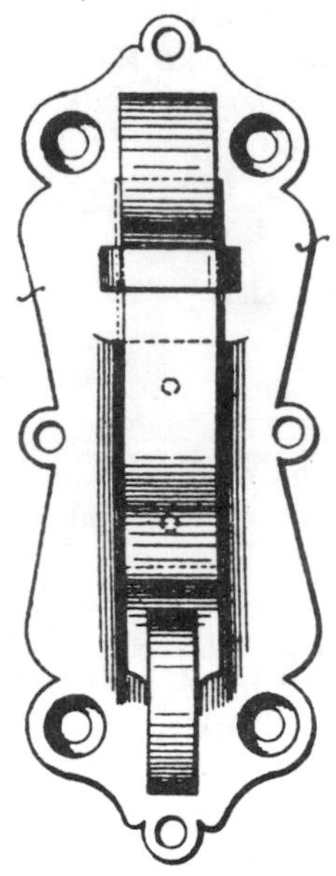

different coloured lights at the flick of a switch, it having been demonstrated that different colours were efficacious in treating different diseases (336/1901, for example).

The widespread use of electrical treatment is understandable — even at the turn of the century electricity was not in general use as a means of providing power in private houses, and still a source of wonder in public places. At the Paris Exhibition of 1900, where it featured in so many of the pavilions, electricity was 'A new force . . . a mysterious power . . . that transforms and recreates everything . . . The Fairy Electricity was born of the heavens, like true Kings. The public laughed at the words "Danger of Death" written on the pylons — it knew that electricity cured everything, even the "neuroses" fashionable at the time.' [32]

Germs

The word 'germ' as the 'seed of a disease' seems to have come into use at the beginning of the nineteenth century. However, it was Pasteur who laid the foundations of bacteriology, identifying, for example, the mechanism of fermentation, and the causes of such diseases as anthrax, diphtheria and rabies. The public thus became aware that germs, or microbes, were everywhere, just waiting to strike. An obvious seat for their lurking was that of the water closet, and many weapons for combating them thereon were patented.

We have rolls of disposable paper (with an appropriate hole) to be pulled across the seat before use (16029/1902); the collapsible celluloid cover to be carried in the pocket (18897/1902) — what happened to the germs which had crawled on to the underside?; the ventilated pan (27817/1904); and the seat which is itself a pad of tear-off sheets (11734/1905). However, perhaps the most elegant solution for class-conscious germs was the extra seat which could be locked in the 'up' position, only the head of the household, or the senior partners in the firm, being allowed keys (24194/1901).

Another communally-used device close to the person was, of course, the telephone. Here again, we find rolls of tear-off sheets to be pulled down over the mouthpiece (25948/1902, 9683/1904 & 12341/1905); or a gas heater to burn up the germs (21574/1903); or replaceable inserts soaked in antiseptic (13597/1905). Perhaps the most advanced was the device which made use of the telephone's magneto-electric generator to produce ozone, which was then distributed all about the instrument by means of plumbing (2488/1901).

Concern about the communion cup was not confined to the Edwardian period, but, though ecclesiastical periodicals still carry correspondence on the topic from time to time, none of the hygienic systems propounded — using the inevitable rolls of paper, for example (15399/1903) — is in use today, as far as we know.

Finally, we feel some sympathy for the inventor who feared that his daily newspaper might attract germs if it were thrown on the ground by a lazy delivery boy, and therefore designed means of clipping it to his door (10590/1905). Presumably he had no letter-box, in case the germs crept through it.

Artificial insemination

As we have remarked elsewhere, there is a tendency for a person to think that an invention dates from about the time that he himself became aware of it. Knowledge of the invention may be delayed if prudery is an ingredient, and it may come as a surprise to many that our year of publication (1979) marks the bicentenary of the Italian Abbé Lazare Spallanzani's successful artificial insemination of a hunting bitch with the semen of a spaniel.

Our inventors show an interest in the subject, for there is an insulated chamber for transporting semen (16903/1902). The same inventor may have subscribed to the theory that masturbation destroys potency, for he was co-inventor of a device for preventing self-abuse in horses (48/1904). Should the animal attempt to abuse itself, it receives an electric shock. We have thought deeply about this topic: whatever horses do, they don't do it lying down, for the invention embodies a switch to break the circuit should the horse do so.

Perpetual motion

Perpetual motion — getting something for nothing, in effect — clearly appeals to our baser natures. There are schemes for perpetual motion machines to be found in the earliest literature, and today there can hardly be an enquiring child who has not suggested connecting an electric motor to a dynamo so that it can produce the power to drive itself.

It is easy for us, with our knowledge of physical laws, to laugh at the seekers for perpetual motion. But just as the search for the philosopher's stone helped to lay the foundation of modern chemistry, so was the search for perpetual motion responsible for its being proved an impossible aim. However, knowing that something is impossible is, to some, merely a challenge, and some perpetual motion machine designs are so ingenious that it takes considerable analytical probing to discover why they won't work. It is not enough merely to *know* that they won't work: we have to know *why*.

The principle we mentioned above — that of a

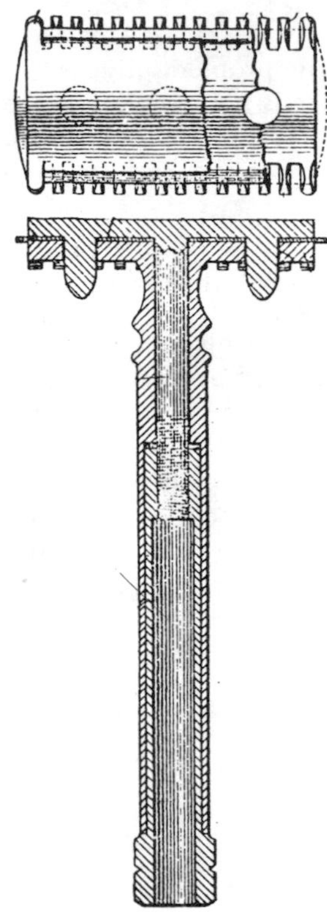

motor driving a dynamo — is an old one in modern dress. A favourite design is the water-wheel which pumps its own water (26571/1903 & 5723/1904): the former has energy to spare, for it is specifically stated that a power take-off may be provided to drive a dynamo. This is typical of the necessary optimism of the perpetual motionist, epitomised by a Victorian patentee in his phrase: 'To prevent any dangerous increase of speed, a brake-shoe can be tightened against the water-wheel shaft.'

In the five years we are studying, 46 perpetual motion machines were patented: we show 11. Two (11318 & 19822/1901) embody hydrostatics; two use a fan caused to rotate by the motion of the vehicle to assist in driving the vehicle (6086/1902 & 25353/1905); another makes use of the weight of the rider of a velocipede to compress air to drive said velocipede (6702/1902); two use falling weights to raise other weights which fall in their turn (26163/1904 & 7789/1905), the latter being especially noteworthy as it demonstrates the perpetual motionists' principle of making a design ever more complex in an attempt to overcome the failure of simplicity; a buoyancy motor (15746/1905); and an interesting machine using the expansion and contraction of the magical quicksilver to provide an overbalancing wheel (17600/1905) — if this rotates, it will be thanks to energy from without, possibly an early attempt to use solar power.

In 1911, the American Patent Office decided that a working model of a perpetual motion machine would have to be produced before its application could be processed. Although this may have saved inventors the government's fees, there is no evidence that it deterred them from spending as much time and money as ever on model-making.

In England, what the Patent Office kindly dubbed 'self-driving motors' were acceptable until 1932, when the ground of objection to an application 'that the invention is contrary to well-established natural laws' was introduced. But to perpetual motionists, such attempts by authority to save them from themselves must be seen merely as short-sighted obstructionism.

A final note: readers who wonder at our inclusion of the station indicator (12768/1904) should consider the 'overbalancing wheel' type of perpetual motion machine, and reflect that the indicator does not turn of its own volition.

Advancing technology

As we indicated above, another of our selection criteria was technological advancement. In *New and Improved* [4], R Baker, of the Science Reference Library, lists and describes 'inventors and inventions that have changed the modern world'. His selection, rightly within his context, examines inventions not necessarily first in their field, but the first to be workable. From his list we include Hornby's Meccano (587/1901); Sturmey's bicycle gears (16221/1901); Booth's vacuum cleaner (17433/1901); Ostwald's nitric acid manufacturing process (698/1902); Hooley's roadmaker (7796/1902); Anderson's puffed wheat (13353/1902); Pitney's postal franking machine (21234/1902); Lanchester's disc brake (26407/1902); Gillette's safety razor (28763/1902); the Cooper-Hewitt Company's fluorescent light (3444/1903); Burger's thermos flask (4421/1904); the Wright Brothers' glider (6732/1904); Fleming's radio valve (24850/1904); and Anchütz-Kaempfe's gyro-stabiliser (6359/1905). We have not illustrated all of these, as the complexity of the illustrations is often meaningless without a more lengthy description than we are able to give here.

Apart from this selection, there are other patents in our period which presage technological advances, but which for one reason or another were ahead of their time: here they are.

Ader's vessel adapted to slide on the surface of water (2844/1904) is built in the shape of an aeroplane — body, transverse wings and tail — and is supported on the water by a cushion of compressed air. This is clearly the principle of the hovercraft, for which the world had to wait another half-century. Since one of the difficulties of making a hovercraft is the design of the flexible skirt, Ader had an incomplete perception of the demands of the idea.

Lee's improvement in telephonic communication (7676/1904) would have made every power station into a modern Tower of Babel by connecting every consumer served with electricity to additional apparatus enabling all to communicate with the station. The idea of using existing wiring for communication is attractive, and a system (now illegal, we are told) was marketed some twenty years ago by Labgear Ltd for inter-office communication.

Hülsmeyer's Hertzian-wave projecting and receiving apparatus to give warning of the presence of metallic bodies (13170 & 25608/1904) clearly foresaw the possibility of what we now call radar. In his first patent, Hülsmeyer suggested that the projector be turned 'to each point of the compass in turn, and so arranged that the officer in charge knows at once the direction from which the warning comes.' In his second patent, he suggests tilting the projector 'in a vertical plane until the reflected Hertzian effect is a maximum; from the angle the distance may be calculated.' Since he was looking for metallic bodies at sea, Hülsmeyer must have been equating his Hertzian waves with

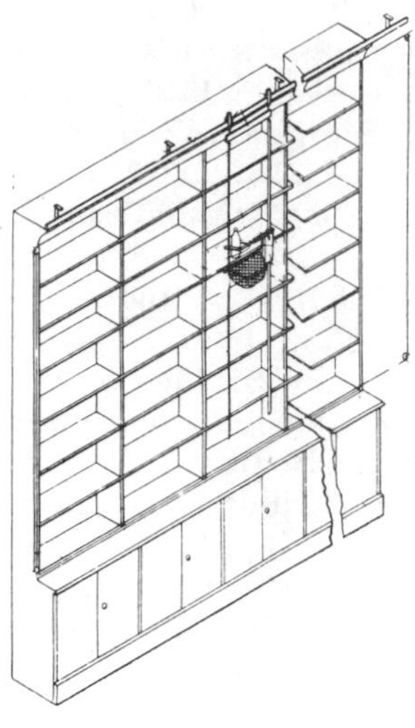

projectiles fired from a gun. However, it seems unlikely that the wavelengths demanded by radar could have been known to, or generated by, Hülsmeyer, so the principle remained an idea awaiting its technology for another three decades.

Tremaine's coffee package (24151/1904) was but a step away from the tea or coffee bag, produced for caterers in the USA in the 1920s. It was not until the mid-1930s that tea-bags were generally available for domestic use — in America — and the UK had to wait until 1952.

Lanchester's mechanically propelled road vehicle (7949/1905) uses a heavy flywheel to store energy in an easily available form. In fact, the secret is not just to use a heavy flywheel, but to spin it very fast, an impossibility until modern composite materials were developed. Today, there are many experimental vehicles using Lanchester's principle in one way or another.

Ward's appliance for removing articles from shelves (22587/1905) was an Australian storekeeper's answer to the problem of access to high storage. Its principle is now well known in vast warehouses, where forklift trucks perform according to Ward's dream — with varying degrees of success.

Nothing new under the sun

Observant readers may care to pick out inventions from our selection which seem to have come round again. The improved marking instrument (11070/1905) has recently been marketed: a pen fitted with a light for writing in dark places, aimed at the executive who has everything, including the need to sign debit slips in dimly-lit restaurants. And we have just found a 1975 US patent for a solar-powered sola topee — containing a 'personal self-contained fan unit' — Prince Hozoor Meerza's hat (19015/1902) brought up to date.

A short history of the British patent system

Early Letters Patent

A grant of Letters Patent — a patent — is essentially a monopoly granted by the Crown for a limited period to enable the petitioner to exploit some invention or process. In return for this privilege, he must disclose his method so that those skilled in the art can use it after the expiration of the monopoly period.

A number of isolated grants are recorded long before any organised patent system developed anywhere. For example, in 1236 Henry III confirmed a grant made by the Mayor of Bordeaux to Bonafusus de Sancta Columba, whereby he and his fellows alone in Bordeaux were permitted to make 'cloths of divers colours after the manner of the Flemings, the French or the English', for a term of 15 years. In 1421, a three-year privilege was granted to the Italian architect Filippo Brunelleschi, by the Signoria of Florence, enabling him to use a method of transporting heavy blocks of marble by water, which would 'operate at any time, and at a lower cost than formerly.' And again in Italy, in 1444, Antonius Manni de Francia received a 20-year grant to construct 'waterless cornmills'.

The first English patent was probably that granted to the Flemish-born John of Utynam by Henry VI, on 3 April 1449, for making coloured glass. This was also a 20-year grant, and John's most memorable commission was the production of coloured glass for Eton College (founded by Henry VI).

It was nearly a century before the next English patent grant, again for coloured glass, to one Henry Smyth in 1552. Between 1561 and 1590, Queen Elizabeth I granted some 50 patents; however, although 'the intention of the Crown was sound', many monopolists abused their privileges, leading to a protest in the Commons in 1601. Reform was promised, but nonetheless many restricted patents were granted by James I (and Sixth), until further public outcry constrained him, in 1610, to issue a declaration that he would grant patents only for 'projects of new invention so that they be not contrary to law, nor mischievous to the State by raising the prices of commodities at home, or hurt of trade, or generally inconvenient.'

Here was the introduction of the notion of public interest into the patent system, incorporated in the Statute of Monopolies (1624), wherein it was rendered illegal to grant a monopoly except 'for the term of fourteen years or under hereafter to be made of the sole working or making of any manner of new manufacture within the Realm to the true and first inventor'. The term of 14 years was determined as it was twice the length of the standard apprenticeship, so that others could be instructed in the art.

A notable early patent — the second in the Chronological Index — was that granted to Nicholas Hillyard (1617), 'His Majestie's servant and principal drawer for the small potraits & imborser of His Majestie's medaillies of gould . . .'

During the reign of Queen Anne (1702-14), it became the practice to file an 'instrument in writing' with an application for Letters Patent, giving a description — or specification — of how the article was to be made or the process carried out. Such a description is obviously important; if a monopoly is to be granted, its confines must be clearly stated, and Lord Mansfield's judgement (1778), that a patent was void on the grounds of the insufficiency of its specification, left the importance of that document in no doubt. Nowadays when we speak of 'a patent', we are almost always referring to the specification.

Development of the modern system

It has been suggested that the British patent system helped to create the country's industrial supremacy, which existed at the time of the Great Exhibition of 1851. Be that as it may, the system was far from perfect, and in 1851 a Select Committee was set up to investigate its shortcomings. Evidence presented to the Committee suggested that the high cost of obtaining a grant of Letters Patent was forcing many inventors to 'work their inventions secretly', and other serious defects were revealed: there was no immediate security for the petitioner on lodgement of his petition, and there was no system for the would-be patentee to search in order to establish his originality. It was also pointed out that an application had to pass through ten stages, to seven offices, and twice needed the Sovereign's signature. As a result of these enquiries, the Patent Law Amendment Act came into force on 1 October 1852, and the Patent Office opened its doors on 29 December that year. Professor Bennet Woodcroft FRS, who had been a prominent member of the Select Committee, was at the helm, and within two years had published a *Chronological Index of Patents 1617-1852,* with associated cross-reference systems.

As far as the inventor was concerned, however, the most important effect of the Act was to reduce the cost of protection — which was now immediate — from £300 (for the Three Kingdoms) to £25. This caused an upsurge in the number of applications, so that the number filed annually soon surpassed the total for the previous century.

One of the criticisms of the 1852 Act was that there was no examination of applications for their novelty, though such an examination would have been difficult without the reference system which Bennet Woodcroft was painstakingly building. It was also said that the reduced fee was yet too high, and the next major reform — the Act of 1883 — reduced it still further, to £4. As might be expected, this resulted in another sharp increase in the amount of work which the Patent Office staff was called upon to handle: in 1882 there had been 6241 applications and 4337 grants; in 1884 there were 17110 applications and 9118 grants. This increased workload introduced severe delays into the process of obtaining a patent, so the Patent Office staff was increased accordingly.

Examination for novelty was not, however, introduced until the next Act — that of 1902 — came into force on 1 January 1905. Then, the 50-year search was instituted as part of the procedure, 'not so much to improve good inventions as to exclude bad ones' — our readers should seek their own definitions of 'good' and 'bad' in this context. Suffice it to say that the 1902 Act made little difference to the numbers of applications after it came into force, though the Comptroller-General reported an 8 per cent decrease in applications in 1905 compared with the previous year, 'possibly due to the anxiety of inventors to lodge applications before the 1902 Act came into force, and thus avoid not only the official search fee for novelty prescribed by that Act, but also the payment of a sealing fee before a patent could be granted.' [3]

Carrying out the 50-year search was possible only because of the thorough systems which Bennet Woodcroft had set up, and searching was greatly facilitated when the new Patent Office library and reading room was opened to the public on 9 January 1902. It is a magnificent building, decorated in the *art nouveau* style, a fitting setting for our researches. Long may it remain!

Patent agents

As the modern patent system came into being in the 1850s, so did certain people become 'practised in the arts of petitioning the Crown for the grant of a Patent'. Such people — often consulting engineers — were able to sell their expert knowledge to inventors, and the profession of patent agent was born.

Today, a patent agent is one of the most specialised of specialists, for not only must he understand patent law and matters technical, he must also have a 'capability with language to enable him to apply the protection of the law to the technical inventions of which he is the custodian'. He is not concerned solely with the arts of searching the literature and drafting and submitting applications: he has to understand foreign patent practice, and licensing and litigation, not to mention handling British applications on behalf of overseas patentees.

Apart from the advances in technology which obviously demand more knowledge on the part of the agent, his task is not very different now from what it was in 1891, when the Chartered Institute of Patent Agents came into being with

236 members at the year's end. At the end of 1977, there were 1124 Chartered Patent Agents — a select profession indeed.

Edwardian patent agents
It is no more necessary to employ an agent when applying for a patent than it is to employ a solicitor when drawing up a will; in both cases, however, the specialist can save the amateur a great deal of trouble. Many of our inventors saw the advantage of using a patent agent, and we append (page 000) a 'league table' of their names. Readers in the trade will note that many well known today appear in the list, and many members of the firms descended from Edwardian times may be amused to see what their predecessors were doing at the turn of the century.

Rate of decay of protection
After a patent has been granted, renewal fees have to be paid annually from the fourth year to keep it in force. We are often asked how many of our inventors kept their patents in force: we cannot give details for each, but we can indicate the rate of decay, which remains remarkably constant from year to year.

Annual Averages 1901 — 1905

Applications received 28379	*% of original remaining*
In force to end of	
4th year 15076	53.1
5th year 5250	18.5
6th year 3786	13.3
7th year 2933	10.3
8th year 2331	8.2
9th year 1923	6.8
10th year 1599	5.6
11th year 1332	4.7
12th year 1126	4.0
13th year 927	3.3
14th year 693	2.4

Conventions

We now come to the body of our book: the abstracts from the patent specifications. Following the **Title** and the description, we give whatever is available from the following:
(Patent number/year)
Name(s) of inventor(s), *Profession(s)* or *description(s)*
Address(es) of inventor(s)

(Agent)

Sometimes a provisional specification is filed to give instant protection, followed by a complete specification; sometimes the original application is complete. We do not indicate this unless the applicant changed agents in mid-stream; in such cases we give their names for the provisional (P) and the complete (C) applications. The absence of an agent's name implies that the patentee did not employ one.

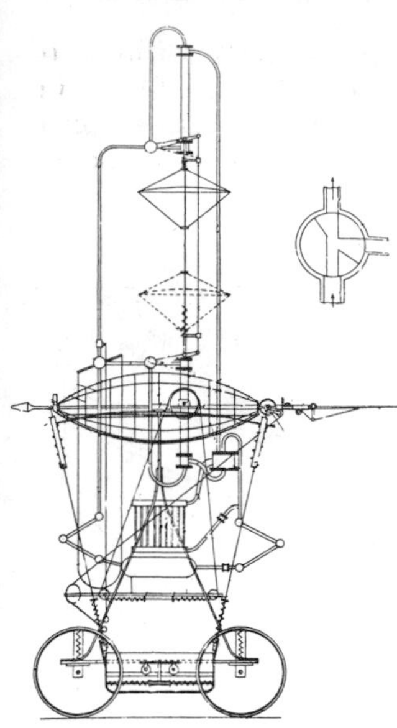

Blunt's Improved Flying Machine is made to imitate the flight of birds.

It is driven by steam — propelled by the downward stroke of the propellers tilting the rear of the aeroplane upwards; gravity causes the machine to move in an inclined plane downwards and gather way enough to carry it upwards when the rear of the aeroplane is depressed. The propellers may also serve to force the machine forwards by inclining them downwards towards the front.

The steering is partly effected by the aeronaut putting his weight on one or other of the cords from which the car hangs.

346/1901
Arthur Henry Philip Blunt *Civil Engineer & Surveyor*
98 Albert Street, Regents Park, London NW, England

Wagener's Improved Spoon has its bowl arranged at right angles to the handle so that the edibles can be brought to the mouth in a facilitated way, necessary turning of the hand when introducing the spoon into the mouth being rendered superfluous. The angular arrangement of the mouth portion also enables a better dividing or cutting up of cake or, as the case may be, fruit and the like.

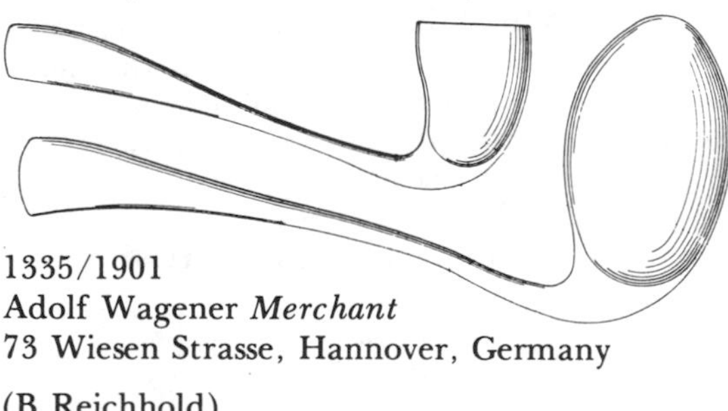

1335/1901
Adolf Wagener *Merchant*
73 Wiesen Strasse, Hannover, Germany

(B Reichhold)

Hornby's Improved Toy or Educational Device for Children and Young People relates to arrangements by which children can construct mechanical objects, buildings etc from independent pieces. The equipment may include a file, screwdriver and pliers for working the pieces. The pieces serve for the construction of bridges, tunnels, stations, signals, signal boxes, hoists and buildings in general, as well as cranes and railway lines as shown in the illustration.

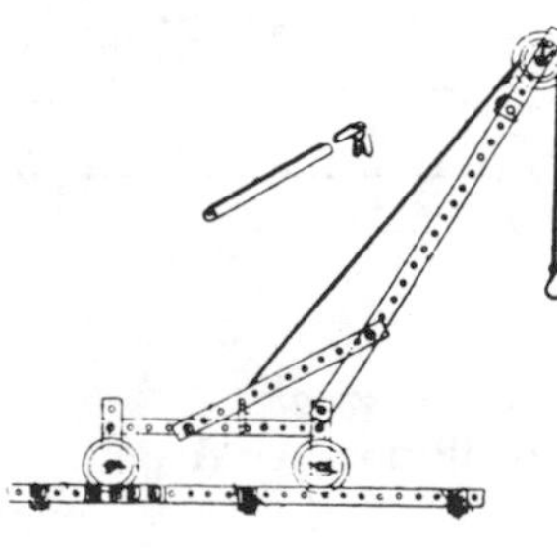

587/1901
Frank Hornby *Manager*
10 Elmbank Road, Sefton Park, Liverpool, England

(W P Thompson & Co)

The Ottos' Improved Light Bath facilitates the treatment of disease, for experiments in the science of cures effected by light have demonstrated that various forms of diseases require to be treated with different coloured lights, the cure being effected quicker, or only being effected at all when, instead of ordinary white light, red or blue for example are used.

The greater institutions therefore always keep light bath apparatus on hand for white light and other apparatus for the coloured lights beside. Smaller institutions are neither able to afford, nor have space to store several apparatus, and changing bulbs requires care and occasions loss of time. The present apparatus obviates these problems, with the added advantage that the colour of the light can be changed while the patient is in the bath, it being unnecessary for him to leave the same. This is effected by means of providing banks of lamps of various colours, wired so that, by means of the switch lever, the patient can be treated with light of appropriate colour.

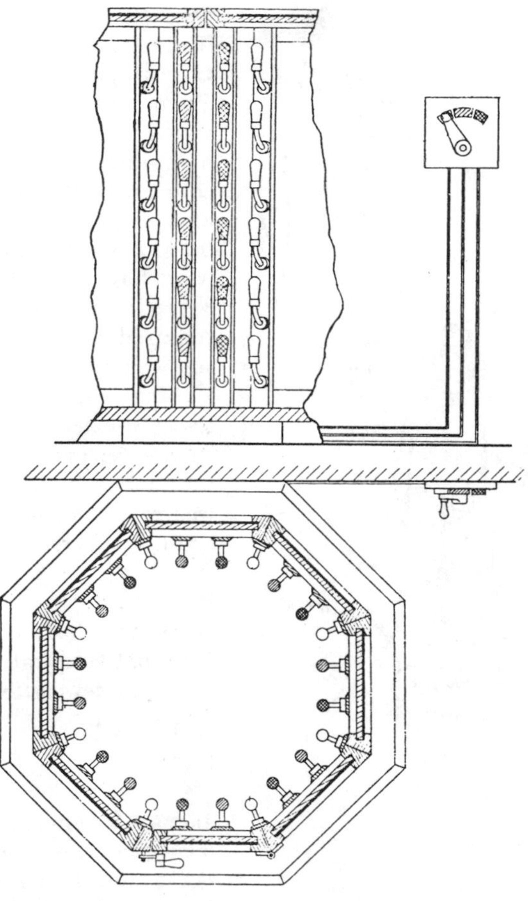

336/1901
Robert and Karl Otto *Merchants*
22A Luisenstrasse, Berlin, Germany

(Wheatley & Mackenzie)

1901

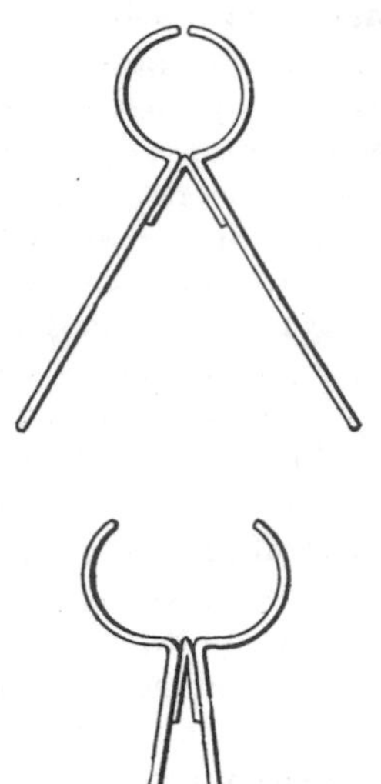

May's Improved Clip or Holder for Eggs obviates the necessity for having to wait for a boiled egg to cool before it can be transferred from the spoon with which it has been removed from the boiling water to the egg cup, without burning the fingers.

The device consists of a spring clip having curved parts to encircle or grip the egg and hold it until it is placed in the cup. To remove the egg from boiling water the handle is gripped, opening the forked and curved holder and allowing it to be passed over the egg. On relieving the pressure upon the handle the egg is retained in the holder and can then be removed and placed in the cup by compressing the holder.

761/1901
Arthur Henry May *Soldier*
Ghorporie Barracks, Poona, India

(Geo H Rayner)

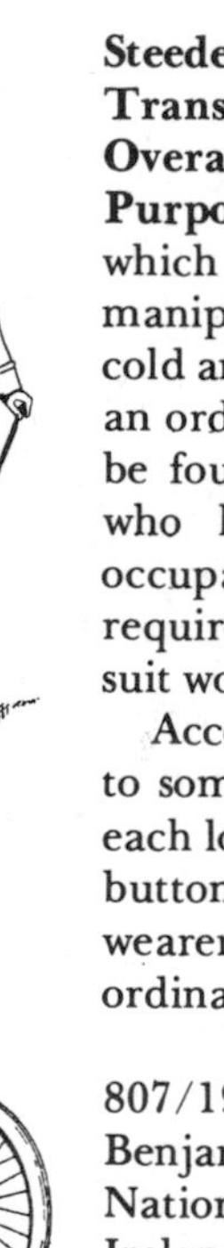

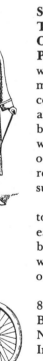

Steede's Overcoat which can be readily Transformed so as to Form a New-shaped Overall suitable for Cycling, Riding or other Purposes provides an adaptation to an overcoat which can, by simple and expeditious manipulation, render protection to the legs in cold and boisterous weather and also be used like an ordinary overcoat if the wearer wishes. It will be found especially useful for wear by persons who have to ride daily to their places of occupation, when the comfort of an overcoat is required, but where the usual riding or cycling suit would be out of place.

Accordingly, the overcoat is divided (at will) to some extent at the back and is so made that each lower part can be conveniently fastened, by buttons or otherwise, round each leg of the wearer to form a better protection than an ordinary overcoat.

807/1901
Benjamin Hosford Steede *Physician*
National Hospital for Consumption for Ireland
Ireland, Newcastle, County Wicklow, Ireland

(P R J Willis)

Jackson's Brake for Wheeled Vehicles overcomes the disadvantages of the usual brakes. For example, a frictional brake applied to the pneumatic or rubber tyre of a comparatively heavy vehicle proceeding at a good pace is highly destructive to the tyre, and a band brake upon the hub of the wheel has a tendency to strain the wheel and wrench the tyre. Accordingly, this invention provides a skid or shoe arranged so that, by means of a system of levers, it may be passed underneath the tread of the wheel and so raise the wheel from contact with the ground.

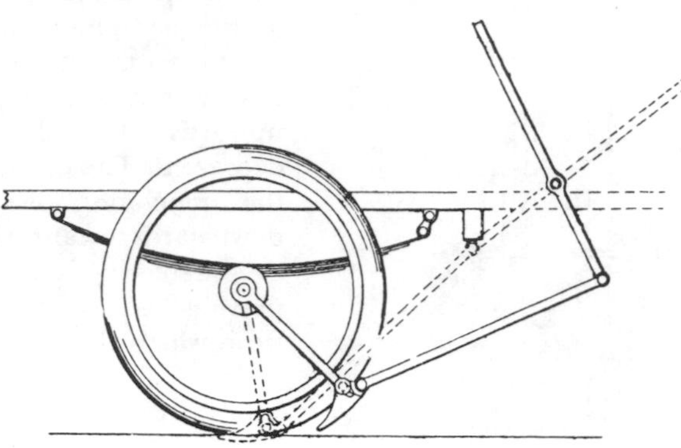

858/1901
Ralph Jackson *Engineer*
Moss Lane, Altringham, Cheshire, England

(J A Coubrough)

Luria's Invalid or Surgical Bedstead is provided with various fittings to facilitate surgical and medical operations, to regulate the temperature, and to suit the convenience of the patient. It is fully adjustable, with leg and arm rests, mattress and pillows, and can be filled or inflated with hot water or air by means of pipes. There is a water heater near the bedstead, a tank under the mattress which can be used for heating or as a bath, and an ice-tank over the bed for cooling. An adjustable support may be used for apparatus or books etc.

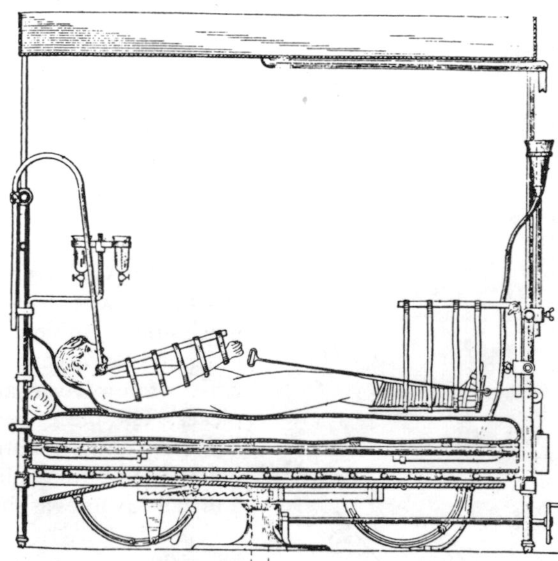

670/1901
Adolfo Luria *Physician and Surgeon*
100 State Street, Chicago, Illinois, USA

(A M & W M Clark)

Nicola's Improved Flycatcher Adapted to be Used for Advertising Purposes may also be arranged as a calendar in such a manner that each sheet represents a day of the week so that the person using the flycatcher, after having torn off the sheets for a week, is compelled to read the advertisements printed thereon.

6818/1901
Louise Nicola, *née* Kappel
43 Ebendorferstrasse, Magdeburg West, Germany

(Herbert Haddan & Co)

Fagan's Improved Physical Culture Apparatus consists in a framework of harness or tape or other bands, fitting round the chest, back, shoulders, neck and legs, such bands bearing rings to which elastic cords or other suitable springs are attached, the other ends being attached to the hands or feet, or to both, by hooks or otherwise. The apparatus is donned in the same manner as a waistcoat and pair of trousers except that the two are in one.

The user throws the arms upwards and downwards, for example, and moves the legs backwards and forwards from the knee or thigh. Such apparatus is very portable, and can even be carried in the pocket.

8009/1901
Bernard James Fagan *Actor*
5 Green Street, Leicester Square, London, England

(Browne & Co)

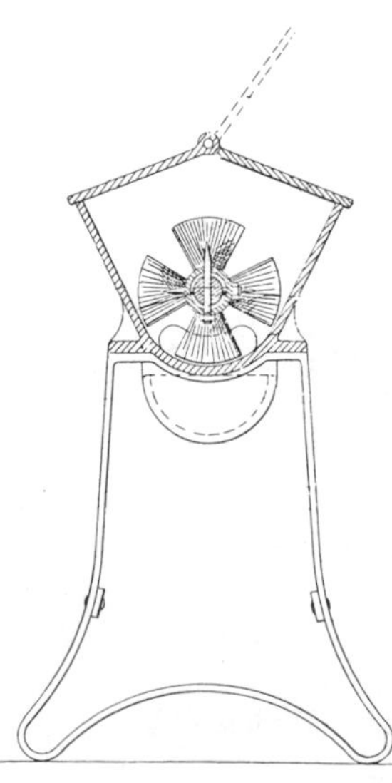

The Littles' Mechanical Sand Box for Golfers is designed to protect the sand or other material from excessive rain or heat and to deliver any desired quantity or the proper consistency for making the sand tee which is still preferred by the majority of golfers on account of its being workable into any desired shape or size, and is always at hand when required.

The box contains a screw conveyor which, when turned, will cause sand to be deposited through a hole onto the ground. Although it is of simple construction and not likely to get out of order, should the device become disabled the box can be used as an ordinary sand box without danger or inconvenience.

1216/1901
Gilbert Little & Archibald Little *Engineers*
New Conveyor Co Ltd, Brook Street, Smethwick, Staffordshire, England

(Charles Bosworth Ketley)

Aurich's Improved Method of Storing and Displaying Souvenirs and Articles of Interest is designed to allow a collector of souvenirs and memorabilia to view the said articles at any time without necessitating a tedious search in boxes and the like where such souvenirs used to be stored heretofore.

The collecting stand may be of any suitable material and shape and be provided with recesses, holes, hooks, screws, pins, and eyelets for receiving the shafts of flags, pennants, standards, staffs and shafts all suitably decorated: military persons will obviously select regimental colours; gentlemen generally the favourite colours of ladies.

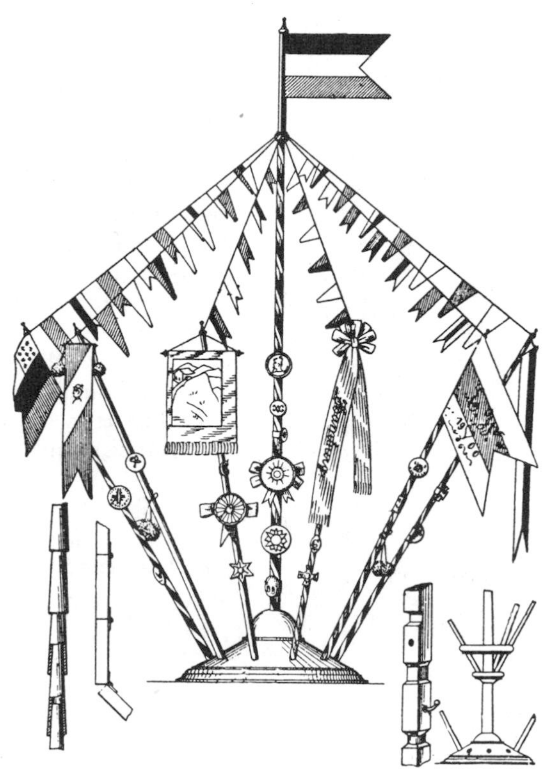

1333/1901
Alban Aurich *Manufacturer*
Hartmannsdorf, nr Chemnitz, Germany

(B Reichhold)

Kelly's Improved Apparatus for Jumping is a trampoline concealed in the case of a grand piano.

1840/1901
Woody Berry Kelly *Gymnast*
257 West Campbell Street, Bath Street, Glasgow, Scotland

(Hughes & Young)

1901

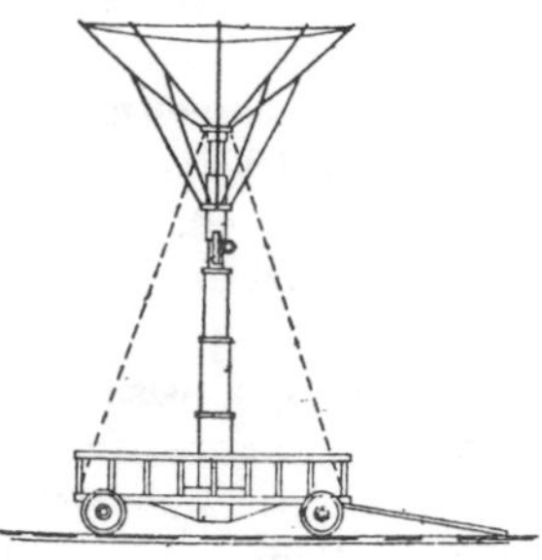

à Brassard's Improved Portable Fire-escape consists of a jumping sheet supported on a telescopic stand on a wheeled base — rather like an inverse and inverted umbrella — which when arranged ready for use can be very widely extended, but during transport only requires very little room.

There will be no hesitation about jumping into this sheet extended in close proximity to the user as is the case with fire escapes of other constructions.

1909/1901
Franz à Brassard *Merchant*
33 Augusta Strasse, Aachen, Germany

(W P Thompson)

Turnbull's Self-Emptying Spittoon for the Floors of Railway Carriages provides a receptacle wherein refuse such as cigar-ends may be readily dropped by smokers, such spittoon being automatically emptied onto the permanent way when nearly full and closed again automatically after such discharge.

It comprises a lidded sheet-brass pan at the bottom of which is a hole one inch in diameter fitted with another lid so as to keep closed when no refuse is in the pan but self-opens when the pan is nearly full and then closes automatically.

8429/1901
John Henry Turnbull *Carriage Fitter*
24 Chandless Street, Gateshead, Co Durham, England

(Hughes, Son & Co)

Wiswall's Improved Dusting Cap for Personal Wear is such as is worn by ladies while doing light housework, particularly dusting, and is a dainty and attractive cap which is at the same time durable and inexpensive. It is made from a rectangular piece of light fabric, preferably a man's handkerchief with a highly ornamental coloured border, and an elastic cord to make it grip the head. A bow of tape or ribbon may be affixed to the front part at the middle. This method of construction is inexpensive and the product dainty, serviceable, and convenient for application and removal.

2033/1901
Marion Wallace Wiswall *Manufacturer*
40 Mount Pleasant Avenue, Boston, Massachusetts, USA

(Herbert Haddan & Co)

Paterson and Reford's Combination Walking-stick and Advertising Lap-card-table is especially designed for the use of travellers and others on long rail-road journeys wishing to drive away the monotony by indulging in a game of cards to pass time away pleasantly. A hollow walking stick is so constructed as to contain a concealed spring roller carrying a fabric which, when unrolled, may be used as a lap card-table or puzzle and game-board and advertising medium.

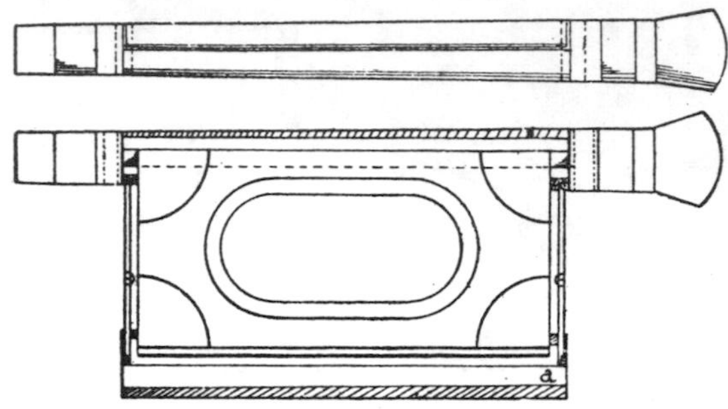

1531/1901
Archibald Hendry Paterson *Cabinet Maker*
Muirhouse Street, SS Glasgow, Scotland

John Hamilton Reford *Professional Brewer*
24 Havelock Street, Byres Road, Glasgow, Scotland

Clark's Improved Sun-screen for the Heads of Horses and other Animals is particularly adapted to provide for the free passage of air between the animal's head and the protecting arrangement, or "screen". The "screen" consists of an awning of canvas or other material on a frame, which either rests on the animal's head or is held by rods suitably attached to the collar and saddle.

The frame may be made to fold up when not in use. The "screen" does not interfere with the contour or appearance of the animal and moreover does not hide its head.

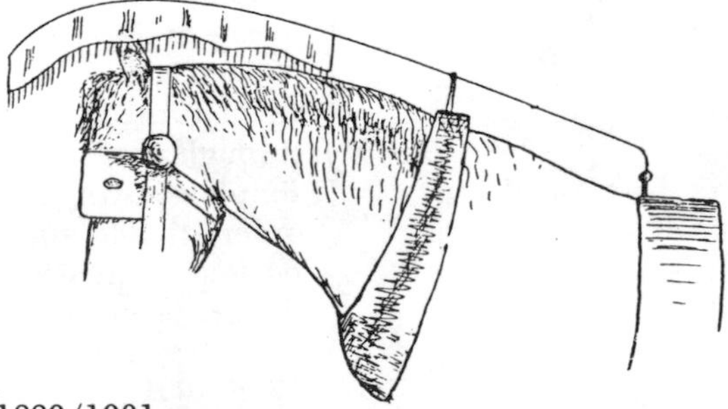

1889/1901
William Samuel Clark *Grocer and Provision Merchant*
77 Oxford Street, Swansea, Wales

(Hughes & Young)

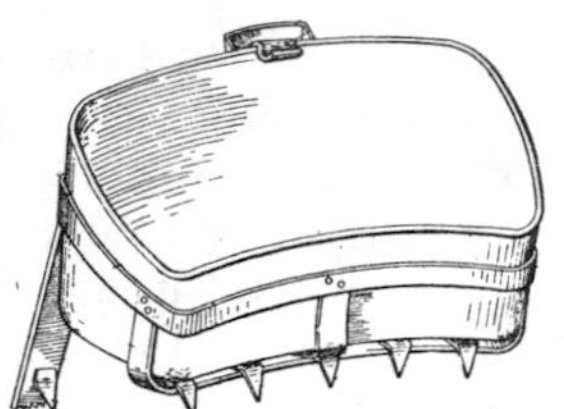

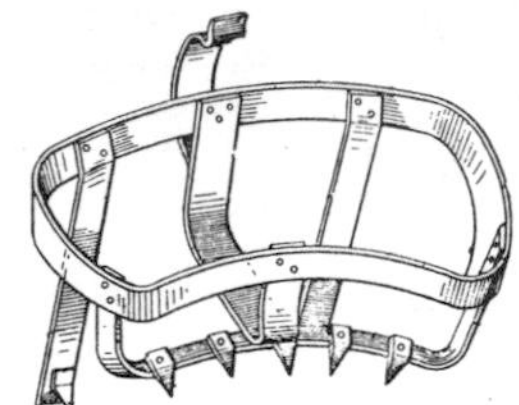

Wolfskill's Improved Bed Pan Holder is a basket-shaped frame designed to prevent the bed pan slipping backwardly from the patient when in use, thus defeating the object for which it is intended. The basket or frame into which the pan is slipped and clipped is preferably built from strips of sheet metal and has a row of prongs, riveted to a transverse member, which engage with the bedclothes.

2299/1901
Susan Cooper Wolfskill *Farmer and Fruit Grower*
Winters, California, USA

(Wheatley & Mackenzie)

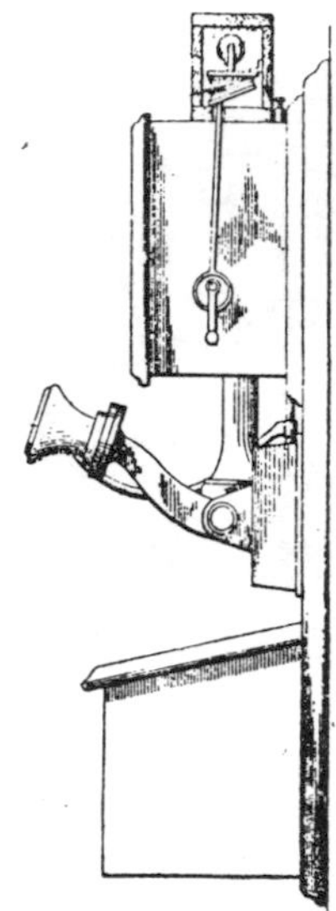

Axtell's Apparatus for Disinfecting the Transmitters and Receivers of Telephones meets the well-known and understood necessity for the application of some efficient and at the same time simple means of achieving same, to wit, by using the current obtained from the "magneto" forming a portion of all ordinary telephones to produce ozonised air, the antiseptic properties of which are also well known. Turning the handle of the magneto operates a small bellows or pump, which passes air through the ozoniser and thence to various parts of the telephone.

2488/1901
Cyrus Fletcher Axtell *Mechanical Engineer*
22 West 113th Street, New York City, USA

(A M & W M Clark)

Penotti's Improved Shower Bath overcomes the problems found with other shower baths; that the bather not only has to hold a chain if the water is to flow, being thus unable to use his hands for other purposes but, moreover, that there are no proper means of mixing the hot and cold water so that there is a risk of the bather being scalded.

The first problem is overcome by arranging the floor upon which the bather stands to move under his weight, so turning on the water supply as required.

The second problem is overcome by having a three-way cock giving "cold", "mixed" and "hot" positions.

3674/1901
Giovanni Penotti *Mechanic*
Via Lagrange, No 24, Turin, Italy

(Abel & Imray)

Gillet's Appliance for Enabling a Performer to Appear to Stand upon One Finger is an expedient for artistic purposes; the standing upon one finger is only apparent, and the use of the appliance deludes the public.

The appliance consists in a piece of iron grasped in the hand with a protuberance extending down behind the finger upon which the artist apparently stands. Anyone attempting to employ the artifice must of course have great assurance and dexterity in standing upon one hand and one arm and making thereon his productions.

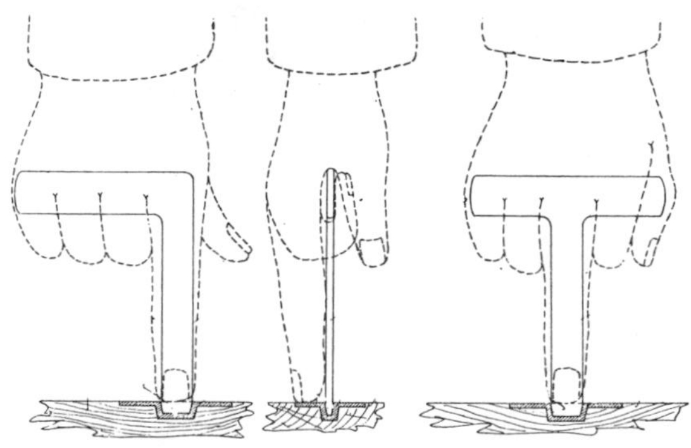

2021/1901
Fred Gillet *Artist*
Scala Théatre, Cologne, Prussia

(B Brockhues)

Altermann's Improved Arm Rest for Violin Players prevents an incorrect position of the instrument being assumed by beginners who, instead of holding the instrument up and to the right, generally have the habit of holding the instrument down and also sidewise. Accordingly, the pupil is provided with a belt to which is attached a telescopic support for the arm; a chain prevents the arm from moving too far to the left.

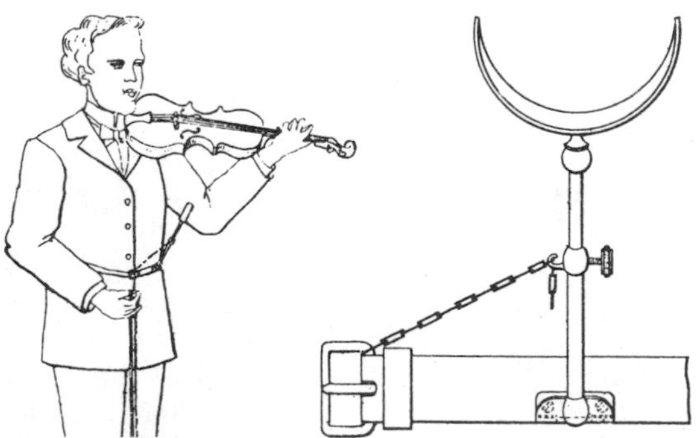

3564/1901
Ezor Altermann *Violinist*
50 Boulevard Malesherbe, Paris, France

(Haseltine, Lake & Co)

1901

Crôtte's Process of Preserving Beer discharges into it small quantities of an antiseptic by means of an electric current of high tension. This renders the organic or germ life of the beer harmless so that it will keep, without deterioration, for a great length of time.

The antiseptic may be boric acid or iron perchloride in a solid or liquid form, and is contained in a copper tube, silvered externally, which acts as one of the electrodes. The other is also preferably of copper and is mounted in a non-conducting frame; it may be inserted in a sponge soaked in the antiseptic.

The high tension current causes infinitesimal quantities of the antiseptic to pass through the copper tube by cataphorical action. After that, in order to further insure the destruction of organic life in the beer, a current of low tension (110-150 volts) is passed for five to ten minutes.

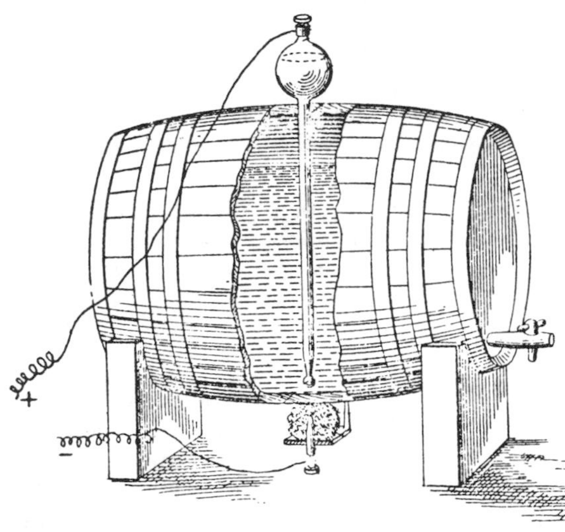

2690/1901
Francisque Crôtte *Electrician*
66 Central Park West, Manhattan, New York, USA

(Tongue & Birkbeck)

Jaffe's Rocking-chair-bed is a convertible item of furniture of box-like construction. It is made at the top like an armchair and stands on four tiny legs provided with castors to allow easy movement from place to place. It is also provided with adjustable rockers hinged underneath, one each side. These are turned up when not in use and carry projecting bars which serve as supports for a footboard which can be slid under the chair.

3765/1901
Herman Jaffe *Hebrew Teacher and Amateur Mechanic*
40 Fashion Street, Spitalfields, London E, England

Möller's Improved Chalice avoids the problems of the ordinary metal cups hitherto used, where many lips consecutively touch the metal brim. It is constructed so that a freshly filled and untouched cup is presented to each communicant. It has a ring of cups which can be filled from a central reservoir. A rotating plate with a hole in it allows the cups to be filled one at a time by means of the pressure of wine within the container.

3052/1901

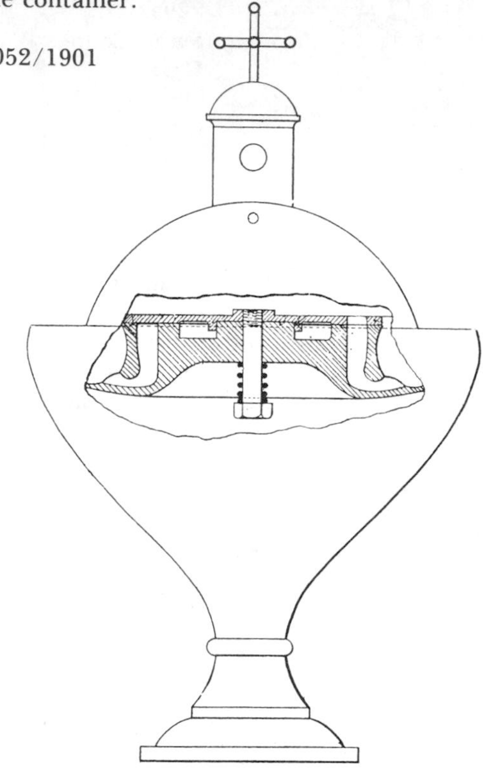

Möller's Further Improved Chalice has a ring of cups around a central reservoir, into which wine is forced by screwing up the base, the spout having been pointed in the appropriate direction. Leakage past the piston collects in a recess.

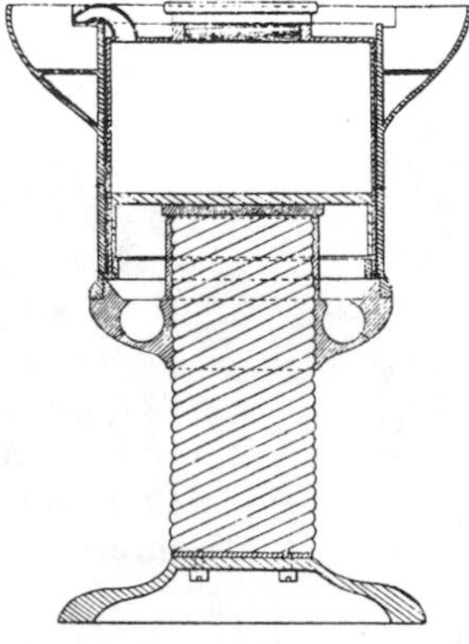

7648/1901
Thöger Christian Theodor Möller *Merchant*
12 Pilestraede, Copenhagen, Denmark

(Edward Evans & Co)

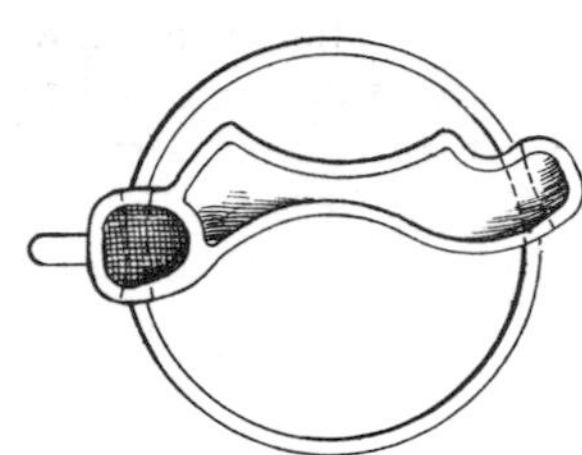

Krimer and Andrew's Improved Moustache Guards for Drinking Purposes may be detachably mounted on a cup or on the bowl of a spoon, and supersede the existing practice of making fixed guards.

In one form the guard forms part of the handle of a spoon; and, when applied to a cup, the spoon is reversed and is held in position on the mouth of the cup by a stop.

The guard may be plain, open, or ornamented, and an indent, groove or projection to serve as a stop can be formed on the spoon, which, in combination with the pressure exerted by the thumb of the user, will hold the combined spoon and guard when laid on the rim of the cup in the required position so as to permit drinking without the slightest difficulty whilst preventing the liquid coming in contact with the moustache.

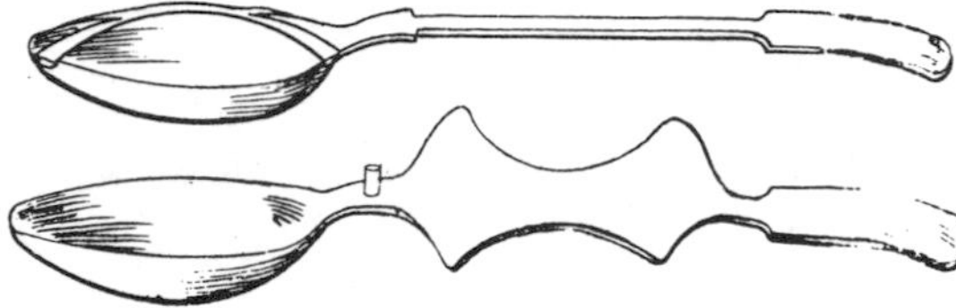

3871/1901
Moses Lazar Krimer *Merchant*
22 Springdale Road, Stoke Newington, London N, England

William Munn Andrew *Draughtsman*
16 Brook Street, Kennington Road, London SE, England

(Benjamin T King)

Marriott and Rigby's Further Improved Folding Fork for Use at Picnics has the particular advantage that it cannot possibly double up by accident when in use.

This is effected by forming ears at the end of the stem which project and catch against the sides of the handle when the fork is open, and so prevent it from closing except in the reverse direction.

4102/1901
Charles Frederick Marriott *Rubber Moulder*
18 Sandy Lane, Aston, Birmingham, England

William Bernard Rigby *Fitter*
52 Sandy Lane, Aston, Birmingham, England

(Lewis Wm Gould)

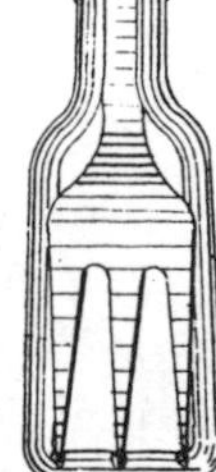

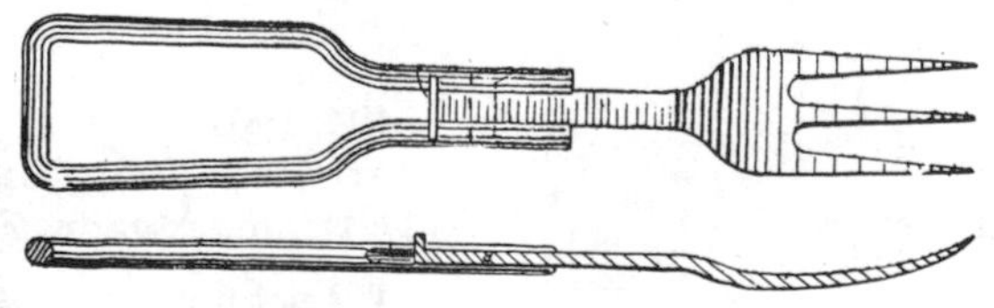

Nicholl's Improved Flying Machine consists essentially in providing one or more articulated members so constructed and arranged that they can be caused to vibrate by the action of the wind and thereby supply or assist in supplying the necessary motive power for propelling the machine. The machine is constructed of a framework with a roof of bamboo, metal etc with a propellor at the rear. The propellor is operated by wind and a spring after the machine has attained a certain momentum, or by hand or by a motor. Spring loaded legs deaden the shock on landing. When not required to carry a person, the machine may be held captive after the manner of a kite, and steered or controlled by a person or persons on the ground.

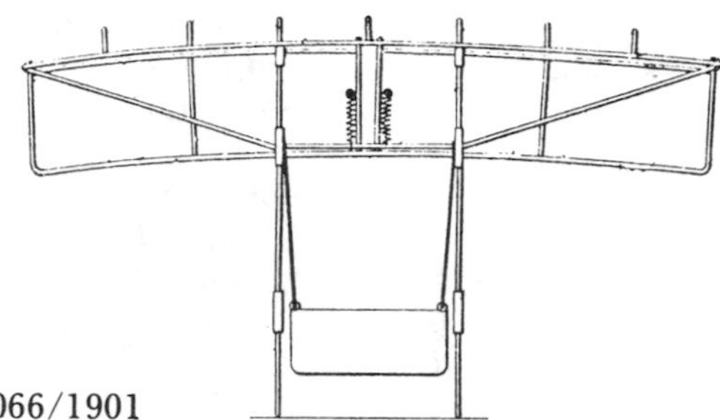

1066/1901
George Nicholl *Mathematical Instrument Maker*
153 High Holborn, London, England

(Haseltine, Lake & Co)

Scheffer's Improved Weighing Apparatus is a device adapted to bedsteads or perambulators so that the exact nett weight of children and adults may be determined, which is not indicated in using ordinary weighing devices involving the inconvenience that the weight of the dresses worn by the persons which are weighed must be deducted from the total weight indicated by the balance, which in most cases is possible only by rough estimation. On the contrary, when the apparatus is used, the person to be weighed is lying in the bed and consequently the weight of the dresses, which are all taken off (leaving out of the question shirts and other light body-linen or bed-dresses) does not substantially alter the weighing results.

Accordingly, the bedstead or perambulator body is mounted on a weighing machine, and the latter is set to zero. The person to be weighed then climbs into the bed and his exact nett weight so determined.

4517/1901
Ludwig Scheffer *Gentleman*
51 Wörthstrasse, München-Haidhausen, Germany

(Edward, Evans & Co)

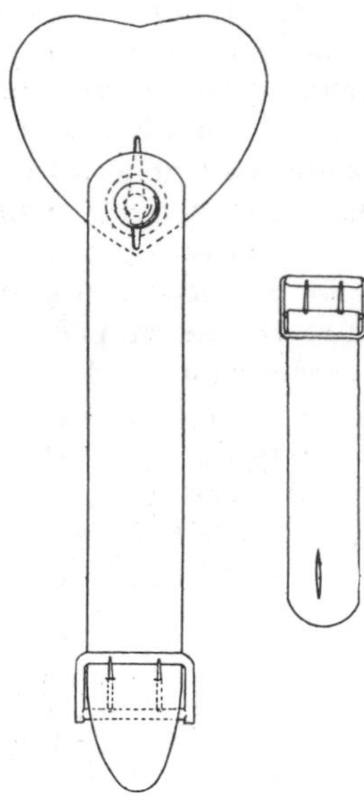

Roberts's Improved Means for Holding the Backs of Shirts in Position provides for each shirt a means of adjustment, not in substitution for, but in conjunction with, and as auxiliary to, braces, so preventing the possibility of the formation and continuance of the usual hunch or protuberance between the shoulders of the wearer, which spoils the fit or set of his waistcoat and coat whilst imparting a humpbacked appearance to himself.

The invention comprises a length of strong linen or other suitable material, attached to the shirt about four inches above the top of the trouser line, and furnished with a vertical button hole, to which is buttoned a strap which is buckled to a suitable contrivance fastened to the trousers or underpants, thus imparting sufficient tension to compel the back of the shirt to assume and retain a smooth and uncrumpled condition.

4660/1901
Thomas Roberts *Manufacturer*
34 Victoria Street, Westminster, London, England

(Browne & Co)

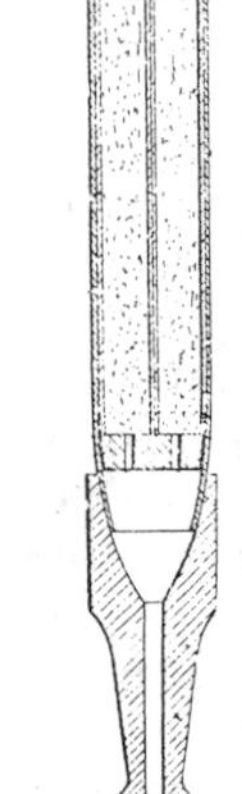

Ferré's Improved Appliance for Generating Gases and Vapours suitable for Surgical and other Purposes is particularly to be used where the gases resulting from the combination of chemical elements are too unstable to enable them to be kept, or where there is an advantage in utilising them at the very moment of their generation, their action having, at such moment, a greater efficacy.

Accordingly, the invention comprises two impermeable receptacles containing the materials, over which streams of air are passed. Having been allowed to mingle, these will give rise to a compound required in the laboratory, or in industry, or for inhalation.

The device may take the form of a cigar, in which case the two impermeable receptacles consist of paper, textile material etc filled with hydrophile cotton and plugged at the ends. If one receptacle contains cotton soaked in hydrochloric acid, and the other cotton soaked in an aqueous solution of carbonate of ammonia, drawing air through the mouthpiece will produce an absolutely neutral and inoffensive white vapour giving a perfect imitation of smoke.

The appliance thus forms a toy or a "surprise" as well as an inhaler.

5050/1901
Henry Ferré *Pharmacist*
18 Rue Mogador, Paris, France

(Ernest de Pass)

Hughes's Improved Driving Mechanism for such Velocipedes as are Driven by Muscular Power provides an alternative means of propulsion to that used in the modern bicycle, which outrages the most fundamental notions of animal mechanics and indeed of common sense — the rotary crank drive, which has superseded the other drive motions all over the world and whose principal fault is that it is not adapted to be economically driven by a downward force of muscular origin, such as the cyclist in his ordinary position is pre-eminently fitted to exert.

In Hughes's improved driving mechanism this defect is remedied by a pair of pedals swinging up and down which apply force to the driving wheel alternately; a cord passing over pulleys causes the depression of one pedal to make its fellow rise.

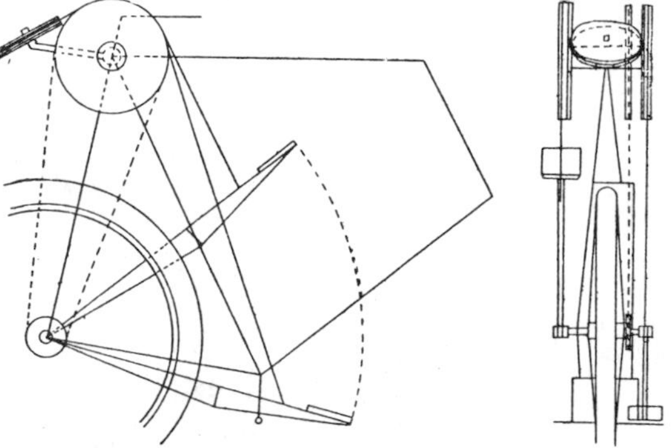

4759/1901
Robert Frederick Hughes *Inventor*
16 Westmorland Street, Marylebone, London, England

Cohr's Drinking-horn to be used with the Chalice prevents contagion whilst not changing the communion service as the sacramental wine is received directly in the mouth of the communicant, each communicant providing his own drinking-horn.

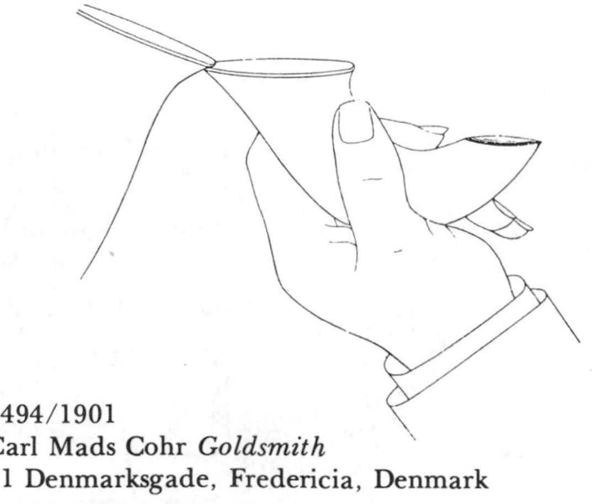

5494/1901
Carl Mads Cohr *Goldsmith*
11 Denmarksgade, Fredericia, Denmark

(B Reichhold)

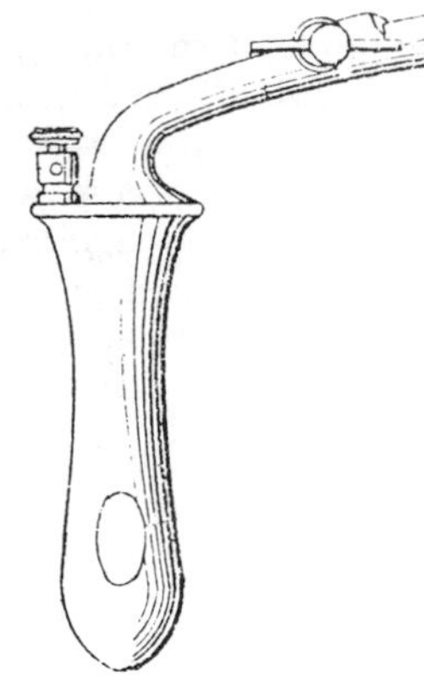

Mosher's Improved Electric Surgical Dilators provide means for the application of electric currents to the canals or discharge-passages of the body in such manner that the current will pass through and act upon only such parts as require electric treatment. A rectal dilator, for example, may have rubber or other insulating arms attached to pivoted metal arms with an expanding screw. Each arm has a metal electrode with a terminal to receive the battery connections.

The arrangement is such that contact is not made with the sphincter muscles. Dilators with more than two blades may have the extra blades placed in multiple connection, or not connected.

5220/1901
Charles Leonard Mosher *Physician*
Chatham, County of Columbia, New York State, USA

(Boult, Wade & Kilburn)

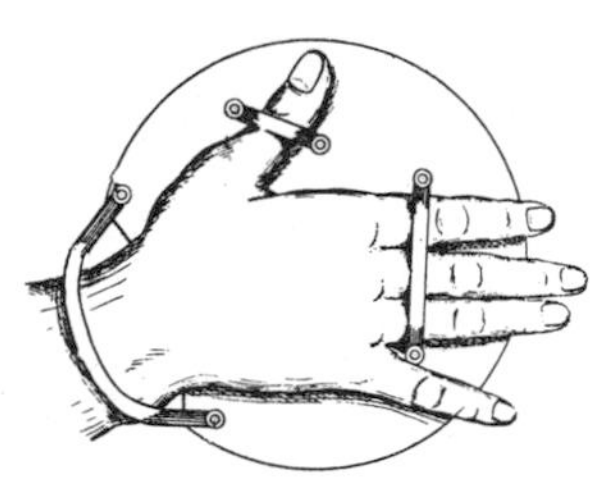

Coad's Improved Appliance for Aiding Swimming is a disc-like plate with straps to fasten it to the hand. It provides a means for affording considerable assistance to a swimmer to enable him to propel himself through the water with greater ease and at increased speed than heretofore, with no special skill being required.

6265/1901
James Coad *Marine Engineer*
Commercial Road, Hayle, Cornwall, England

(Hughes & Young)

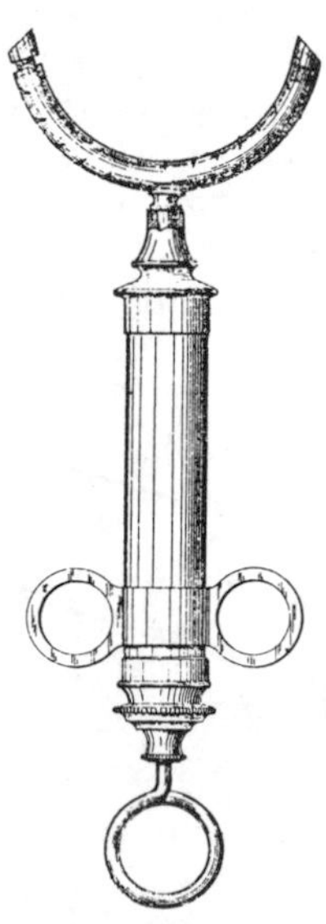

Schneider's Improved Appliance for Cleansing the Teeth dislodges particles of food remaining in the hollows and defective or carious parts of the dentition without the use of the customary tooth-brushes and tooth-picks which are injurious since they destroy the enamel. The device consists of a syringe fitted with a tube curved to correspond with the teeth, and furnished with holes through which mouth-wash, disinfecting liquid or the like is forced under pressure. By moving the piston back and forth the liquid can be used again and again until the teeth are clean. Different arrangements of holes and slots may be used, and the ends of the curved tube may be removed for cleaning.

7004/1901
Adam Schneider *Manufacturer*
101 Bernauer Strasse, Berlin, Germany

(Ernest de Pass)

Jones's Improved Course or Track for Cycling is a construction which is designed for use in cycle races or performances, on music hall and other stages and places of public amusement and in other places of limited area. It can be readily erected and removed as required and, being partly open or with interstices between the whole or a portion of the floor of the track so as to enable the spectators to see through it, is especially adapted for the above purposes.

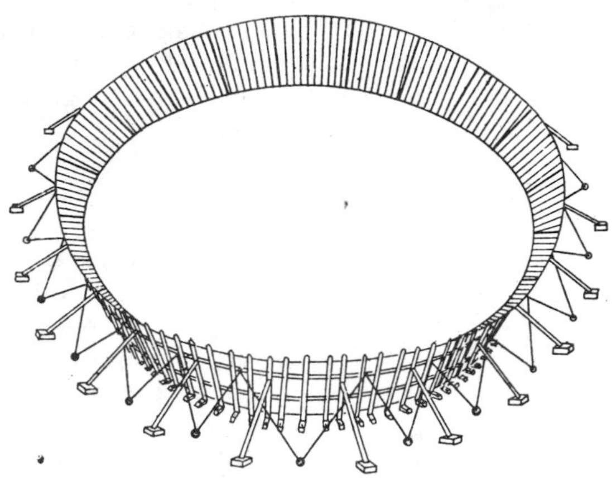

6126/1901
Charles Henry Jones *Professional Cyclist*
105 Truro Road, Wood Green, Middlesex, England

(J C Jackson)

Jopp's Improved Cycle Signalling Device enables a rider to both ring a bell and, at his will, to sound a whistle or horn. The arrangement is ingeniously devised so that on lightly pressing a friction wheel on the rotating wheel of the bicycle the whistle or trumpet comes into operation, while on pressing it hard a bell rings.

Notwithstanding the many advantages of a signal device of this kind, it is in no wise more complicated than an ordinary cycle signal apparatus capable of emitting only one signal.

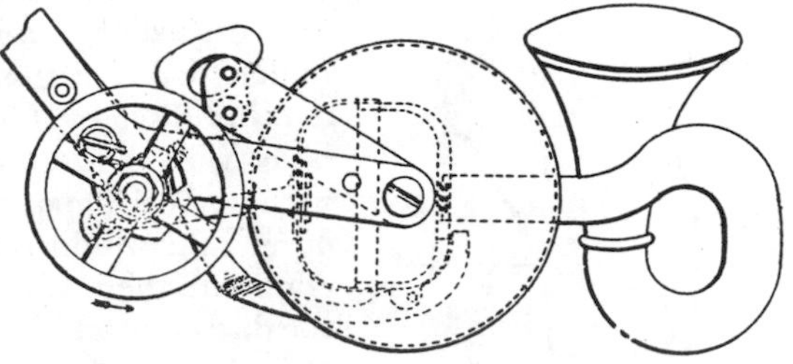

4697/1901
Andreas Jopp *Manufacturer*
Mehlis, Thuringen, Germany

(W P Thompson & Co)

1901

Pearce's Apparatus for Providing Shelters or Hiding Places for Fish is a kite-like board anchored so as to float near the bank and provide artificial cover for trout so that they can have some retreat when the weeds growing in the river are cut short. It obviates the defects of previous shelters which have been attached to the bottom of the stream and hence become inoperative through silting up.

1762/1901
William George Pearce *Baronet*
Cordell & Chilton Lodge, Hungerford, Berkshire, England

(Abel & Imray)

Scott's Apparatus for Treating Diseases by Antiseptic Vapours or Fluids provides an improved means whereby these can be applied under pressure and at a suitable temperature for the treatment of human or animal diseases.

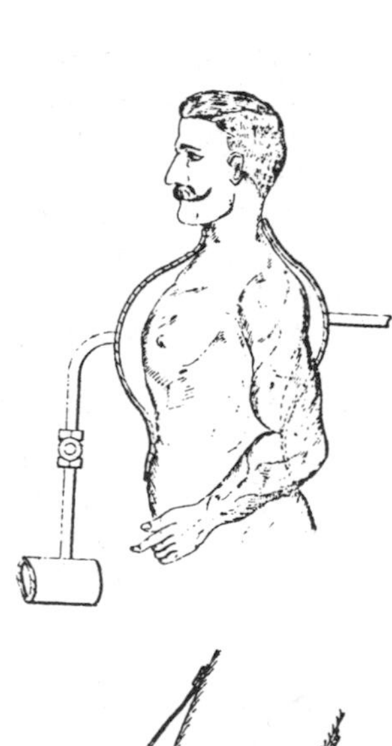

The system of employing antiseptic fluid for the destruction of micro-organisms may be explained by the operation of divers, who occasionally work in an air pressure of 80lbs to the square inch above atmospheric. This very clearly demonstrates that the atoms composing the atmosphere pass into every part and molecule of the body forming an equilibrium of pressure, for at this enormous pressure not the first cell of the brain nor the most delicate fibre or structure is broken or crushed. Of course, in treating persons under this system it is not necessary to employ such enormously high pressures as divers are subjected to, only a few pounds above atmospheric being sufficient pressure to permeate the body with inconceivable millions of antiseptic atoms.

In disinfecting a person it is absolutely essential in all cases to gradually raise the pressure and gradually release it. The body may be charged any number of times as desirable with new antiseptic atoms. While probably 75 per cent of the diseases from which man suffers are caused by micro-organisms, it will be understood that this system may also be applied to other diseases, for instance by permeating the body with a fluid which by combination with pathological deposits in the tissues of lime and other salts would render them soluble and thus eliminate them from the system, and with them the disease of which they were the cause.

In this apparatus the patient is enclosed in an airtight chamber. The air is drawn off, and the temperature is raised, and at the same time antiseptic vapour is admitted thereto. The pressure of the vapour is then raised above the atmospheric, to cause it to permeate the body, being forced through the pores of the patient's skin, or through the wounds or affected parts. The purifying vapour so reaches the internal organs and effectively acts at the seat of disease.

Air or oxygen is supplied with the vapour to permit respiration, or another appliance may be provided to permit the patient to breathe atmospheric air.

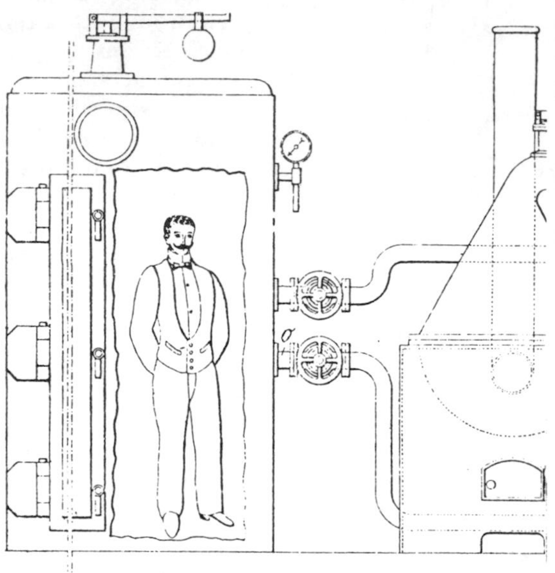

7056/1901
Robert Scott *Gentleman*
8 Graingerville North, Westgate Road, Newcastle-upon-Tyne, England

(A F Spooner)

Morin's Improved Nest for Hens provides a construction designed to discharge a freshly-laid egg from beneath a hen into a separate compartment, so that the egg can be removed without disturbing the occupant of the nest. It also prevents egg-eating hens from eating the eggs and provides a soft and yielding bottom to minimise the danger of breakage when the eggs are discharged.

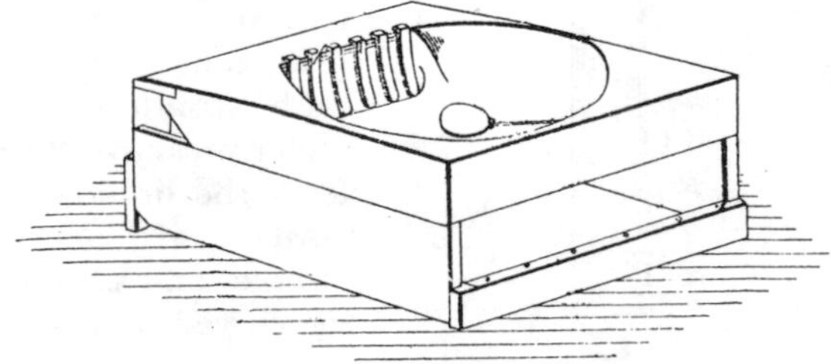

14761/1901
Louis Paul Morin *Contractor*
St Hyacinthe, Quebec, Canada

(Haseltine, Lake & Co)

Rigby's Improved Velocipede obviates the shortcomings of machines already in use, in that such machines require two or three wheels to support them, while Rigby's requires but one. The greatest innovation is in the balancing-bar which supports the saddle. The shape of this bar admits of considerable variation, and indeed is immaterial so long as it fulfills its purpose of enabling the rider to balance. The machine is provided with small wheels which prevent the rider from falling either backwards or forwards; two wheels may be placed a short distance apart on these axles so that the machine will stand alone when not in motion.

The action of the machine is, roughly speaking, as follows – when the rider mounts (the back small wheel being on the ground) and begins to pedal, the action of pedalling and the weight of the rider lifts the small wheels off the ground.

By modifications in the position of the pedals and shape of the saddle supports, the machine may be constructed for the use of ladies.

8038/1901
Marshall Rigby *Commercial Clerk*
The Mill House, Stanthorne, Middlewich, Cheshire, England

(John G Wilson & Co)

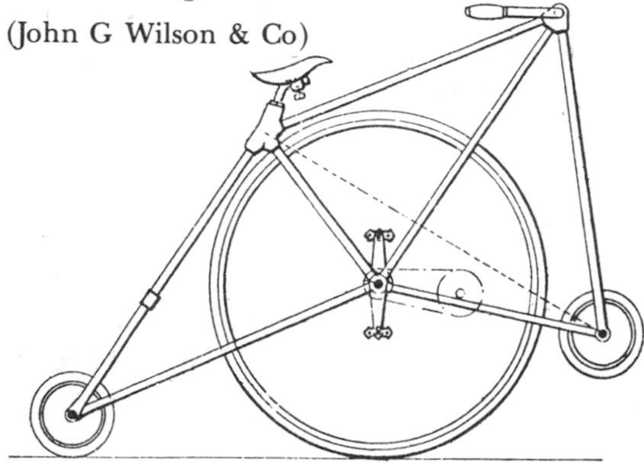

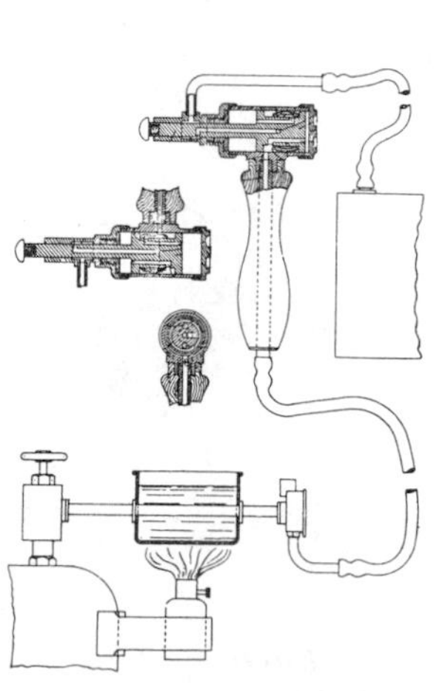

Sartori's Device for Mechanical Skin Treatment or Massage is a self-acting appliance designed to treat any part of the body of the patient by a far more rapid and efficient kneading action than has been possible before. The kneading action is produced by a knob or mallet reciprocated by a compressed-gas engine. The knob may act either directly on the skin of the body or via an elastic plate which is thereby rapidly vibrated and the shocks distributed so as to intensify the kneading, pounding action of the usual treatment by hand.

8726/1901
Graziano Sartori *Chemist*
22 Kreuzberg-Strasse, Berlin, Germany

(Wheatley & Mackenzie)

Böhm's Improved Umbrella provides a relatively increased interior head-space, is capable of relatively closer rolling in closed condition and is generally superior in point of utility and general efficiency.

This is achieved by providing the umbrella with brace-rods above the ribs and an umbrella-stick in two parts, the upper part carrying the ribs and runners of the brace-rods. The lower part of the stick embodies the handle and supplemental brace-rods and the whole may be rolled very closely because practically no part of the stick projects among the folded ribs.

8634/1901
Maria Apollonia Böhm *Spinster*
224 East 126th Street, New York, USA

(Hughes & Young)

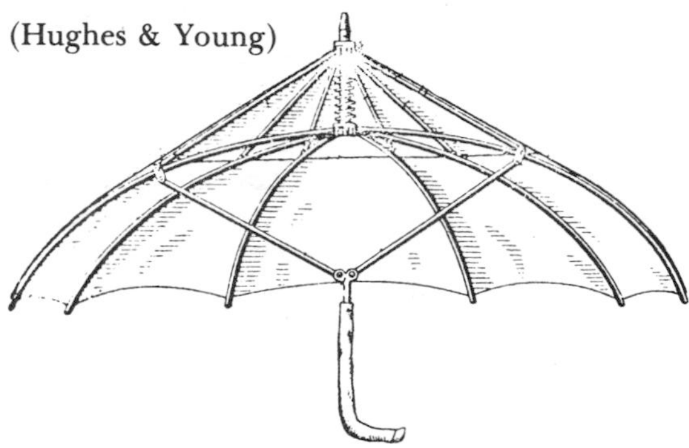

Cotman and Walker's Toy or Device for Word Building and the like is designed to be both instructive and a source of amusement for young children. It consists of prismatic blocks or cubes, preferably of wood or confectionery, prepared with holes to allow of them being threaded on spindles, with any desired number of blocks on each spindle and with letters or figures on their several faces.

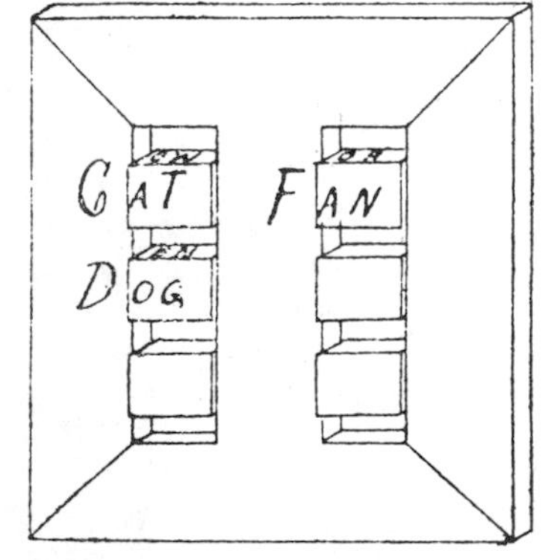

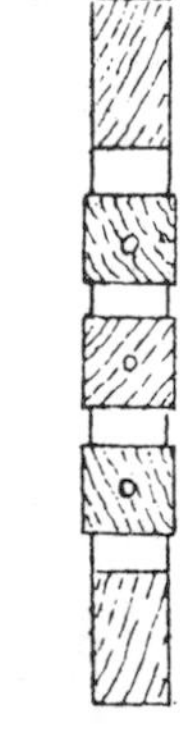

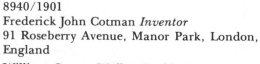

8940/1901
Frederick John Cotman *Inventor*
91 Roseberry Avenue, Manor Park, London, England

William George Walker *Builder*
20 Wordsworth Avenue, East Ham, London, England

1901

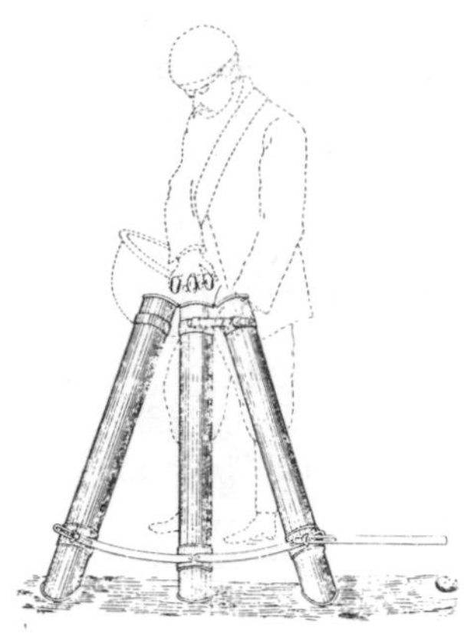

The Jervises' Improved Means of Planting Potatoes and Seeds provides a simple and efficient apparatus for planting these at regular intervals and in more than one row simultaneously. The apparatus may be designed for hand use or arranged in connection with a vehicle for use with horse power; nonetheless, however it is used a saving in labour and time is effected.

In using the hand apparatus, the labourer rests the tubes in the trench and feeds the potatoes or the seed one into each upper end thereof from a suitable receptacle secured to his person. He then moves the tubes to the next position, guided by a gauge attached to the tubes, and feeds again, and so on.

8975/1901
The Rt Hon Carnegie Parker Jervis, Viscount St Vincent and
Hon Cecil Leonard Jervis
Norton Disney, Newark-on-Trent, Nottinghamshire, England
Hon Ronald Clarges Jervis
Sutton-on-Derwent, Yorkshire, England

(George Downing & Sons)

Taaffe and Backhurst's Improved Hansom Cab is provided with rest bars or supports so that, should the horse stumble or fall, or the shafts break, a wheel mounted in a fork carried by a prop in the centre of the vehicle falls to the ground to keep the vehicle upright.

9109/1901
Albert George Taaffe *Artist*
98 Pevensey Road, Eastbourne, Sussex, England

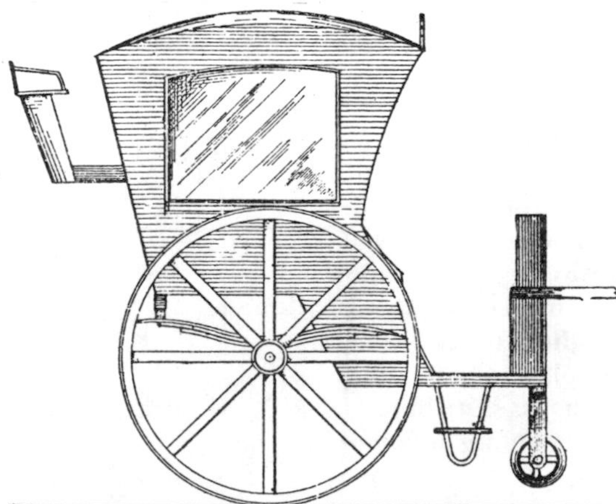

Thomas Bertram Backhurst *Joiner*
26 Rylstone Road, Eastbourne, Sussex, England

(Hughes & Young)

Doré and Evanovitch's Improvement to Motor Vehicles is designed to provide means whereby the retardation of velocity of such vehicles due to the resistance of the air is diminished, and the liability of injury to pedestrians in crowded thoroughfares greatly reduced.

A guard or shield of V or double crescent shape in plan view is secured to the front of the vehicle. The shield can be raised and lowered by suitable gearing. It may be produced upwards so that a glazed frame protects the occupants from wind and dust. This raised portion, when the wind is favourable, can be employed as a "sail" and can be made movable so that it can be set to catch the wind. The guard or shield can be formed artistically so as to materially enhance the appearance of the vehicle to which it is fitted.

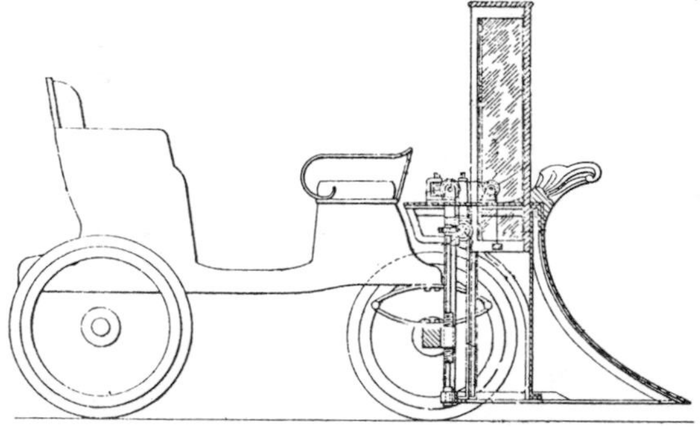

9373/1901
Edwin Stephen Doré
80 King William Street, London, England

Gusser Evanovitch
29 St James Square, Notting Hill, London, England

(G F Redfern & Co)

Cowland's Improved Fish Fork has a pivoted, spring-loaded prong, which may be moved by means of a rod, thus enabling the user to remove bones from the fish.

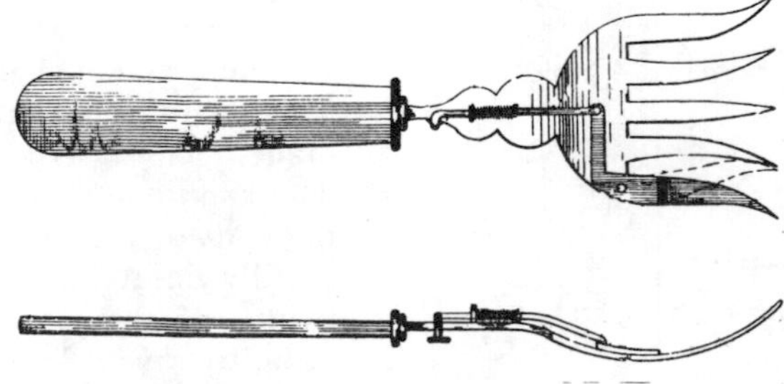

10089/1901
Henry Culver Cowland *Stockbroker's Clerk*
3 Victoria Villas, Maybank Road, South Woodford, Essex, England

(Hughes & Young)

Braatz's Improved Property Quadruped for Stage Performances is constructed so that the front legs of the animal ape those of the person performing, who has the appearance of a rider, and the hind legs are moved mechanically by a suitable arrangement of levers and cords.

In the example shown, a donkey, the front legs are connected cross-wise with the hind legs to get the proper method of walking of an ass, but in the case of a camel they would be connected up in parallel as will be readily understood. A pistol may be supported by a tube of the frame within reach of the performer.

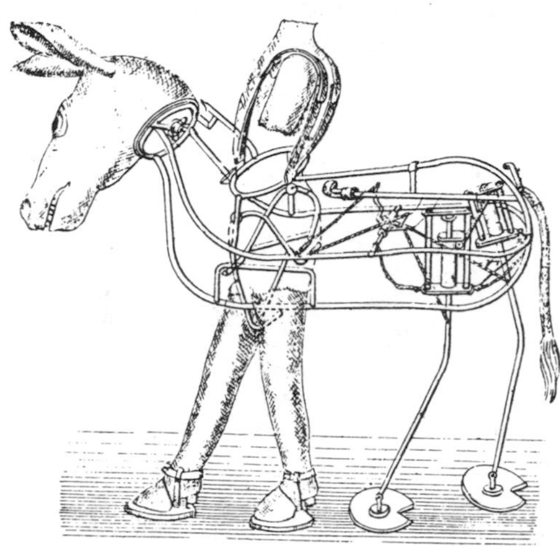

11705/1901
Alexander Braatz *Music Hall Performer*
Schützenhaus, Oranienburg, nr Berlin, Germany

(Marks & Clerk)

Tredinnick's Improved Self-driving Hydraulic Motor consists of a water cistern or reservoir to the bottom of which is connected one or more pipes which pass downwards and then are bent back and pass over or through the walls of the cistern to empty themselves back into the cistern, which empties itself into the pipes as fast as the water is returned into them. The flow of water thus occasioned may be employed for driving machinery. The cistern may be provided with a cover for travelling, as the machine may be made of any size and material that will contain water.

This invention resulted from my discovery of the way by which water can be forced to raise itself above its own level without the aid of any other power behind it. The knowledge that upward flowing rivers are to be found under the surface of the earth convinced me that it was possible for water to raise itself above its own level up here on the surface as well as underground if only we could find out the conditions under which it flowed from the subterranean reservoir. I concluded a fall was necessary on its commencing its journey in order to ensure the moving power to send it up any channel and my experiments proved that conclusion to be correct. The self-driving hydraulic motors raise water above its own level without the aid of any other motive power, are extremely simple in form, and supply perpetually running water. The extreme cheapness of this new way of managing water recommends it to all; the most wonderful results can be obtained at nominal cost.

When lowered into a well or mine the apparatus may be employed to raise water therefrom. By fixing the apparatus at one end of a canal and carrying the pipe to the other end of the canal the water can be turned into a running stream. By fixing the apparatus at both ends of the canal the direction of the stream can be changed at will.

I fancy I have found out the way in which the "ancients" managed water as a motive power to aid them in carrying out their big undertakings, and they possibly made their motors or apparatus of bricks and mortar or blocks of stone hewn into shape and fitted together.

11318/1901
Emma Harrison Tredinnick *Teacher of Music and Drawing*
34 Cambridge Grove, Ilfracombe, North Devon, England

Hazard's Improved Trouser Suspender will hold the trousers well up in front and not unduly elevate them at the rear. It will become loose and slack when the wearer sits down, stoops or bends over and thereby avoid strain on the trousers while the wearer is in these positions, prevent wear on the seat, bagging at the knees and pulling off the buttons and will at all times give the wearer ease and comfort.

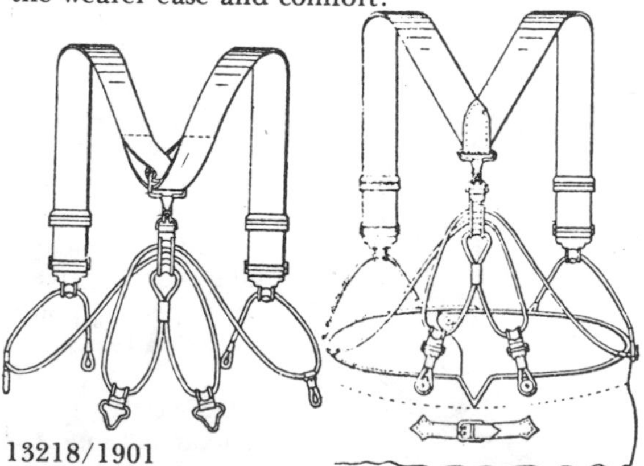

13218/1901
Henry Thomas Hazard *Lawyer*
Los Angeles, California, USA

(W P Thompson & Co)

1901

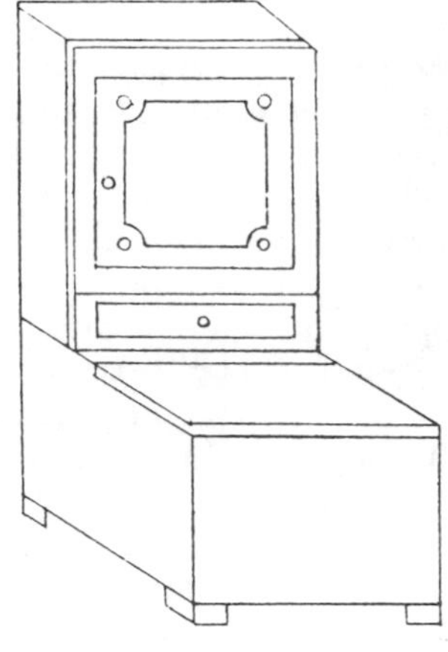

Luther's Improved Cupboard for Kitchen and House is equally well fitted for both the simplest and most elegant household. It can be used for different purposes; it is destined for coals and fuel in general, but can also be used for storing books, bottles or any other objects if placed in the sitting room. The bottom case rests on rollers and is of large size, and its lid can be opened in any direction required; in it a large quantity of coal may be stored. The upper case has a drawer in which utensils for lighting the fire — such as matches and fire-lighters — and utensils for cleaning the fireplace may be kept. The cupboard may also be used for storing briquettes or compressed fuel and when the lid is shut it can be used as a chair.

10597/1901
Martin Luther *Master-Carpenter*
11 Tilsiterstrasse, Berlin, Germany

(F G Harrington)

Lüngen's Improved Water Spraying Device is a perforated ring furnished with handles for easy manipulation. It is attached to a flexible hose connection and easily passed over the whole body in order that the water jets may act uniformly and in all directions on the same.

10674/1901
Wilhelm Lüngen *Merchant*
Köln-Deutz, Germany

(Boult, Wade & Kilburn)

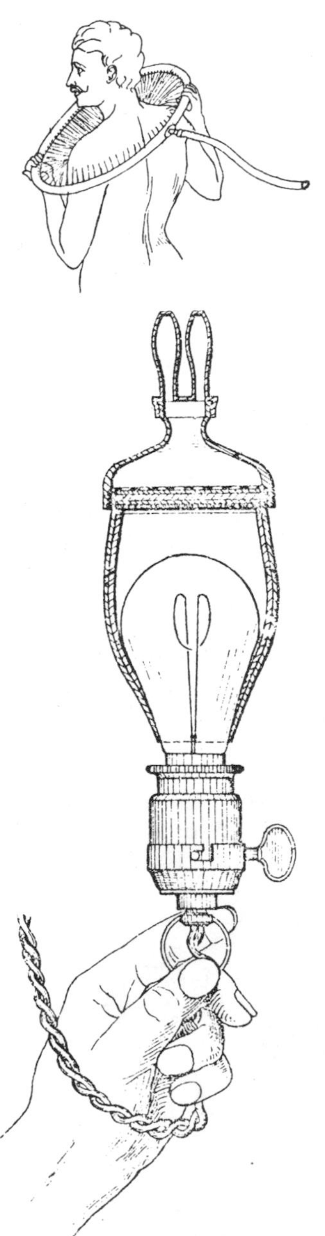

Kautz's Improved Inhaler differs from other apparatus hitherto known in respect of there being no disagreeable development of gases or smell of spirit and no danger of burning or explosion, while at the same time a steady and uniform development of heat is produced. It permits of the inhalation of solid vapourising substances which have recently come extensively into use in medicine, such as benzoic acid, menthol, camphor, sal ammoniac and the like. The appliance comprises a casing carrying a perforated glass plate on which is placed wool to receive the substance to be inhaled, adapted to fit over an electric lamp or other form of heater. A cap screwed on to the top of the casing is adapted to receive a tube or tubes through which the vapours are inhaled by the mouth or nostrils of the patient.

10327/1901
Theodor Kautz *Apothecary*
212 Bahnhofstrasse, Bad Reichenhall, Germany

(Cheeseborough & Royston)

Tyrell's Improved Revolving Automatic Self-adjusting Targets for Shooting Galleries have two exceedingly realistic figures normally maintained by gravity in a horizontal position, which are permitted to revolve about an oblique axis when either of them is struck, thereby giving the effect of their chasing one another around a pole.

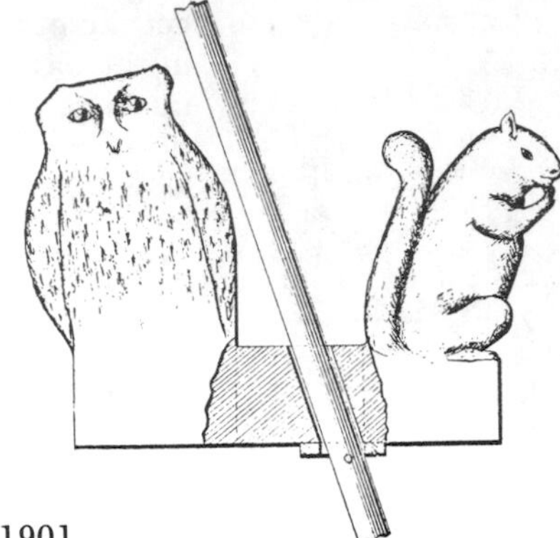

11416/1901
George Tyrell *Gentleman*
1012 Main Street, Fort Worth, Texas, USA

(J P Bayly)

Valley's Improved Bicycle employs an arrangement whereby the steering and guiding capability of the machine is increased. The axles of both the wheels are movable in a horizontal plane and are compulsorily connected with one another in such manner as to be always simultaneously operated; the method of coupling can be such that the front and rear wheels are differentlt deflected.

This arrangement enables the cyclist to move in a circular course of much smaller diameter with the same diversion or movement of the handle bar, and facilitates steering without using the hands which is of very great importance particularly for military purposes.

11693/1901
Gustaf Valley *Engineer*
65 Gustaf Adolfstorg, Malmö, Sweden

(W P Thompson & Co)

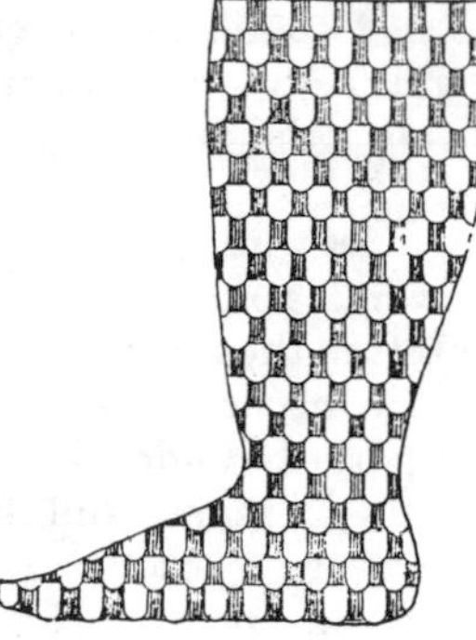

Classen's Improved Swimming Socks may be drawn on to the foot and leg in the manner of an ordinary stocking, and are furnished with a plurality of webs or flies which open in the backward movement of the foot and close in the forward movement thereof.

12216/1901
Carl Classen *Inventor*
336 East 19th Street, New York, USA

(Wheatley & Mackenzie)

Schlüter's New and Improved means for Packing Tea, Coffee, Spice and other Tropical Products is especially destined for serving in the export of the grocery goods above enumerated and enables the consumer or purchaser to obtain colonial goods originally packed in the country of origin without the contents having been interfered with. The goods are enclosed in a bag or metal case which is placed in a box made of bamboo cane or the like, fitted with a cork or plug which is then sealed.

The disused boxes may be converted into various articles; by cutting one diagonally, and fitting handles, a pair of grocers' scoops may be made; by cutting longitudinally, stud boxes, pen rests etc may be produced; slots cut in half the case enable it to be used as a letter rack.

12162/1901
August Hermann Schlüter *Merchant*
6 Gellert Strasse, Hannover, Germany

(A Archd Sharp)

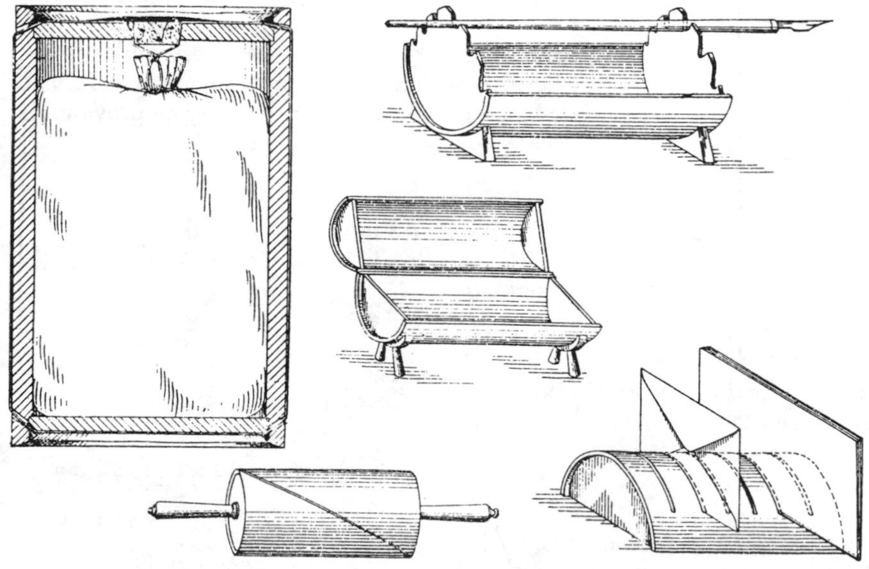

Drury's Improved Clothing for Infants, Children and Adults facilitates dressing and undressing by eliminating many unnecessary fastenings and is also suitable for use in hot countries since it is not double at the front.

Infants' long clothing for indoor wear, the petticoat and day gown, are made as one combined article as shown, so that the infant can be dressed or undressed while in a recumbent position, the time spent being only a few minutes and no pins or other dangerous fastenings being employed, so that it is not subjected to the discomfort, if not positive suffering, necessary with the common or conventional style of garment.

For enabling young boys or girls to dress and undress themselves easily, the garments of a set, frocks, pinafores, petticoats, drawers, scarves and helmets, are constructed to open some to the right and some to the left of the front and provided with overlaps, which enable fewer fastenings to suffice. These garments are styled

"dress myself" clothes, because young children are able to do without any assistance in fastening their articles, which is not the case with present day garments which fasten behind.

A day set of clothing for female invalids may be constructed on similar principles, and includes a vest which may also act as a cholera protection belt, bodice stays with lateral openings at the front to facilitate breathing, and a tea gown which may be used over a night gown.

These patent fasteners may also be employed in making a shirt suitable for climbing, walking, or driving in motor cars as it overlaps at the front and will only come apart with a strong pull. A man's set comprising a shirt and smoking jacket may also be made in such a manner as to facilitate dressing and undressing as previously described with reference to women's and infants' garments.

10119/1901
Agnes Louisa Drury *Authoress, and Widow of the late Charles Garling Drury*
Thames View, Broom Water, Teddington, Middlesex, England

(W Lloyd Wise)

1901

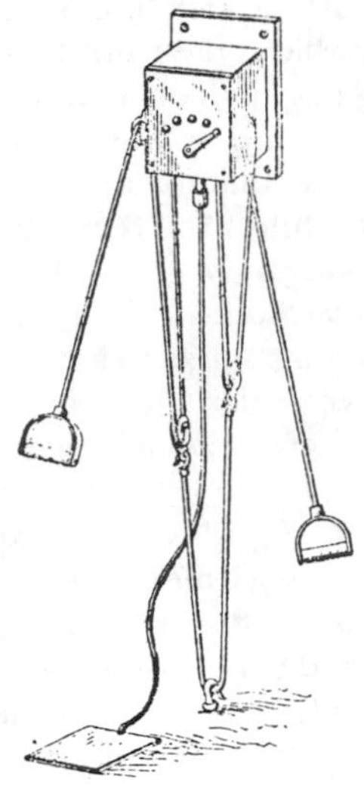

Courtney's Improved Exercising Machine is a pulling apparatus arranged to administer a practically continuous current of electricity to the user. The action of pulling the handles (which perform the alternative office of conducting electricity to the hands) generates electricity during both forward and backward movements. The user may also stand on a plate to receive current through his legs. Starwheels acting as circuit breakers serve to provide a simple means for increasing or diminishing the strength of the current passed through the operator's body.

9610/1901
Albert Williamson Courtney *Manufacturer of Medical & Electrical Specialities*
147 Niagara Street, Buffalo, New York, USA

(Haseltine, Lake & Co)

Tulloch and Handcock's Improved Cricket Ball enables the bowler to attain any desired variety of bowling effects, one of the said effects being what is termed by cricketers as "breaking" or "twisting". For this purpose, the ball is made unsymmetrical, or loaded eccentrically with lead, mercury or other fixed or movable substance, or with ridged or grooved surfaces, or with projections or depressions. Such a ball is especially applicable for batting practice and will be found especially useful for enabling the batsman to gain experience with different kinds of bowling.

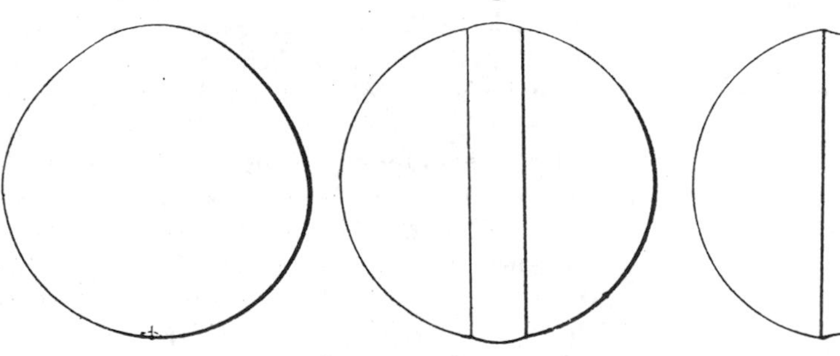

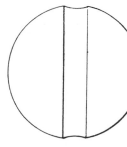

16361/1901
Major Hector Tulloch CB
The Hollies, Bickley, Kent, England

Henry Willam Handcock *Electrical Engineer*
5 Albion Road, Sutton, Surrey, England

(Haseltine, Lake & Co)

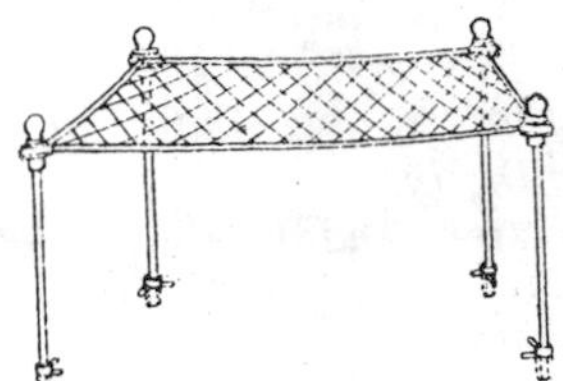

Johnston's Improved Bed meets the problem that, in all houses, beds occupy when not in use, not only a ridiculous amount of space but prevent to a great extent the rooms being used, and in small houses this loss of space leads to considerable inconvenience. It would therefore be a great benefit if the bed could be entirely removed in the daytime and the room converted into a sitting room, and this is the object of the present invention.

This new bed comprises a net with an eyelet at each corner which slips over a post. Four such posts are plugged into sockets in the floor. When the position the bed has to occupy has been selected, the carpet is temporarily folded back, the site of the sockets carefully measured and the flooring cut out to receive the sockets. The carpet is then replaced and slits cut in it above the socket holes. When it is desired to erect the bed, the carpet at these holes is lifted by the point of a finger and the tubes are dropped into the sockets and the hammock or net placed over the tubes. When not in use the net containing the bedding can be rolled up and put in a cupboard, leaving the floor entirely clear.

17049/1901
George Bogie *Commercial Traveller*
22 St Michael's Road, Bowes Park, Middlesex, England *for* Thomas Ruddiman Johnston
18 Hiyoshicho Kiobashi Ku, Tokio, Japan

Jones's Improved Apparatus for Cycling, Athletic and other analogous Performances is designed for use in a circus or on a music-hall stage or other limited area and so constructed that it can be readily erected and removed as required and any performance taking place inside it can be readily seen by persons outside it. The apparatus consists of a hollow globe, made in segments of metal work or other suitable material, which can be bolted together to form a rigid structure. It may turn on pivots, or be held stationary. The track or position for the performers may be on the inside or outside, and when on the inside doors etc may be provided to allow the performers to enter.

17328/1901
Florence Ann Laurence Jones *Married Lady*
105 Truro Road, Wood Green, Middlesex, England

(F W Golny)

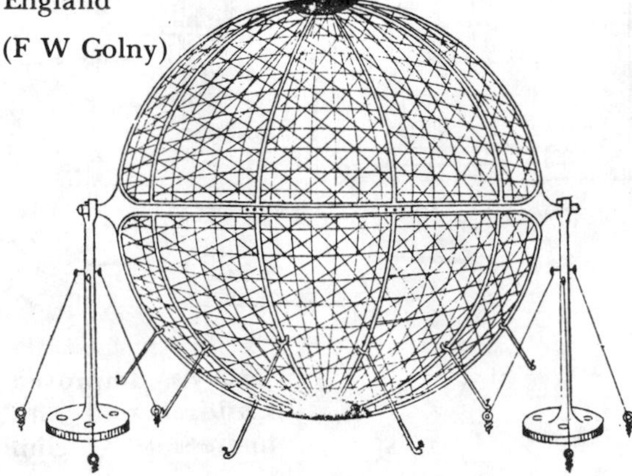

1901

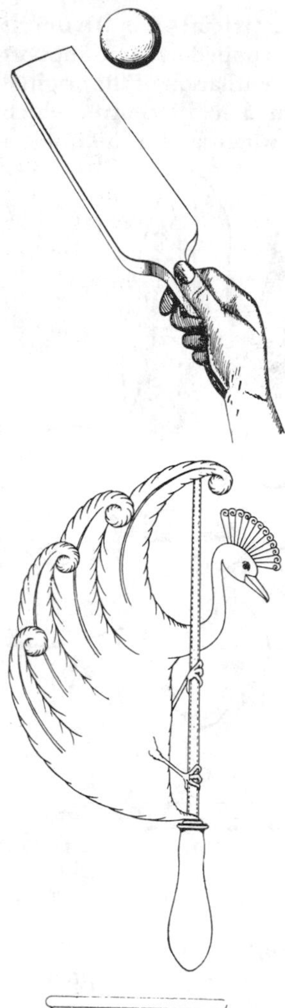

Wythe's Improved Detonating Toy is designed to produce a series of sharp detonations without replenishment of the detonatable material. Balls, staves, bats etc are coated with materials which explode locally so as to produce a detonating noise when two of them are struck together. One may be coated with an oxidising agent such as chlorate of potash, the other with amorphous phosphorus or other oxidisable material.

The area of action being small, detonation may be repeatedly produced with the same articles.

15452/1901
William H Wythe *Inventor*
Bloomfield, New Jersey, USA

Rizzi's Coronation Revolving Fan may be made of paper, linen, feathers, wire etc in any ornamental design and mounted on a tube which rotates on a handle. The fan is rotated by actuating the handle thereby giving a pleasant and continuous motion of air. It is particularly pointed out that it is on the revolving action of the fan caused by the hand that the protection of the patent law is specially required.

14052/1901
Ferrante Rizzi *Professor; Skin Specialist*
4 Onslow Place, London SW, England

Kerr's Improved Means of Exhibiting Advertisements causes the movement imparted to a toilet roll as a leaf is withdrawn for use to actuate means of bringing a fresh advertisement into view.

12167/1901
Allen Coulter Kerr *His Britannic Majesty's Consul at Santiago, Chili*
Woodside, Upper Norwood, Surrey, England

(Haseltine, Lake & Co)

Jennings's Improvement in Illuminated Designs for Advertising and other like Purposes is applicable for the purpose of decorating buildings at the time of public rejoicing, and for illuminated shop fascias, notice-boards, tablets and so forth. It consists of tubes, containing rarefied air or other gases, and shaped to represent a letter or letters or designs, which are rendered luminous by being connected with an inductorium, Ruhmkorff coil, or other suitable electrical apparatus. By using different kinds of glass, and by enclosing the tubes in others containing, say, sulphate of quinine, various striking effects may be obtained.

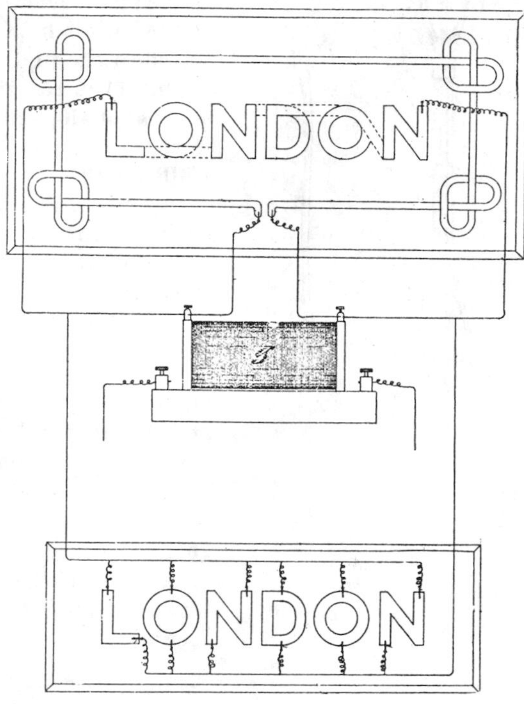

9993/1901
William Oscar Jennings *MRCS England, MD Paris*
74 Avenue Marceau, Paris, France

(W H Beck)

Vasseur and Darby's Improved Aërial Machine is propelled by electricity supplied by a generator to a track along which the cables restraining the machine run.

The aërostat is designed for the transport of passengers, packages, articles of merchandise or any other kind of freight from one place to another.

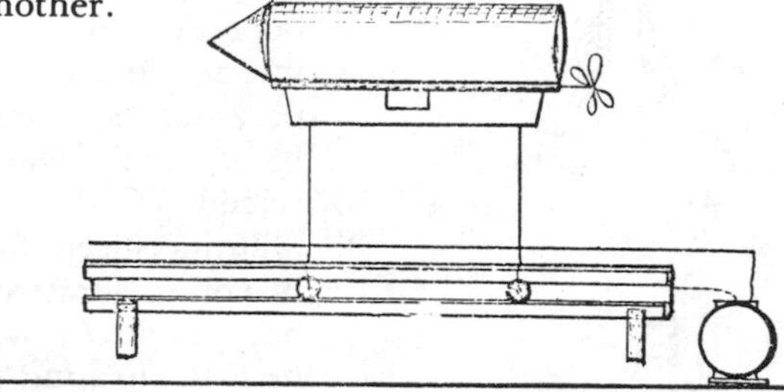

14311/1901
Manuel Vasseur *Agent*
4 Place du Théatre Français, Paris, France

William Evans Darby *Doctor of Divinity*
47 New Bond Street, London EC, England

(W P Thompson & Co)

1901

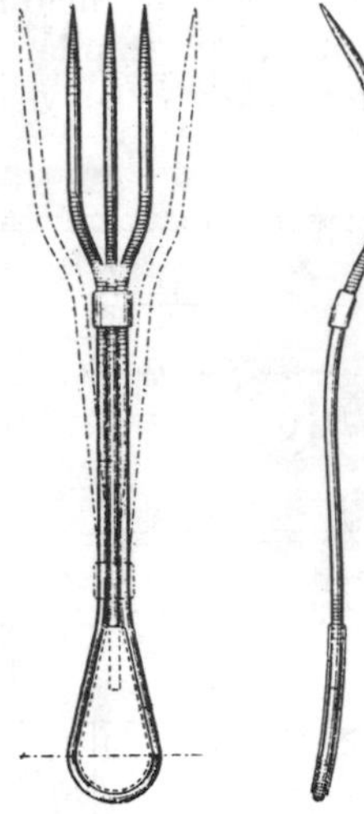

Kowalski and Borowski's Improved Hygienic Table Fork for domestic purposes is so constructed that a very great resisting power is given to it, while at the same time it is very easy to cleanse. The tines are made of separate pointed wires held by a clip. Said tines can be parted by sliding the clip, thus allowing the fork to be cleaned and greatly facilitating the removal of all particles of food adhering thereto.

17401/1901
Felix Ladislas Kowalski
Michel Borowski
Lodz, Poland, in the Russian Empire

(George Downing & Son)

Sturmey's Variable Speed Gears for Bicycles and other Machinery provides a three-speed epicyclic gear completely enclosed in the rear hub of the cycle.

16221/1901
John James Henry Sturmey *Journalist*
Middleborough Road, Coventry, Warwickshire, England

(Douglas Leechman)

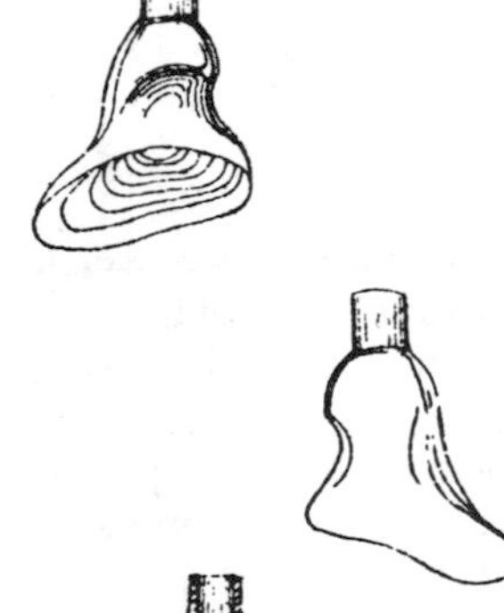

Payn's Improved Magnetic Device to Restore and Aid the Hearing provides a simple and effective device which may be inserted into the ear and retained therein by its shape, and which operates to vitalise or remedially affect the tissues of the elementary organs of the ear for the purpose of improving the hearing or even restoring it in cases where no structural injury has occurred, and for relieving and curing painful or annoying affections of the ear whether the hearing has been affected or not.

It consists of a pair of hollow permanent magnets. One is inserted in each ear, with opposite poles towards each other. When inserted in the ears, these magnets act upon the paramagnetic particles or constituents of the blood and tissue and vitalise or otherwise remedially affect them.

It has been found that by the use of the invention curative effects have in no great time been secured, and, because the devices are hollow, hearing takes place in the natural manner without the intervention of artificial aids.

16423/1901
Samuel Giles Payn Jr *Magnetic Expert*
611 Broadway, Albany, New York, USA

(Wheatley & Mackenzie)

Vickers's Improved Artificial Ear Drum is made cup-shaped with crimped edges to support it in position and allow ventilation of the original drum. A cord is attached to its centre, which passes through a tube, which acts as an inserter.

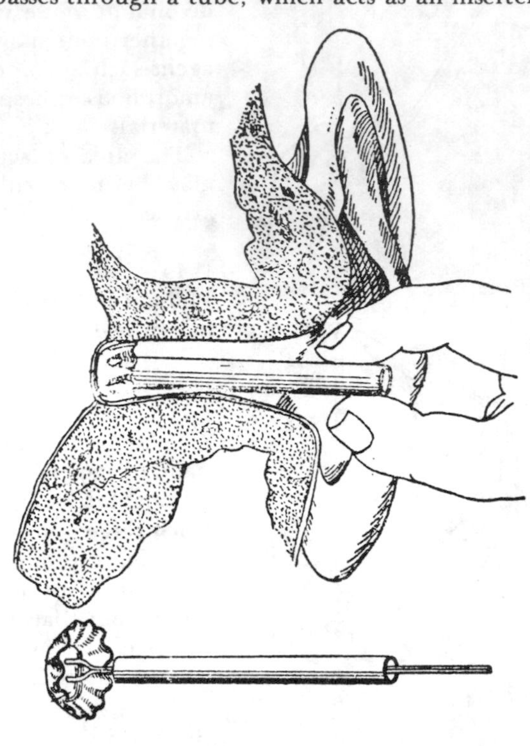

15809/1901
Laura Helena Vickers *Specialist*
1511 North 55th Street, West Philadelphia, Pennsylvania, USA

Goddon's Improved Artifical Ear Drum and Inserter is a disc made of thin woven silk or cotton, membraneous tissue etc and coated with an antiseptic composition of wax and oil. A silken cord is attached to its centre and a flexible rubber tube is provided for insertion. This is removed when the disc is in position, leaving the silken cord exposed to the exterior and able to convey the sound waves to the artificial drum. By the latter's intensified vibration the sounds are conveyed in increased volume across the natural drum to the auditory nerves.

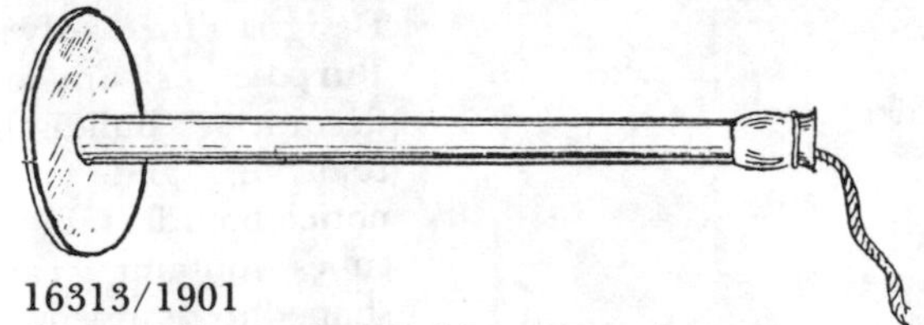

16313/1901
George Alexander Goddon *Manufacturing Chemist*
19 Beulah Road, Thornton Heath, Surrey, England

Meadows's Improved Shelter for the Watchman or Officers on the Bridge of a Steamship consists in a movable platform on a swivel attached to the bridge on which the officer in charge stands. A covered scoop is attached to the front of this platform and can be made to face the wind by turning a handle. The wind strikes the scoop and is changed in direction upwards, and the blast so directed upwards is carried over the man's head whilst it acts as an effectual barrier against any wind coming in a direct line with his face.

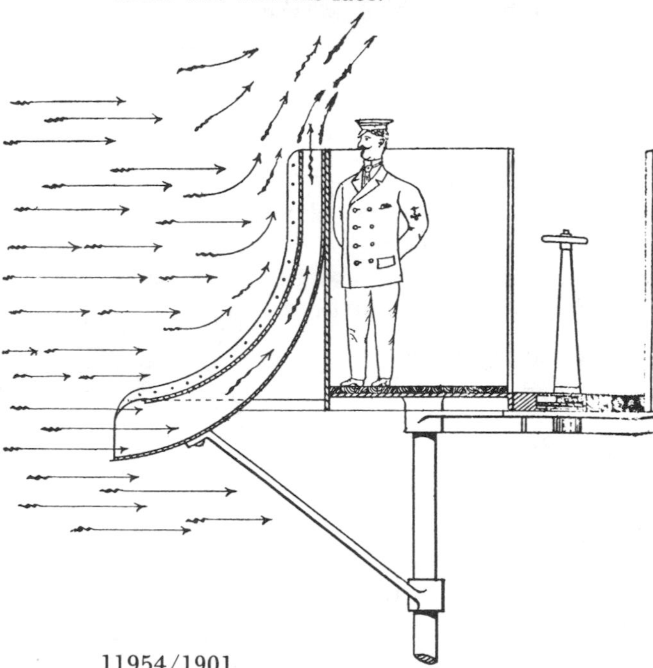

11954/1901
Henry Lloyd Meadows *Clerk of the Crown and Peace for the County of Wexford*
Ballyrane, County Wexford, Ireland

(R Core Gardner)

Batter's Portable Foot and Body Warmer is a device designed to be carried or worn by individuals under conditions where the natural heat of the blood is not sufficient or where there is intense exterior cold against which the person should be guarded. It consists in the employment of pads designed to be worn beneath the feet or hands and connected to a heating system. Tubes filled with spirit or brine heated by a spirit lamp in a case strapped on to the user's back are suggested. The flow of the heated liquid is governed by clockwork, which is in turn preferably governed by an air-compression cylinder.

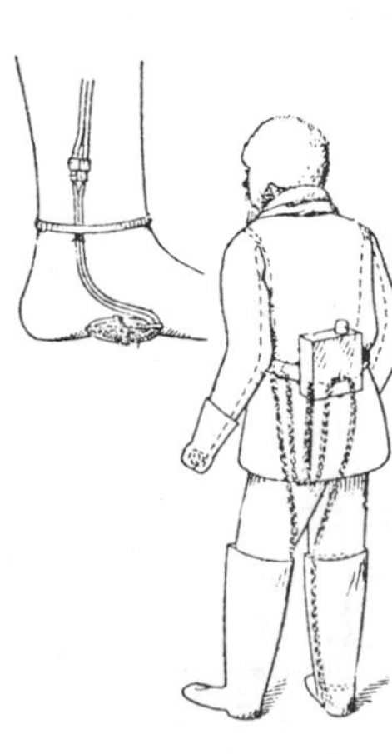

17767/1901
Frank Batter *Engineer*
Tillamook, Oregon, USA

(Gedge & Feeny)

Middleton's Improved Appliance for use in Cleaning Ships' Hulls while they are Afloat comprises a tube of flexible material, at the bottom of which is a kind of diving dress, designed to receive a workman and provided with a glazed window, sleeves, external tool pockets and drawstrings for adapting its length to the size of the operator.

It is lowered into the water and moved around the vessel by cables, so that the workman within can remove barnacles and the like without the trouble and expense of docking the vessel.

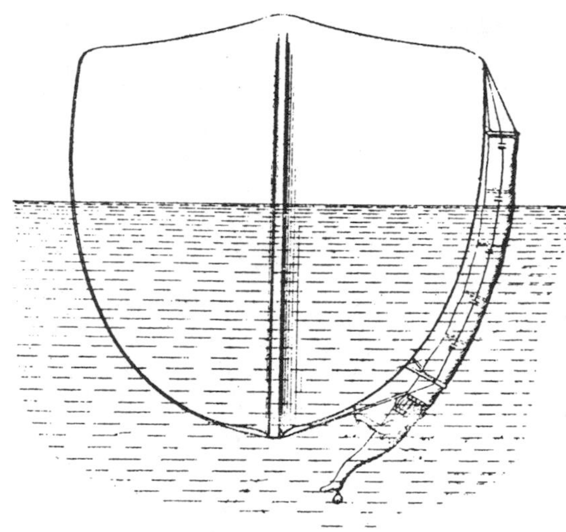

17170/1901
James Calvin Middleton
East Palaka, Florida, USA

(A M & WM Clark)

Fraser's Improved Head-dress Fastener provides a means for securing a lady's hat upon the head by employing two combs fitted inside the hat, provided with springs and manipulated from the outside by means of cords.

17459/1901
William Paton Fraser *Shopkeeper*
219 Easter Road, Edinburgh, Midlothian, Scotland

(Johnsons)

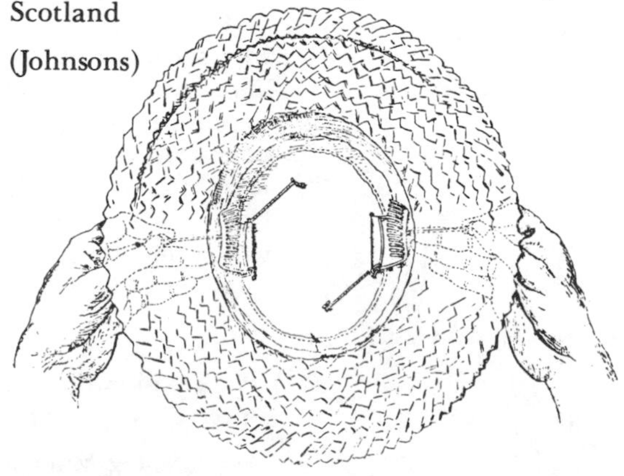

1901

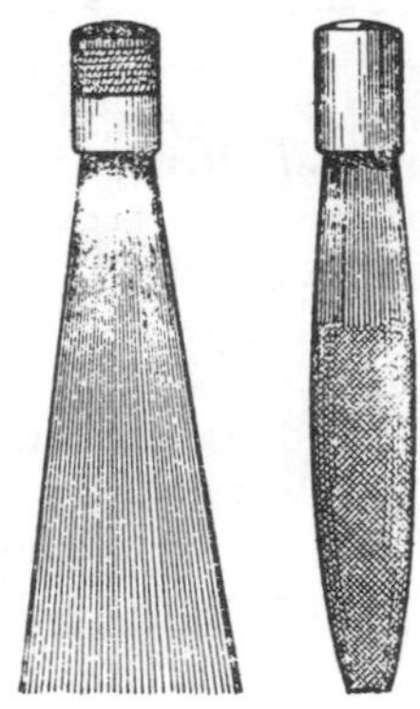

Hesseling's Improved Shaving Brush is of such a quality that, having been used once, it shall then be thrown away to prevent any chance of communicating diseases when barbers use the same brush for different customers. It has the further advantage of being already furnished with the necessary soap, so only a shaving pot is additionally required.

14409/1901
Bernhard Hesseling *Merchant*
33 Corneliusstrasse, Krefeld, Prussia

(Herbert Haddan & Co)

Pease's Improved Receptacle for Holding Coal is lined with rubber or other yielding material to deaden the sound of coal in sick-rooms, and prevents the creation of any noise sufficient to awaken a sleeping person or seriously disturb an invalid.

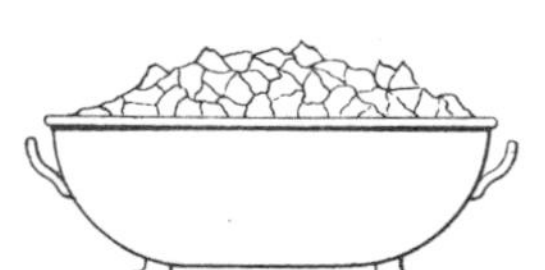

16801/1901
Anna Maria Pease *Professional Nurse*
8 New Cavendish Street, Portland Place, London W, England

(Cassell & Co)

Bernsten's Improved Mouse Trap provides a simple and convenient construction which enables a number of mice to be caught in the same trap.

When the animal jumps on to the pivoted, counterbalanced flap to obtain the bait, its impulsion precipitates it into the receptacle and the flap returns to its original position. To empty the trap a small sliding door is provided at the back, the door being raised when the trap is to be cleared.

16383/1901
William Bernsten *Labourer*
Johnsons Harbour, Falkland Islands, SA

(Geo H Rayner)

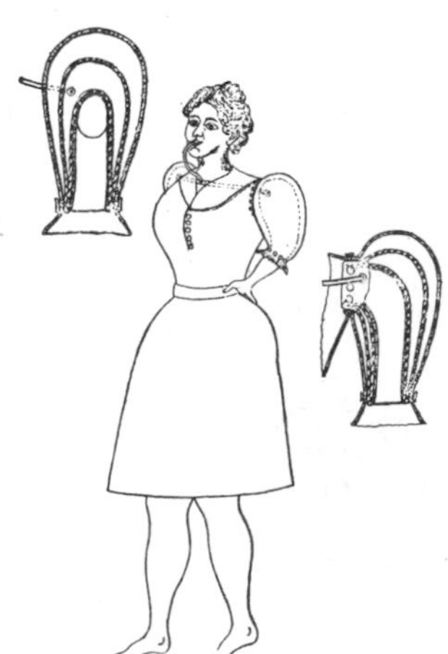

Maccolini's Improved Bathing Dress combines an ordinary bathing costume with a life-saving apparatus which is simple in construction and operation and may also be used as an aid in learning to swim. Each sleeve is provided with a hollow receptacle, secured at the shoulders, and which may be inflated by means of a tube adapted to be held in the mouth.

17113/1901
Icilus William Maccolini *Landscape Contractor*
Inwood, New York, USA

(Wheatley & Mackenzie)

Schulze's Improved Safety Device for Perambulators, Mail Carts and like Vehicles provides a guard which rises automatically as soon as the child gets up in the perambulator and thus prevents it falling out. Should the child rise, it immediately operates a transverse strap so that the catches are released and springs raise the guards; alternatively, the tension on the strap may withdraw bolts from trellises, which may suitably consist of woven cane, and are elevated by a spring arranged beneath the vehicle.

15988/1901
Carl Wilhelm Robert Schulze *Restaurant Keeper*
30 Windmuhtenweg, Leipzig, Germany

(Mewburn, Ellis & Pryor)

Gossmann's Improved Bath and Means for Producing Waves and Executing Gymnastic Exercises therein has a small wheeled carriage in the bath on which the bather sits and pulls against springs so as to cause the carriage to move back and forth and thus produce a powerful wave action.

The carriage may be replaced by a swing.

16130/1901
Henrich Gossmann *Manufacturer*
Wilhelmshöhe, nr Cassell, Germany

(Herbert Haddan & Co)

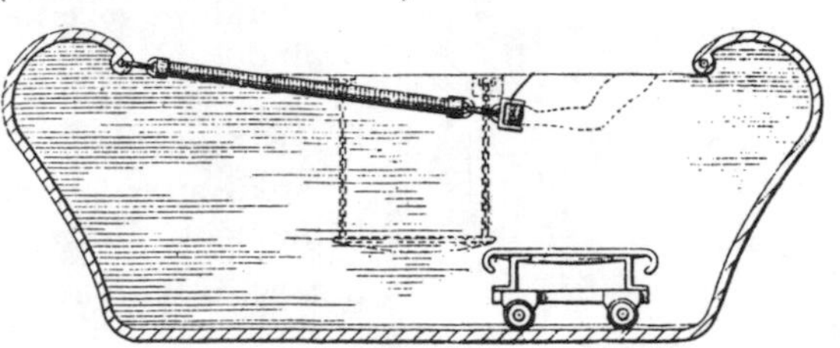

Günther's Improved Face Guard for the use of Cyclists, Motorists and Others consists in a curved celluloid plate mounted in a frame provided with spring ear clips and with a nose-rest carried by an adjustable rod. It is thus possible to adjust the device at any desired distance from the face; it protects the mouth, nose and eyes from the sharp current of air whilst enabling healthy breathing and, the mica-plate being of pale blue or grey, shades or protects the eyes against too strong light.

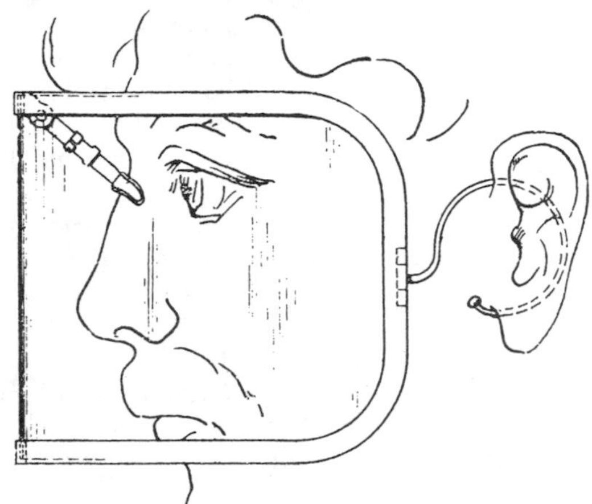

14377/1901
Clemens Günther *Merchant*
10 Kaulbachstrasse, Dresden, Saxony, Germany

(Boult, Wade & Kilburn)

Koppenhagen's Device for Assisting the Swallowing of Pills recognises the fact that many sick people find it perfectly impossible to swallow a pill without first crushing the same between the teeth. As soon as the sick person places a pill on his tongue, it sticks fast to the palate in consequence of a reflective nerve stimulus, and, notwithstanding the greatest masticating or swallowing efforts it cannot be guided to the gullet, so that the patient for good or for evil finds it necessary to take the pill to the front between his teeth, and then only after thoroughly crushing the pill can he move the crushed mass backwards and swallow it down.

The present device is an elongated flask with a grating at the base of the neck. The flask is filled with liquid and the pill is dropped on to the grating. The neck of the flask is then put into the mouth, and pushed down the tongue – tilting back the head enables liquid and pill to be swallowed with no discomfort whatsoever.

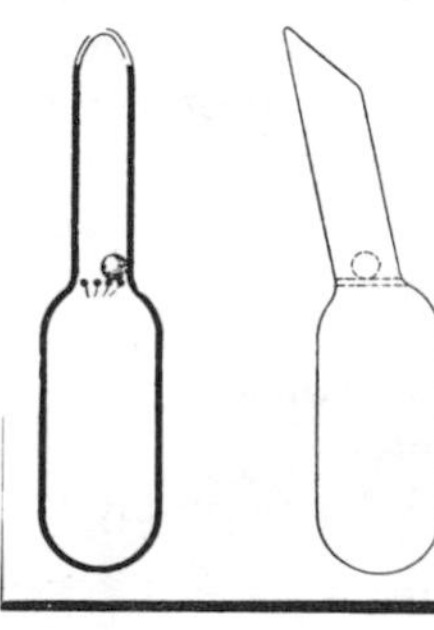

13043/1901
Benno Koppenhagen *Doctor of Medicine*
Unterneubrunn, Saxe-Meiningen, Germany

(W P Thompson & Co)

Kirby's Improved Motor Road Vehicle has a body raised so as to place the occupants beyond the reach of the whirling dust, to widen the scope of observation, and to lodge the machinist apart and more conveniently. To achieve this, the body is mounted on iron uprights, allowing of a machinist's room beneath, with windows. The body is reached by means of a light staircase, the steps of which are covered with white india rubber mat so as to make them less slippery and which is provided with a rope of silk or other material which may be hooked or unhooked at will and serve as a banister.

17418/1901
Emily Frances Kirby *Independent Lady*
24 Army and Navy Mansions, Victoria Street, London SW, England

(H C Fowler)

Wacker's Improved Hat Brush can always be carried in an inconspicuous manner in the hat so that it is at hand for use whenever required. The invention is particularly suitable for bristle brushes or brushes covered with plush or velvet for smoothing silk hats.

17787/1901
Albert Wacker *Manufacturer*
44 Landgraben Strasse, Nuremburg, Germany

(W P Thompson)

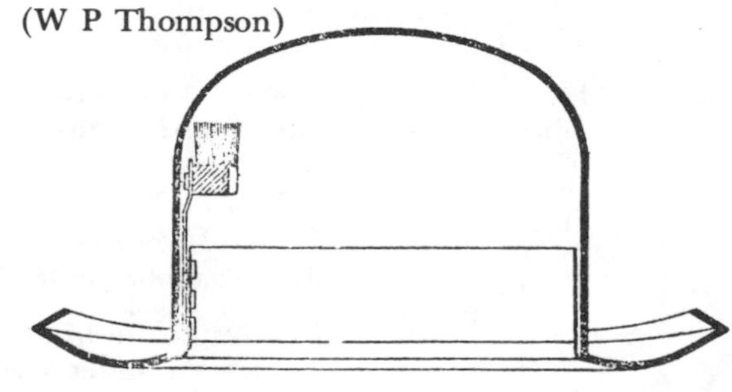

1901

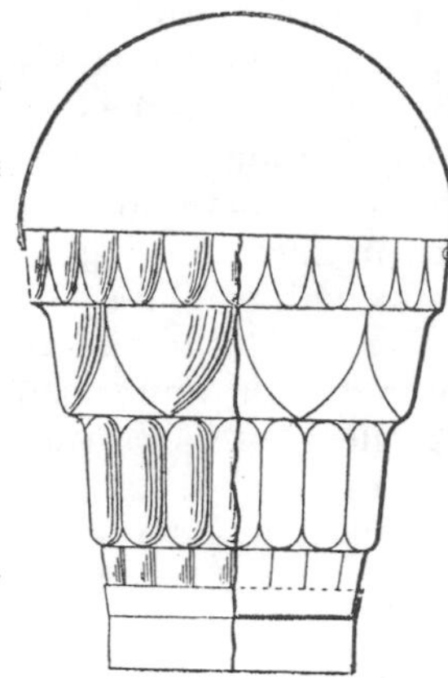

Ewing, Denney & Jolley's Improved Toy Bucket is designed so that the making of sand pies may be rendered more interesting and instructive to the children.

The invention consists in forming the sides or bottom of the bucket with an ornamental pattern in cameo or relief or in forming the buckets in various ornamental shapes, such as human figures, fish or animals.

17788/1901
Jacob Ewing *Printer and Stationer*
12 Euston Grove, Birkenhead, Chester, England

Harry Denney *Salesman*
104 Hawkins Street, Liverpool, England

John Jolley *Salesman*
45 Olney Street, Liverpool, England

(W P Thompson)

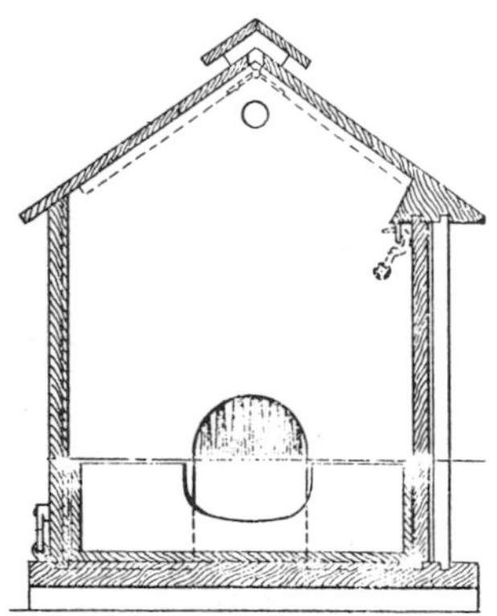

Oram and Prebble's Improved Portable Kennel obviates the disadvantages of conventional kennels in which, consequent upon the permanent form of construction adopted, it is difficult to thoroughly cleanse and dry the interiors, with the result that the kennels become foul in parts and infested with insects and the dogs or other animals, from the dampness of the boards, are liable to attacks of rheumatism and other bodily illness. This kennel can be readily taken to pieces for cleaning and drying and packing for purposes of transit. Inside are two compartments, one with a removable tray, the other with a mat. Bristles are fixed over the top of the main entrance, and over the partition door, so that the dog may brush itself when passing through these openings.

17849/1901
Henry Oram *Estate Agent's Assistant*
24 Anselm Road, Fulham, London, England

Walter Prebble *Estate Agent*
45 Vanstone Place, Walham Green, London, England

(James G Stokes)

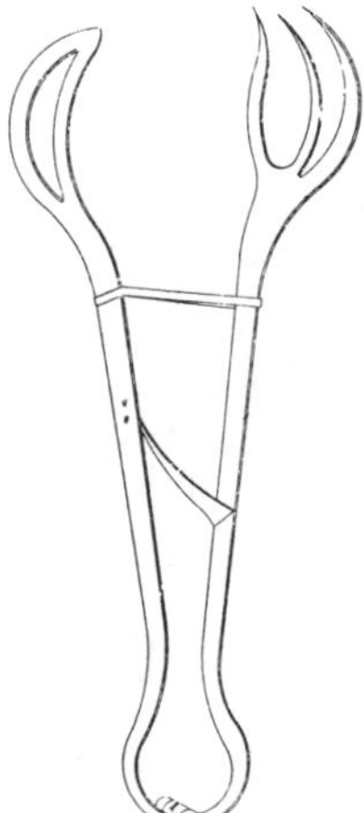

Dodson and Matthews's Combined Egg Lifter and Toasting Fork

22771/1901
Leonard Dodson *Railway Clerk*
9 Newhale Street, Swindon, Wiltshire, England

Charles Alfred Matthews *Mechanic*
10 Deacon Street, Swindon, Wiltshire, England

Stanger's Improved Medical Chair for treating diseases by electro-therapeutic means has an arrangement which admits of the bathing and syringing of the affected parts, so that the most various kinds of remedies can be conducted to the bath. The parts of the chair which support the patient are fitted with cushions which form the electrodes of one pole of an electric circuit, the other being a basin or bath fitted in the seat. The basin is fitted with an airtight cushion or ring at the top and a tube proceeds from its base along which the liquid for treating the patient may be pumped. By suitably adapting the upper edge of the funnel-shaped basin to the form of the body it can be used in the same manner for treating female diseases. The basin may also be connected to a carbonic-acid reservoir for carbonic acid baths, and a spraying device connected to one of the electric circuits may also be used.

18816/1901
Johann Jakob Stanger
Gerbermeister, Ulm, Germany

(W P Thompson & Co)

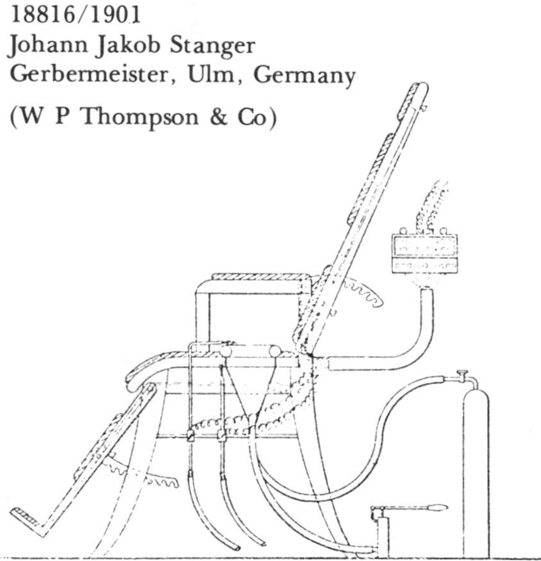

Meyer and Niemann's Improved Animal Trap simplifies and cheapens the construction and renders more serviceable the operation of traps of the kind in which a water receptacle is to receive the trapped animals.

Animals attracted by the bait reach an opening by way of the steps and on stepping on to a roller to get to the bait are immediately precipitated into the water beneath by virtue of the roller rotating instantly on it being touched by the feet of an animal.

25468/1901
Hermine Meyer *Married Woman*
Karl Niemann *Merchant*
Fallingbostel, nr Walstrode, Germany

(Edward Buttner)

Wray's Improved Teapot is so constructed that its spout can be replaced by a new one in case of accident, said spout being screwed into a projection on the pot and possessed of a washer to prevent leakage.

19059/1901
Ernest Wray *Confectioner*
St Anne's Road West, St Anne's-on-the-Sea, Lancashire, England

(G F Redfern & Co)

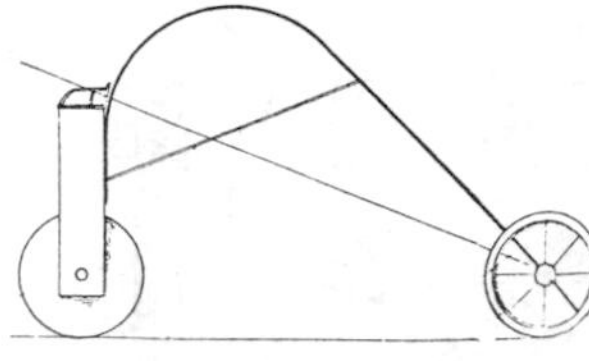

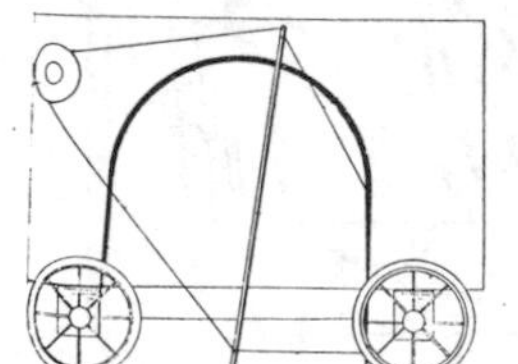

Hanbury's Improved Means for Propelling Vehicles utilises in a novel manner the pent-up energy of a strung bow in conjunction with the distribution of weight. Motive power is obtained by alternately contracting and releasing said bow. In the application of this invention to a cycle, the bow is strung, the saddle placed on it, and the rider, by swinging his weight forward and operating a treadle contracts the bow which is then allowed to extend and push the cycle forward.

The device can also be used for carts: the bow is connected by cords to a pole one of whose ends rests on the ground. When the bow is contracted and released, the pole pushes the cart along.

19344/1901
Ada Isabella Hanbury *Artist*
Ormond House, Poole Road, West Bournemouth, Hampshire, England

(Hughes & Young)

Kühne, Sieves and Neumann's Improved Gymnastic Apparatus is designed for use particularly in a room, the improvements being specially adaptable to that type of apparatus known as the home trainer or exerciser.

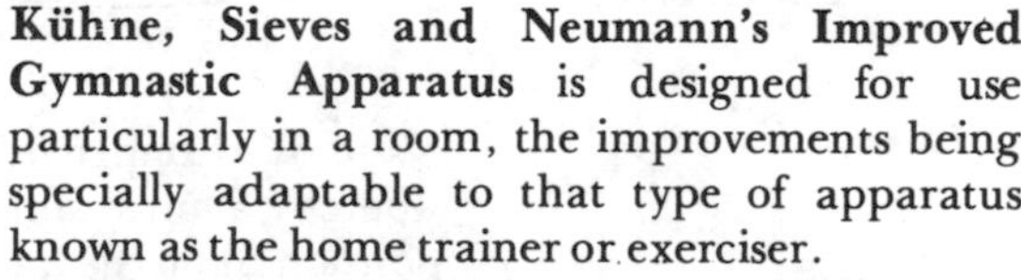

Hitherto, it has been necessary, when a greater strength of apparatus is required, to purchase a complete set of fittings, the others of lesser power thereby being rendered completely useless. This invention makes it possible to regulate the strength of the apparatus so that persons of varying degrees of strength and skill can use it. This is achieved by an arrangement of springs and pulleys, the number of elastic cords attached thereto being variable as desired.

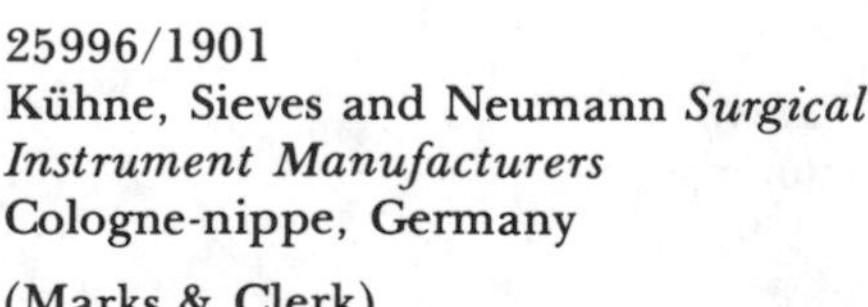

25996/1901
Kühne, Sieves and Neumann *Surgical Instrument Manufacturers*
Cologne-nippe, Germany

(Marks & Clerk)

Addleman's Improved Exercising Apparatus is designed to give as nearly uniform exercise to all the muscles of the body as is practicable, and hence employs a set of cranks for the hands and feet, respectively, of each operator, so arranged that the hand or foot of one side will both be down or up while those of the other side are in the reverse position, so as to give the largest practicable range of difference between them. One can work as hard as he desires and the two operators can work jointly.

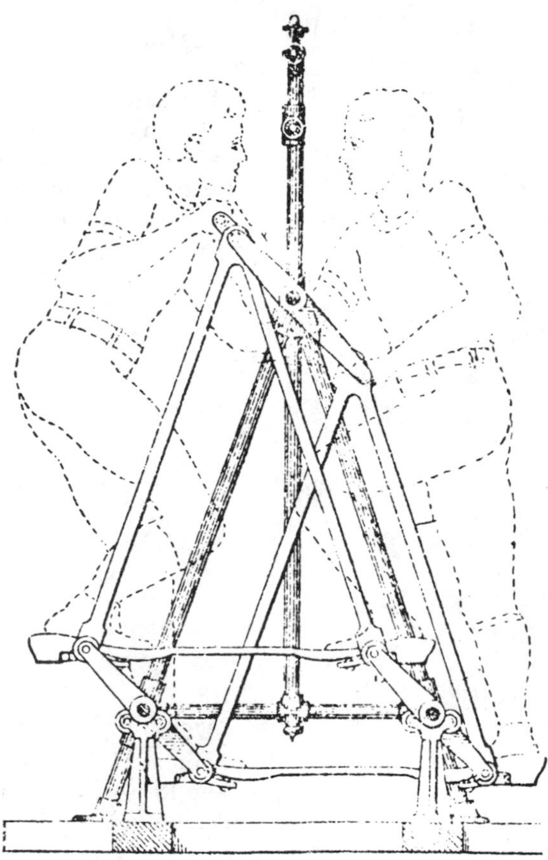

17569/1901
John S Addleman *Inventor*
34 Beuhrer Avenue, Cleveland, Ohio, USA

(Wheatley & Mackenzie)

Procter's Fruit Getting Appliance is a device for removing fruit from inaccessible branches by means of a knife mounted at the end of the handle which also bears a wire loop carrying a shoot of netting to conduct the fruit to the operator.

17757/1901
Thomas Alfred Procter *Engineer and Land Surveyor*
The Poplars, Audley, Newcastle, Staffordshire, England

1901

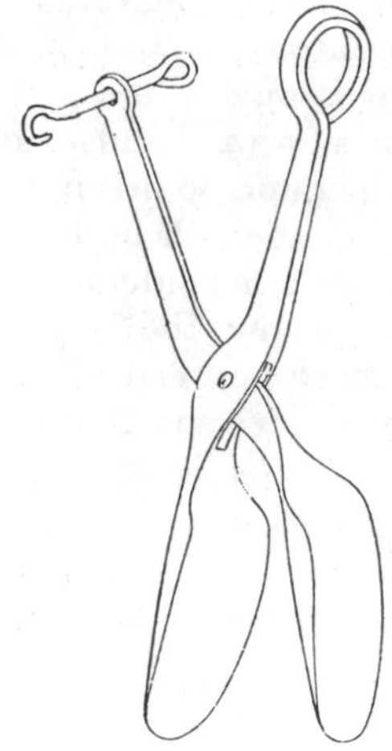

Marshall's Improved Combination Device for Facilitating the Drawing on and Fastening of Boots and Shoes provides a simple, convenient and effective method of achieving this, rendering the use of "tags" on boots unnecessary and also preventing the soiling of the hands. The back of the boot or shoe is gripped securely between the superposed spoon-shaped ends of the device and pulled on in an upwards direction; the button-hook attached is then used in the ordinary manner.

20271/1901

Marshall's Improved Boot Jack renders the removing of boots and shoes a matter of greater ease and despatch then heretofore, being furthermore simple of construction, inexpensive of manufacture, extremely portable and, when closed, lies quite flat and occupies no perceptible space in a bag, portmanteau or the like.

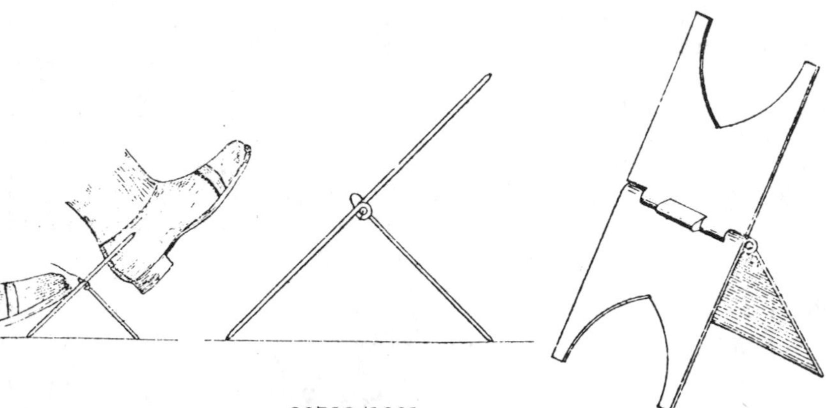

20728/1901
Thomas George Marshall *Clerk in Holy Orders*
The Rectory, Walwyns Castle, Little Haven, Pembrokeshire, Wales

(Hughes & Young)

Klaws's Appliance to aid the Hearing is designed to be worn at lectures, concerts, theatres and meetings. It is designed to fold up compactly and may be carried in a case provided for the purpose. It comprises two shells, or lobes, shaped to fit round the ears and held by rods fastened to a handle which may be provided with means for attaching it to the person. Alternatively, the ear-shells may be provided with hooks to engage with the wearer's hair.

18944/1901
Pauline Antonie Klaws *Married Woman*
454 Collins Street, Melbourne, Victoria, Australia

(John P O'Donnell)

Hooper's Improved Aerating Agitator employs a mode of aërating substances similar to the well-known way of making old-fashioned candy, by continually pulling or stringing out and combining again the mass until by means of the air globules brought in in this manner the same has been increased in bulk to the desired degree.

25995/1901
Beekman David Hooper *Baker*
337 Fulton Street, Jamaica, New York, USA

(Marks & Clerk)

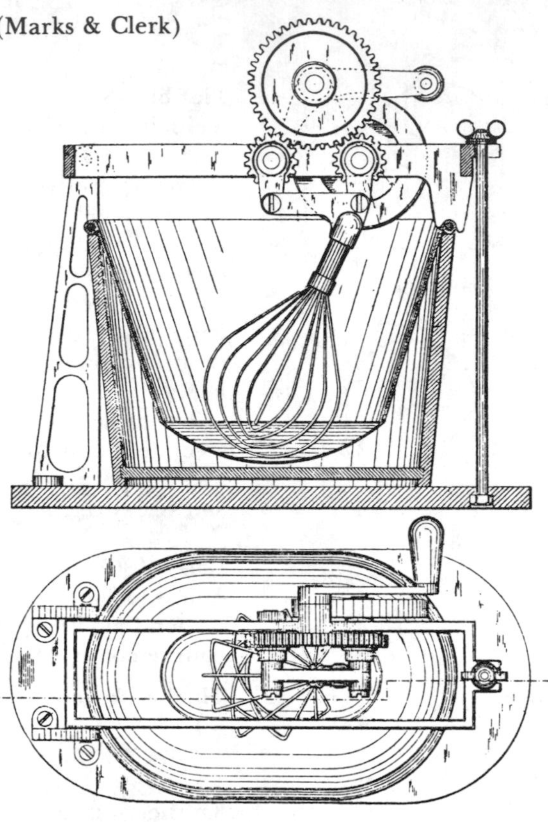

Tsuchii's Improved Hat provides a means whereby a mirror may be carried in a convenient position upon the hat, being held in position by a subsidiary lining sewn to the lining proper of the crown.

20827/1901
Rinkichi Tsuchii *Gentleman*
81 The Chase, Clapham Common, London, England

(Hughes & Young)

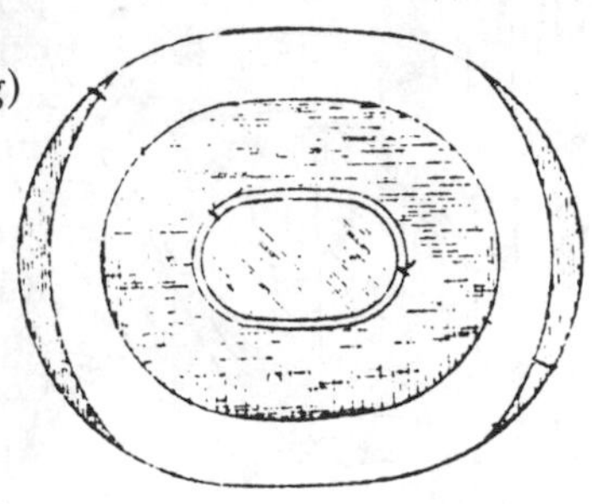

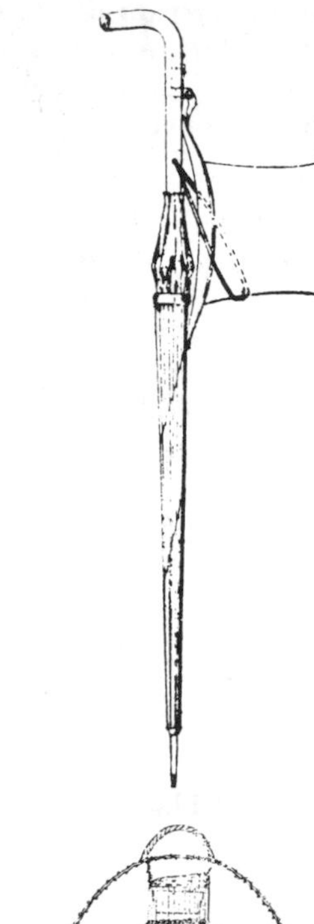

Batchelor's Improved Walking and Umbrella Stick is fitted with a spring rest near the centre which takes in one end of the rim of a hat, whilst the other end of the hat is secured by a clip near the forefinger at the top of the stick, so leaving the hands free. Furthermore, the stick holding the hat can be placed in a corner to serve the purpose of a hat-rack and loss of sticks and umbrellas is prevented since one is liable to go without one's stick but not one's hat. It can also be used to carry the hat when wet, thus preventing rheumatism, and to secure gloves, newspapers and other articles.

22896/1901
William Batchelor *Iron Moulder*
88 Fredrick Street, Portsea, Portsmouth, Hampshire, England.

Schramme's Improved Receptacle to Facilitate the Carriage or Conveyance of Clothes and Other Articles is designed for the convenience of travellers, as clothes and other articles can be packed without previously folding them. It consists of a bell-shaped cover fastened to a baseplate with a wicker form arranged therein on which dresses and outer garments can be arranged. Blouses, bodices and the like can be placed on a second form inside this; umbrellas, boots etc can be packed at the sides, and hats can be placed in the upper part.

20523/1901
Anna Schramme *Married Woman*
44 Friesenstrasse, Hannover, Germany

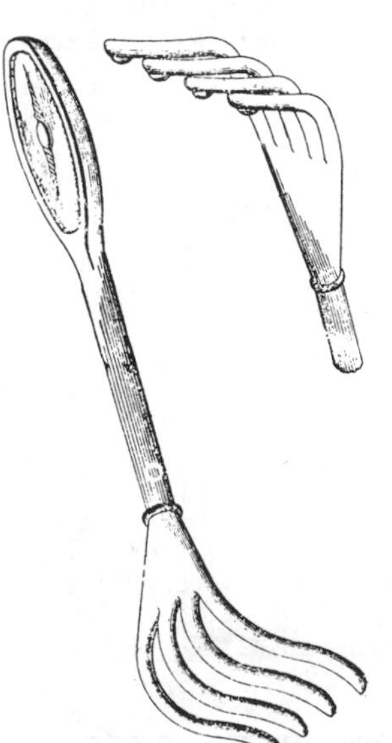

Wirt's Improved Massaging Device is designed to stimulate the scalp and promote the growth of the hair by a gentle rubbing action beneath and at the roots of the hair without tangling the same. The device comprises a hand-like rubbing member and a row of independently yieldable resilient working fingers provided with massaging faces, connected with a fluid reservoir for lotion or tonic. Regular use of this device avoids the hard and harsh action of combs and brushes which have a downward scraping action upon the scalp and thereby often cause injury and so retard rather than promote the growth of hair.

23286/1901
Paul Esterly Wirt *Manufacturer*
Bloomsburg, Columbia, Pasadena, USA

(Boult, Wade & Kilburn)

Sackville's Improved Method of and Device for Improving or Enlarging the Breasts comprises essentially an apparatus for causing, through continued suction, an increased flow of blood to circulate through the mammary glands, thus producing a vigorous growth of tissue and in time a permanent development.

The device consists essentially of a shield of rubber adapted to envelop the breast and make an airtight join therewith. A vacuum is produced by depressing or deforming the shield before its application to the breast, and the edge of the shield is anointed with glycerine or other substance to ensure a perfectly airtight join between it and the skin.

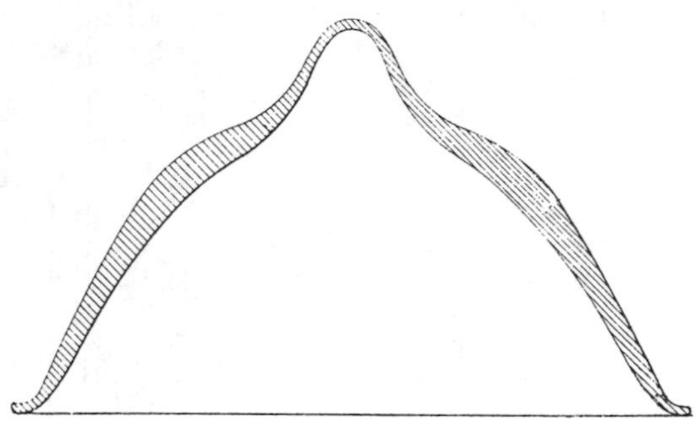

22636/1901
William Stove Sackville *Manufacturer*
15 Girdler's Road, West Kensington, London, England

(Phillipss)

Carter's Toy Animal's Head is a combination of head and springs that can be fixed to any chair, bedstead or the like to form a new toy, which moves both up and down and to and fro when the reins are pulled by the child sitting behind.

23094/1901
Frank Carter *Tailor*
2 St James's Street, Brighton, Sussex

Riddell and Robertson's Improved Jamb of Metal Mantel Register Grates consists in constructing the jambs of open metal work. This may take the form of leaves and stems which formations may cross and recross the opening of the jambs. Behind the said floral or leaf devices may be tiles or marble or other backing so as to throw up the designs of the iron work.

21502/1901
Alexander Riddell &
Robert Robertson *Directors of Watson, Gow & Co Ltd, Ironfounders*
Etna Foundry, Lillybank Road, Glasgow, Scotland

(Johnsons)

Wainwright's Improved Closet Seat recognises that it is frequently desirable that in a closet for general use there should be provided a separate seat for the principal or for members of the family. This invention consists of a lower seat for general use and an upper one for special use which folds down over the lower seat and covers it. When this special seat is not in use, it is folded back and is held by a suitable catch or lock so as to be out of the way of the general seat.

The lock can be opened only by a key in the possession of the person having the right to use the seat.

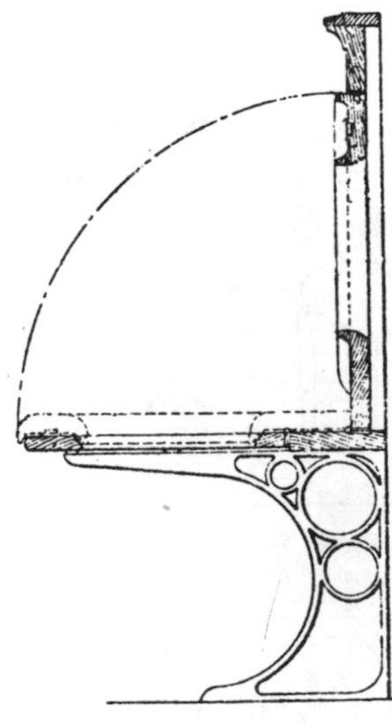

24194/1901
Frank Wainwright *Surveyor*
"St Ives", Hale Grove, Edgware, Middlesex, England.

(Abel & Imray)

Booth's Apparatus for the Extraction of Dust from Carpets and other Materials draws a current of air through the materials treated, and afterwards separates the dust from the air.

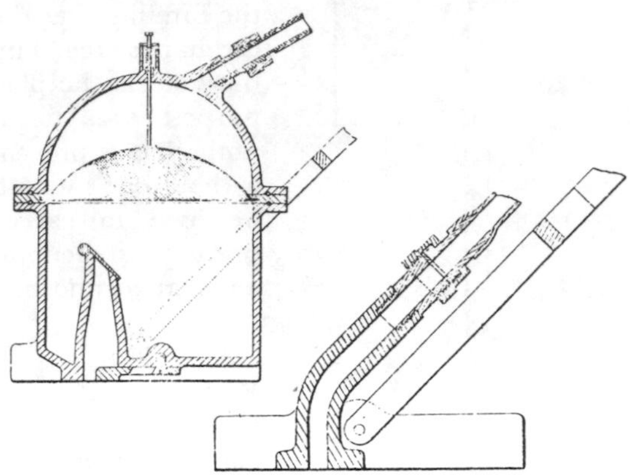

17433/1901
Hubert Cecil Booth *Civil Engineer*
5 Langham Chambers, Portland Place, London, England

(Haseltine, Lake & Co)

Woodhouse's Apparatus for Generating an Inexhaustible Supply of Motive Power without the consumption of Fuel is applicable for driving motor cars, railway vehicles, aërial machines, marine and submarine vessels, insubmergible lifeboats and stationary engines. Air supplied by blowers passes, via dip-pipes under water, or other liquid, into a compressed-air reservoir. The motor or machine top to be driven is supplied with compressed-air from the reservoir, and drives the blowers which feed the device.

19822/1901
Samuel Joseph Woodhouse *Civil Engineer*
29 Wood View Grove, Dewsbury Road, Leeds, England

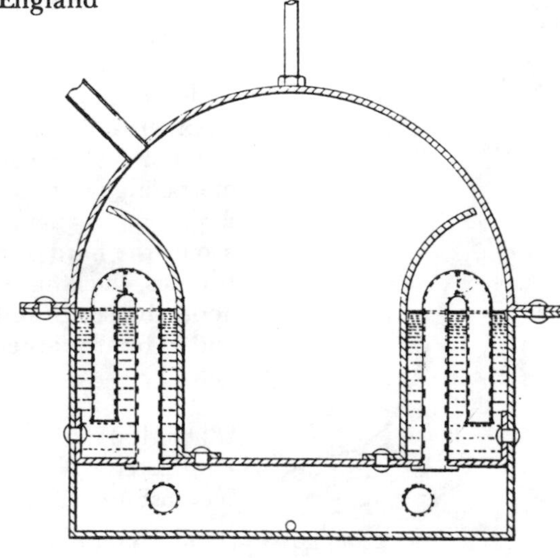

Ready and Ross's Surgical Appliance for Restraining the Movements of Patients is designed to prevent children and other persons who may be suffering from eczema, smallpox, delirium and the like from injuring their faces or scalps by scratching them; at the same time ensuring perfect freedom of the circulation, free room for the play of the muscles and growth of the arm and good ventilation.

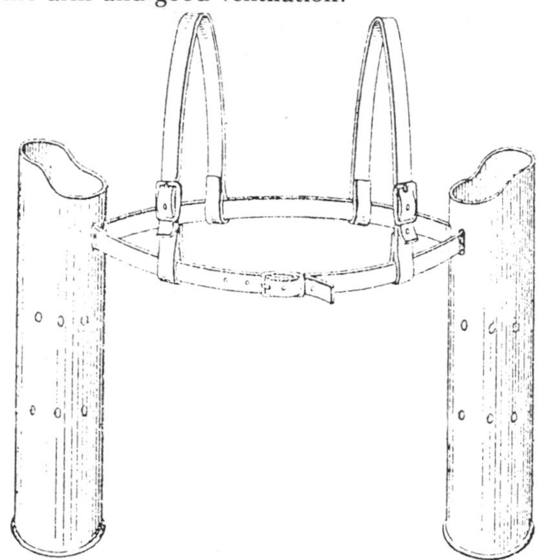

22803/1901
Avery Whipple Ready *Merchant*
156 Fifth Avenue, New York, USA

Edward Hamilton Ross *Merchant*
101 East 25th Street, New York, USA

(Stephen Watkins & Groves)

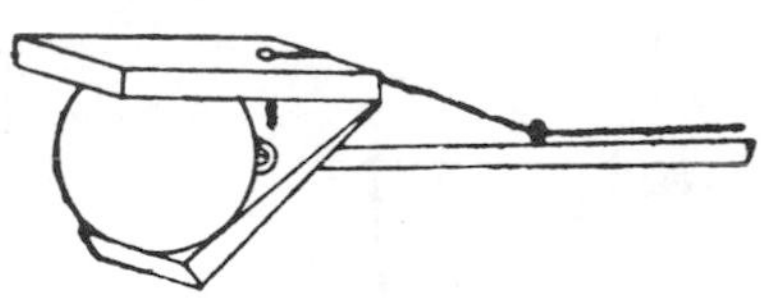

23793/1901
David Walker *Officer of the Inland Revenue*
The Terrace, Pocklington, Yorkshire, England

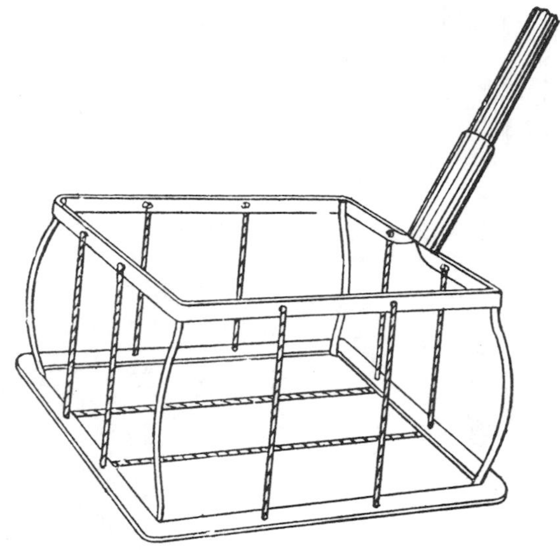

26301/1901
Edward Emrys Roberts *Bachelor of Medicine*
7 Slatey Road, Birkenhead, Cheshire, England

(W P Thompson)

Greenham's, Hammond, van Baerle & Guthrie's, Jeffreys's, Kemp's, King's, Macrae's, Minton's, Ungley's, Walker's and Wrenford's Devices for picking up Ping-Pong Balls are designed to enable balls readily to be picked up in situations where they would otherwise be accessible only with difficulty, under tables or chairs for instance, and to avoid the need for the person employing the device having to stoop to pick up the ball.

Greenham's device takes the form of a bat, skeleton-barred on one side to hold the balls; Macrae's appliance is provided with a tray of a size to hold one or more balls, one ball to pass through its aperture at a time; Ungley's collector is provided with wires conveniently covered with india-rubber to prevent skidding on smooth surfaces.

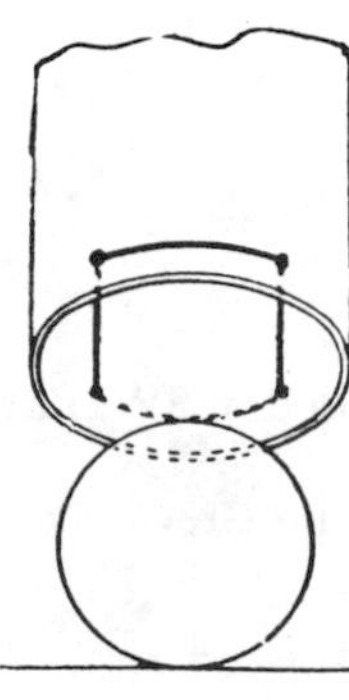

23274/1901
Thomas William Kemp *Ostrich Feather Merchant*
42 & 44 Golden Lane, London, England

(Day, Davies & Hunt)

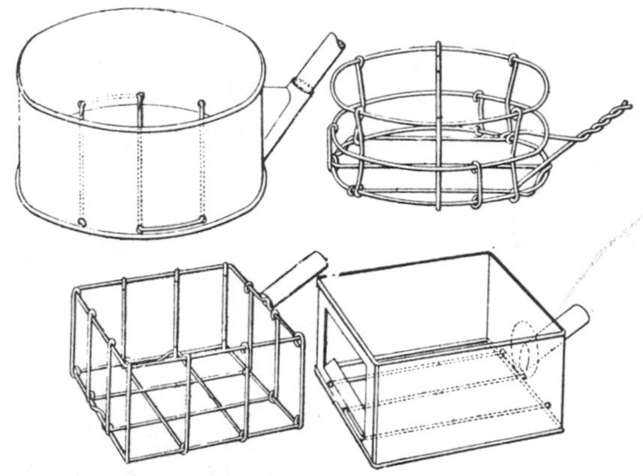

24322/1901
Joseph Hewitt Hammond &
Edward Thomas William van Baerle
t/a Hammond & Baerle, *Dealers in Works of Art*
84 & 85 York Street, St James's Park Station, London, England

Charles Marks Guthrie *Engineer*
Comberton, Wimbledon Park Road, London, England

(Haseltine, Lake & Co)

1902

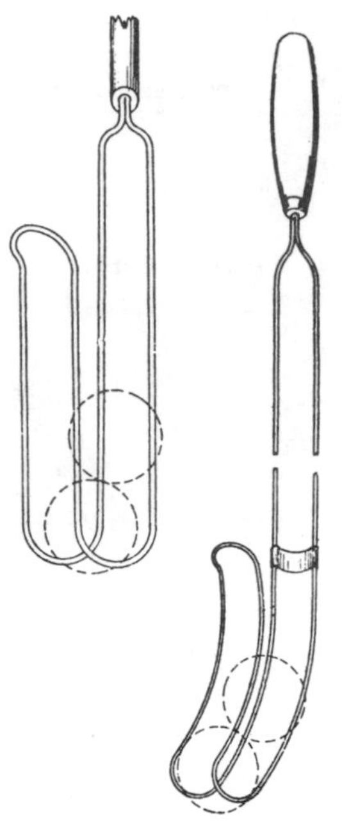
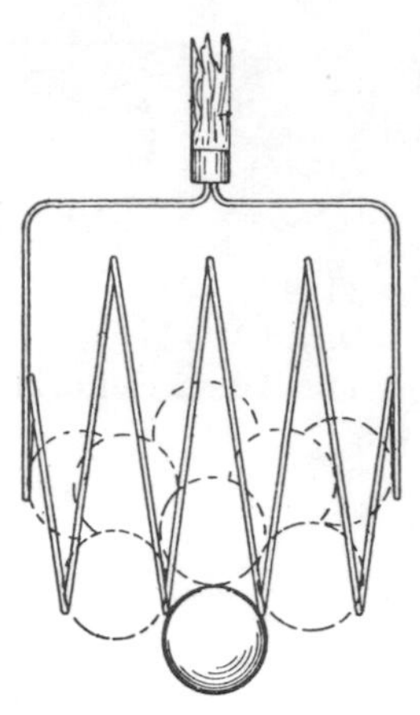
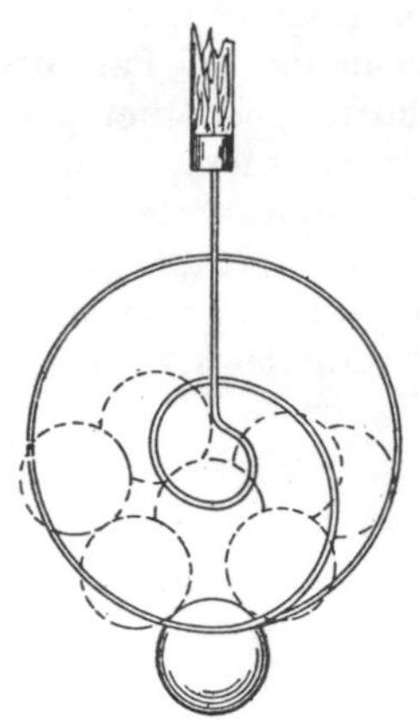

2791/1902
Edward Augustus Jeffreys *Engineer*
Thistlewood, Coppice Road, Moseley, nr Birmingham, England

(Marks & Clerk)

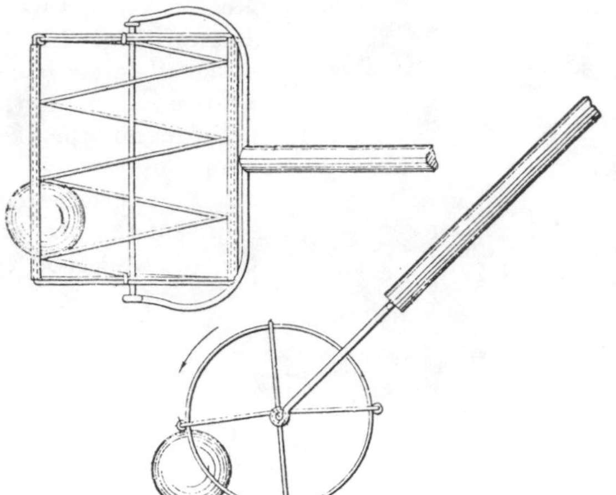

3628/1902
William Thomas Ungley *Wireworker*
8 Garnault Mews, Rosebery Avenue, London, England

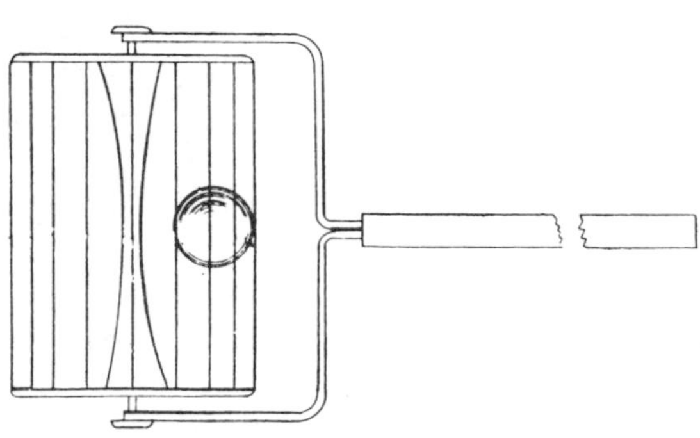

1388/1902
Percy King *Athletic Outfitter*
22 Lord Street, Liverpool, England

(Frederic Prince)

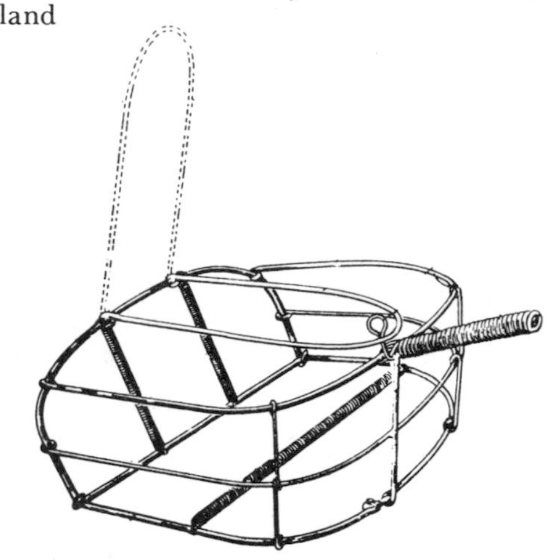

1112/1902
Francis Minton *Solicitor*
17 Philpot Lane, London EC, England

(Boult, Wade & Kilburn)

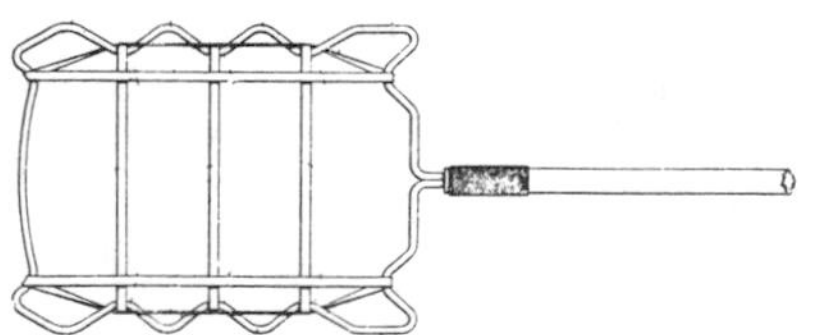

798/1902
William Leonard Wrenford *Gentleman*
59 Bouverie Road West, Folkestone, Kent, England

(A M & WM Clark)

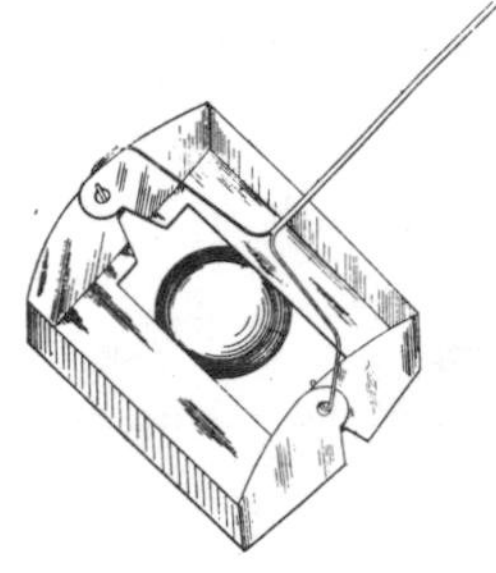

821/1902
Archibald Macdougall Macrae *Indigo Planter*
8 Forest Road, Aberdeen, Scotland

(Johnsons)

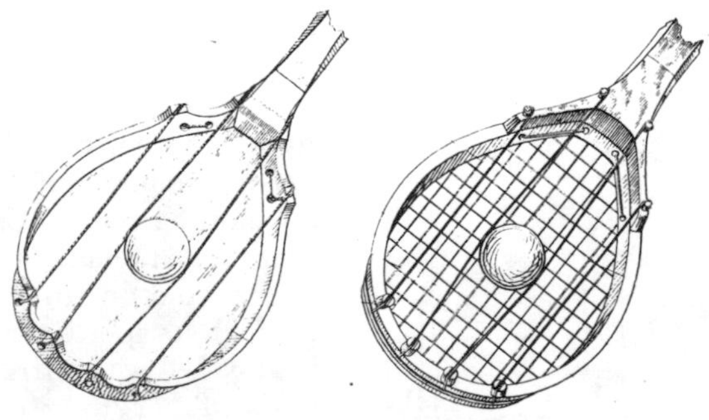

2042/1902
George Greenham *House Agent*
50 Bagleys Lane, Fulham, London SW, England
and later
29 Musgrave Crescent, Walham Green, London SW, England

(Geo Thos Hyde)

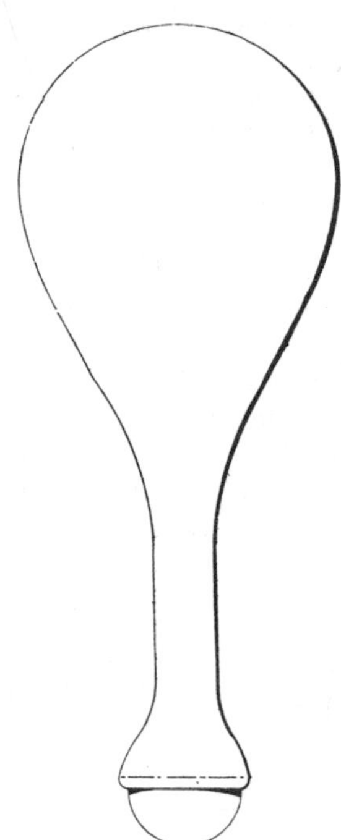

Peck's Improved Ping-pong Racquet Handle enables the racquet or bat, which may be made of wood, vellum or any other suitable material, to be used not only for playing the game but also for picking up the balls and holding them when out of play. To this end, the bat is provided with a cup at the end of the handle.

20452/1902
Edward Samuel Peck *Farmer*
Langton Grove, Eye, Suffolk, England

(Benj T King)

Radcliffe's Improved Table for playing the Game of Ping Pong or Table Tennis is formed with a gully so that balls when falling or thrown into such slot or opening will find their way by gravitation to the before-mentioned trough or tube and thence to the end of the table nearest to the player serving or requiring the balls, avoiding considerable time and trouble in picking up and collecting the balls.

10643/1902
Reginald Heber Radcliffe *Solicitor*
"Bankfields", Waterloo Park, Waterloo, Lancashire, England

(Stanley, Popplewell & Co)

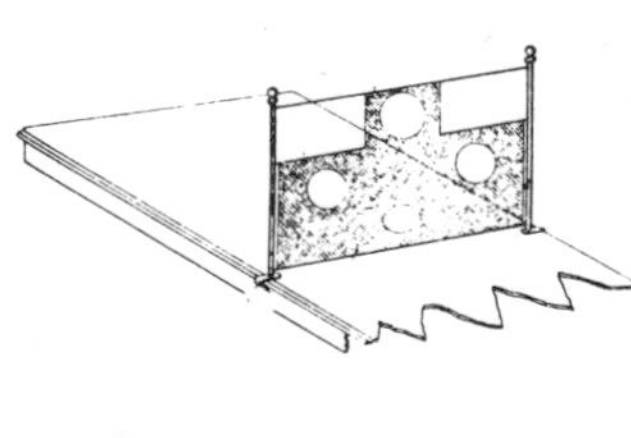

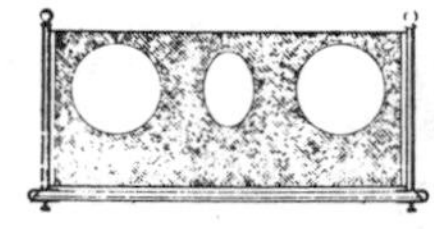

Ingham's New Game comprises a net with holes stretched across the centre of a table. The object is for the player who commences the game to strike a small light ball through one hole in the net, and as the ball bounces or rebounds from the table the player at the opposite end must return it through a different hole, and then the first player must return it through a different opening again. The player who fails to return the ball as it rebounds from the table through the specified hole in the net loses the point.

2516/1902
Herbert William Denton Ingham *Mechanical Draughtsman*
38 Arthur Road, Erdington, juxta Birmingham, England

Graves and Brown's Improved Game embodies figures or objects arranged so as to be received by a pocket when struck by a missile projected from a catapult.

Suggested figures are effigies of notorious criminals and persons opposed to law and order, and these figures are suspended from a gibbet by means of cords tied about their necks so as to hold them in suspension when the supports are knocked out from beneath, representing summary punishment meted out to the victim. A flag may be provided for each figure to designate the character or nationality of the effigy.

1860/1902
Eugene Edwin Graves
William Edwin Brown *Gentlemen*
Black River, New York, USA

(Albert E Ellen)

Borlinghaus's Improved Braces have a special contrivance to remove a disadvantage of usual braces, namely that they hold the pants on a level with the upper edge of the trousers, so compelling them to take up a position, laterally, corresponding more or less with the buttons on the trousers.

By this forced and unnatural position folds are produced which are uncomfortable for the wearer and may unfavourably influence the hang of the trousers.

In these improved braces the pants are held independently of the trousers, a novelty which permits of them being held at various heights and makes them assume a natural position.

1097/1902
Otto Borlinghaus *Merchant*
46 Bonnerthalweg, Bonn, Germany

(Geo Thos Hyde)

1902

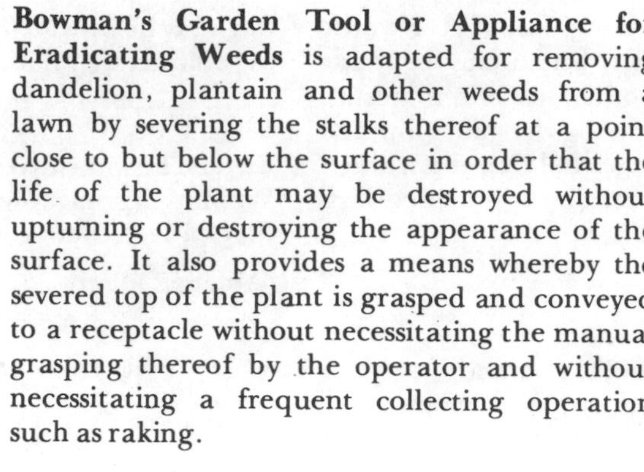

Bowman's Garden Tool or Appliance for Eradicating Weeds is adapted for removing dandelion, plantain and other weeds from a lawn by severing the stalks thereof at a point close to but below the surface in order that the life of the plant may be destroyed without upturning or destroying the appearance of the surface. It also provides a means whereby the severed top of the plant is grasped and conveyed to a receptacle without necessitating the manual grasping thereof by the operator and without necessitating a frequent collecting operation such as raking.

646/1902
Eric G Bowman *Merchant*
Monmouth, Co Warren, Illinois, USA

(Boult, Wade & Kilburn)

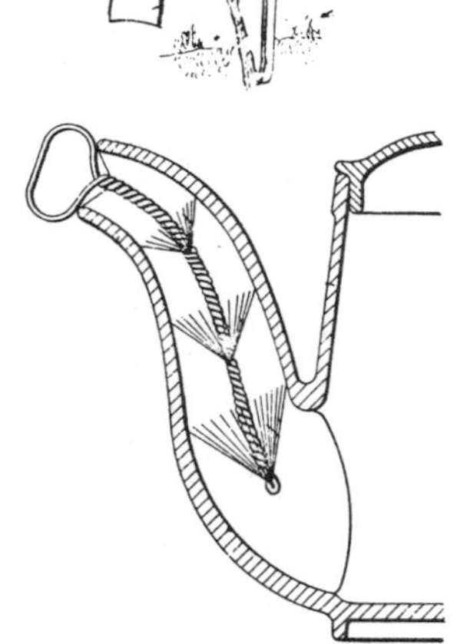

Chambers's Improved Strainer for the Pouring Off Spouts of Teapots prevents the escape of tea-leaves and sediment so that the liquid passes out of the spout in a clear condition.

859/1902
David Chambers *Engineer*
Cottage Grove, San Jose, California, USA

Tilsen's Apparatus for Facilitating the Putting-on of Coats, Mantles and Like Garments is capable of being utilised also for other purposes, such as for enabling an overcoat to be hung up and taken down from a coat hook even when the coat hook is full, for replacing a broken suspensory tab, for hanging up a hat, for keeping a coat collar turned up, for holding up the skirt or train of a dress, and for suspending a muff on the person.

The apparatus consists of a cord with a clip at one end and a ring at the other and is intended to be hung at a suitable spot and a suitable height.

When it is desired to put on a garment with the help of this apparatus, the top left-hand edge of the coat is clipped in the device as shown in the illustration. Then the user, grasping the right hand edge of the coat with his left hand, inserts first his right and then left arm into the sleeve, and then opens the clip to detach the garment from the device.

By using this apparatus the following advantages are gained:

1. The help or services of a second person are dispensed with;
2. The bottom edge of the garment is prevented from dragging on the floor while the first sleeve is being put on;
3. Stretching the garment while putting it on is avoided;
4. There is no liability of the collar being soiled by the person helping to put on the garment, which can otherwise scarcely be prevented.

The apparatus can also be used to hang up a coat on a full rack, in a public room, for example, by passing the cord through the tab of the coat, and clipping the clip to the ring. A gentleman may also utilise the apparatus to keep the collar of an overcoat retained in position when turned up in inclement weather by winding the cord round the collar and then gripping the cord with the jaws of the clip.

448/1902
Georg Tilsen *Retired Army Captain*
74 Kaiser Wilhelmstrasse, Breslau, Germany

(Mewburn, Ellis & Pryor)

Cassels's Improved Lavatory Basin is as free as possible from the danger of conveying infection since it is impossible for one person in washing to use or touch the water previously employed by another. The basin is made with an inlet in the bottom from which rises an inclined jet of water, under the fall of which the hands are placed.

101/1902
William Cassels *Sanitary Engineer*
9 Allan Park, Stirling, Scotland *and later* Killorn Villa, Stirling, Scotland

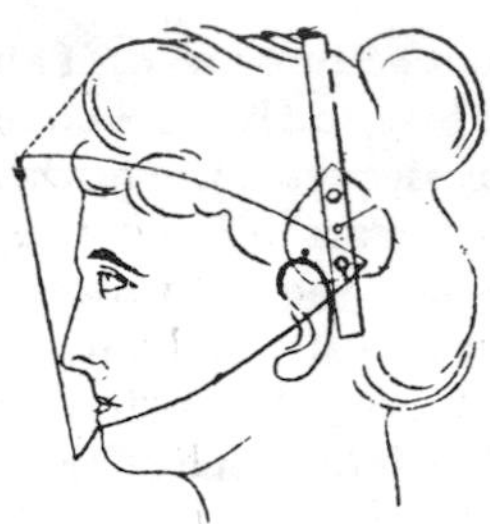

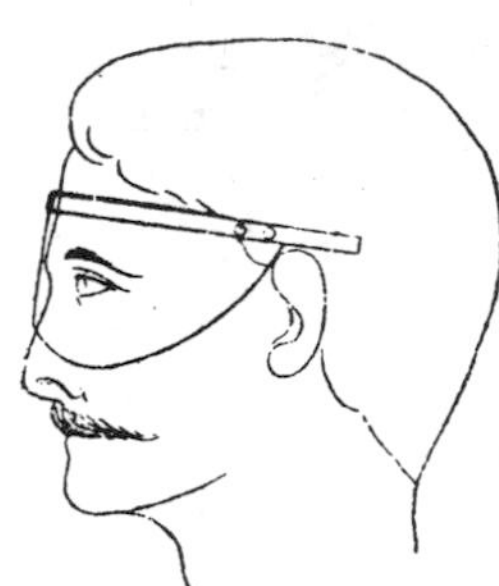

Le Conte's Face Protector, or "Facelière" is intended to act as a shield for the face against the wind, dust and rain. It is to be used by motor-car drivers, horsemen, coachmen, conductors of all kinds of vehicles and all persons exposed to the inclemency of the weather or in all occupations where protection is desired, for example by firemen against the flames, bee keepers obtaining honey from the hives, delicate persons in very cold weather, explorers etc.

It consists of a shield of light transparent material curved into a mask which can be adapted to the head covering of a person of either sex in such a manner as to be adjustable and capable of being raised and lowered at will.

It may be adapted to fit all head coverings, in particular the caps in use by lady automobile drivers called "Napoleon", in which shields may be drawn down in front of the protection and at the back to cover the nape of the neck.

85/1902
Armand Le Conte *Manufacturer*
15 Rue Lafayette, Paris, France

(G F Redfern)

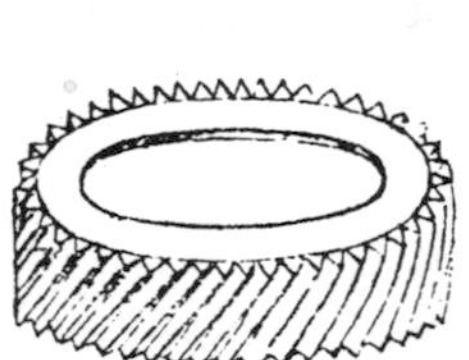

James and Holmes's Improved Match Striking Appliance is designed to fit over such articles as walking sticks, gas brackets, or pipe stems.

100/1902
Horace James *Ironmonger*
119 Devonshire Street, Sheffield, Yorkshire, England

Albert Isaiah Holmes *Boot and Shoe Expert*
56 Club Garden Road, Sheffield, Yorkshire, England

Wilcox's Improved Holder or Suspender for Handkerchiefs and other like Articles is formed of a continuous coil of wire with a dome-shaped top combined with a hook and chain for fastening to dress or belt or wherever desired. By this invention the handkerchief may be readily slipped in between the coils and as readily released and at the same time kept secure. No injury or tearing of the linen or cambric is caused by constant use owing to the wire coils being round and smooth.

2439/1902
Thomas Wilcox (T/A Matthew Wilcox) *Manufacturing Jeweller,*
111 Spencer Street, Birmingham, Warwickshire, England

Mawhood's Improved Navigable Aërostat comprises a balloon of elongated shape, filled with a gas which is lighter than air and fitted with propelling mechanism and steering gear enabling it to be guided in any direction. The craft is driven by a fan situated in the centre of the balloon, the fan being driven by a motor and shaft from below via gearing. An arrangement of weights suspended on rods can be used to effect ascents and descents and in the aërostat illustrated a basket or car is suspended so that it maintains a horizontal position at whatever inclination the balloon assumes.

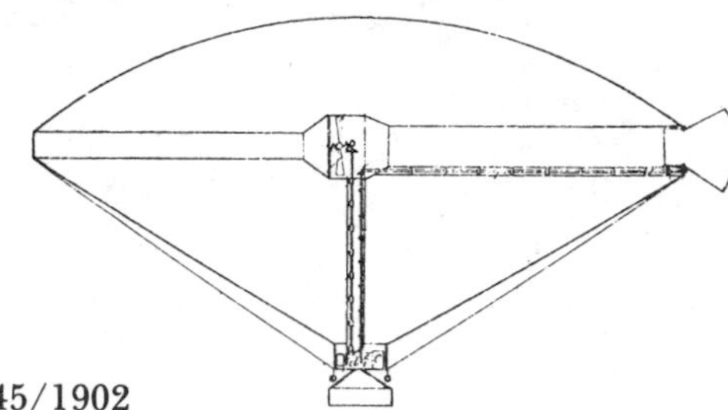

245/1902
Henri Charles Oscar Mawhood *Engineer*
20 Rempart d'Hoboken, Antwerp, Belgium

(Wheatley & Mackenzie)

Gaiger's Improved Head-rest for Perambulators, Children's Mail-carts and like Vehicles will adequately support the head of the child therein when asleep or lying down without the necessity for piling up cushions etc as is at present the case.

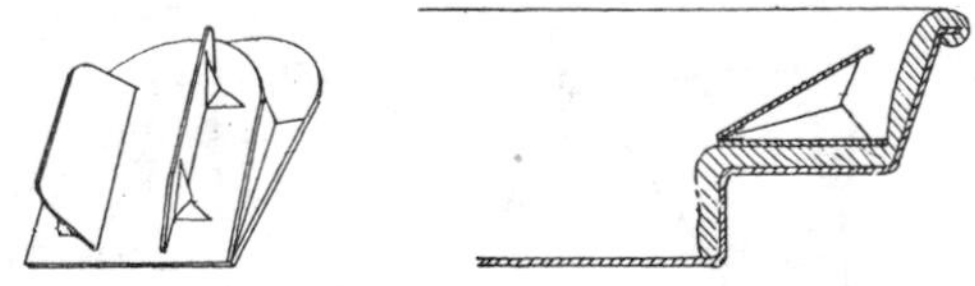

684/1902
Kate Gaiger *Children's Nurse*
Linwood Farm, Ringwood, Hampshire, England

(Albert E Ellen)

Leach's Improved Walking Stick can be reduced to a convenient length for putting in the waistcoat or other pocket when not required, and utilised for the purpose of carrying fluid refreshment such as wines or spirits when open.

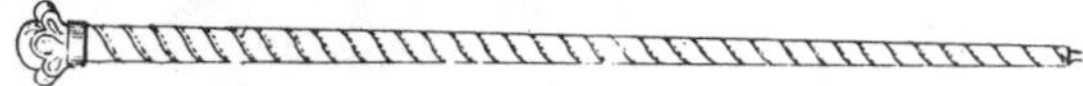

1091/1902
Thomas Leach *Spring Maker*
Altham Cottage, South Parade, Blackpool, Lancashire, England

(George Davies & Son)

1902

Ramm and Behrendt's Mechanism for Cutting and Deflecting the Aircurrent, which is caused by Fast Riding and to Utilize the Power of the Aircurrent as Motive Power to Aid in Propelling a Bicycle consists in a wing or wings mounted on a circular frame and capable of being adjusted according to the direction of the wind and is most advantageously connected to the front wheel of the bicycle through the intervention of an appropriate mechanism.

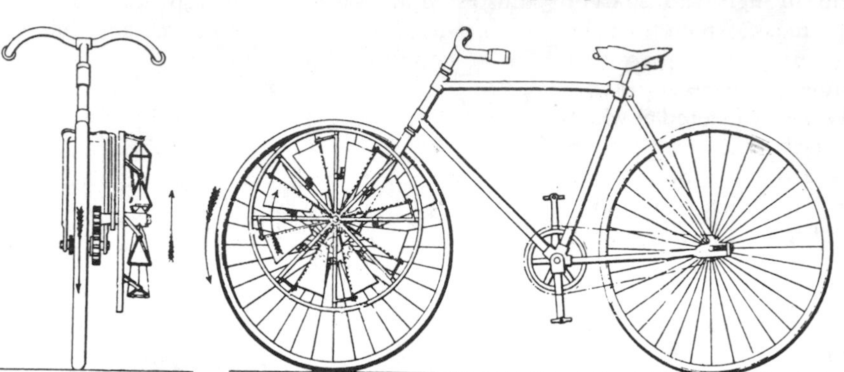

6086/1902
Peter Ramm and Friedrich Behrendt *Porters*
1-3 Kirchstrasse, Bochum, Westfalia, Germany

(Dewitz-Krebs & Co)

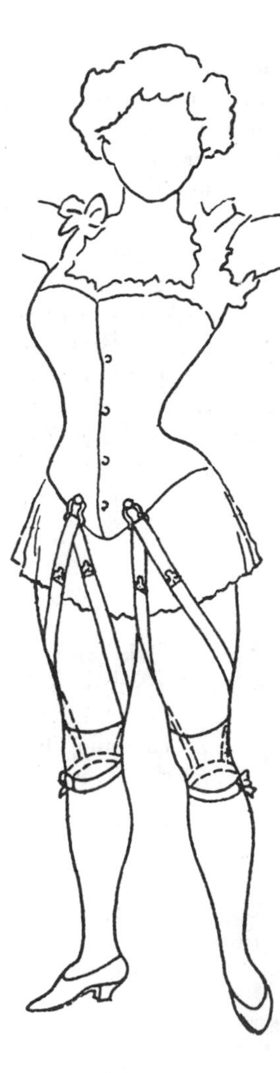

Cayley's Improved Cracker or Bon-bon is for use in a tug of war. It is constructed in what is termed "giant size", being of considerably larger proportions than those already made and, when it is opened, the contents, which may consist of a human being, an animal or an imitation figure such as Father Christmas, immediately fall out.

6186/1902
Frederick William Cayley and
A J Cayley & Son Ltd *Cracker Manufacturers*
Fleur-de-Lys Works, Norwich, Norfolk, England

(T B Browne Ltd)

Hammond's Improved Stocking Suspender causes the stockings to wear longer, gives more comfort to, and greatly improves the figure of, the wearer, particularly if inclined to be stout.

2433/1902
Isabelle Mary Hammond *Married Woman*
St James' Hotel, St James' Street, Derby, England
and later 56 Coldharbour Lane, Brixton, London, England

(W Swindell)

Bernhardt and Salomon's Improved Table Ornament or Appliance for the Use of Smokers combines a casing of any desired form with an electric lighter for lighting cigars. The outside of the casing is adapted for the reception of pictures or advertisements and may also be combined with other useful articles such as ash trays, cigar cutters, hand bells or money boxes.

2858/1902
Max Bernhardt *Merchant*
5 Wallner Theaterstrasse, Berlin, Germany

Hermann Salomon *Merchant*
Villa Thamina, Lausanne, Switzerland

(Haseltine, Lake & Co)

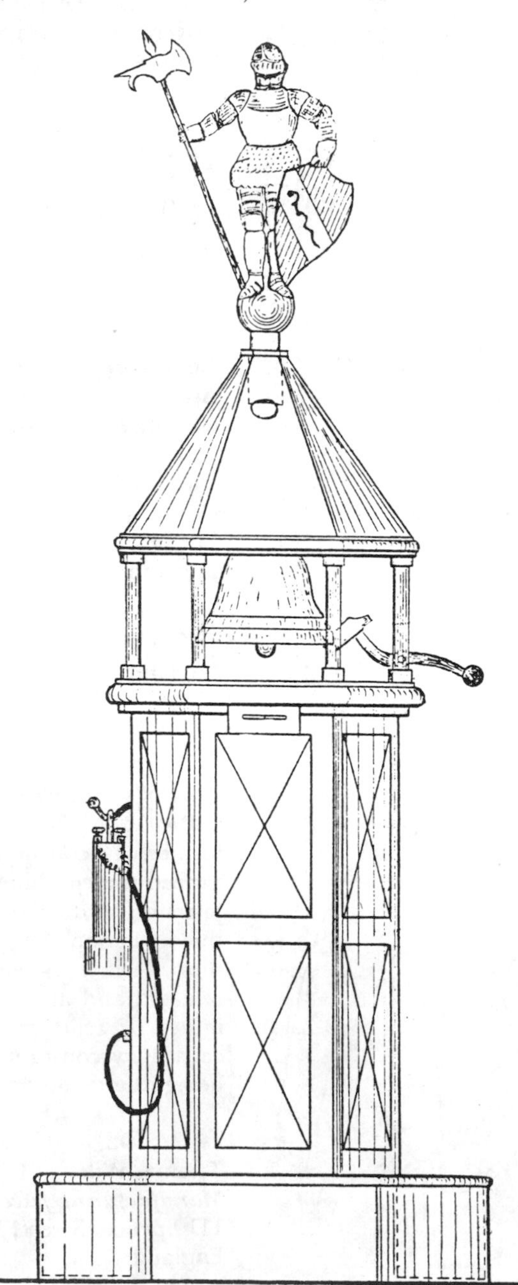

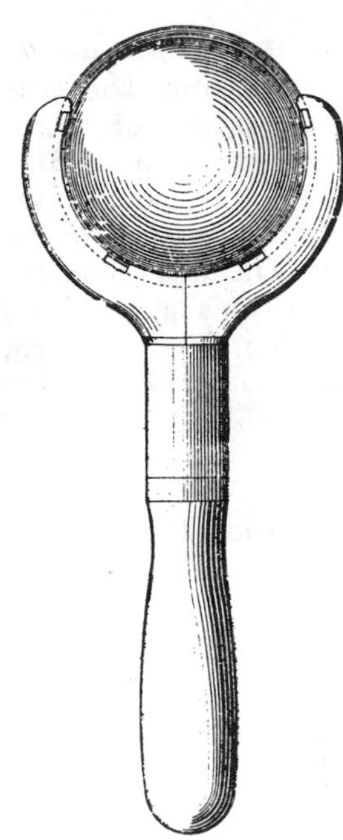

Proben and Fischer's Improved Massage Device comprises a massage roller in which a heavy steel ball weighing from five to ten pounds may be freely manipulated upon the muscles in every direction and whose parts may be readily assembled or disconnected. The massage ball is held within anti-friction rollers projecting from the inner surfaces of curved arms attached to a handle, so that the operator can exert heavy pressure and change the direction of the ball without changing the position of the handle. For a light massage a ball of different weight or of glass or ivory may be substituted.

The handle may also be provided with an opening for an electrode for electric massage treatment.

2167/1902
Charles Ignatius Proben *Physician*
136 East 70th Street, New York, USA

George Fischer Jr *Metal Worker*
947 Home Street, New York, USA

(Wheatley & Mackenzie)

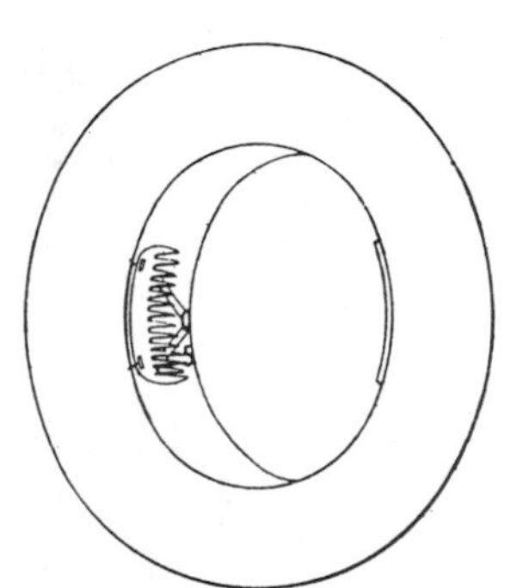

Pomeroy's Improved Hat Fastener overcomes the disadvantages of the hat pins at present in use by females, which are awkward owing to their length, dangerous owing to their sharpened points, and destructive to the hat on account of its being necessary to pierce the hat afresh with them every time it is worn. This invention provides a safe, efficient and comfortable means of attaching the hat securely to the hair, at the same time being adjustable and easily manipulated both in attaching and detaching.

3126/1902
John Pomeroy *Fish Curer*
Catherine Street, North Invercargill, New Zealand

(Boult, Wade & Kilburn)

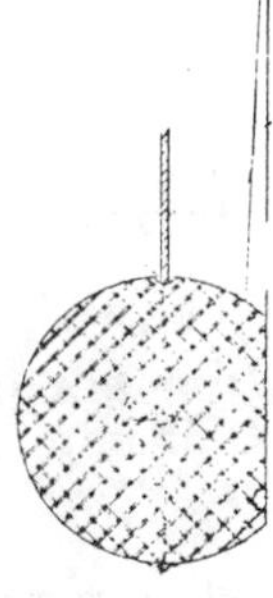

Cheeseborough's Improved Device for Attachment to the Person as a Knitting Ball Holder encloses and retains the knitting ball in a case or receptacle in convenient proximity to the person using it, so that, whilst a thread can be withdrawn in the operation of knitting, the ball is protected from danger of being soiled, stamped on or otherwise damaged.

2621/1902
Joseph Cheeseborough Jr *Hairdresser and Tobacconist*
Surtees Street, West Hartlepool, Co Durham, England

(G J Clarkson)

Sutchall's Device for Use as a Holder or Rest for Cigars or Cigarettes enables them to be smoked as nearly as possible to the very end without the use of a mouthpiece, and is so constructed that it can also be used to hold certain other articles such as a stick of sealing wax.

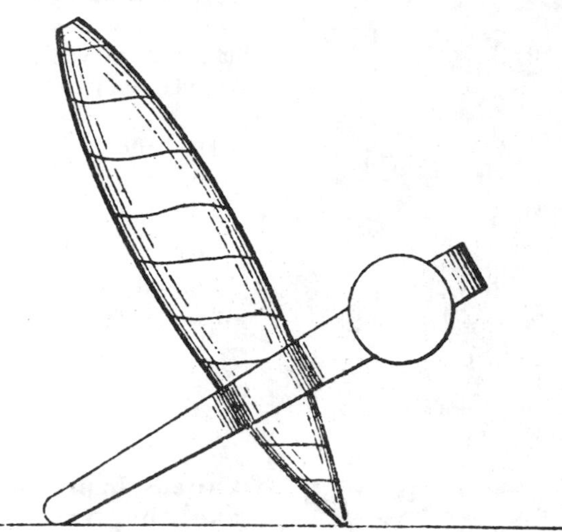

2767/1902
Robert Sutchall *Engineer*
4 Salisbury Terrace, Stockton-on-Tees, Co Durham, England

(G J Clarkson)

Katz's Improved Means for Securing the Contents of Bottles and other Vessels is an arrangement of lockable flaps designed to prevent their surreptitiously being opened by a child under fourteen years of age who may be fetching beer or other liquors from public houses. The device renders it impossible to extract any of the contents except by first unlocking the stopper by means of its proper key, of which there would be two; one key being retained by the publican and the other by the person for whom the child is acting as messenger.

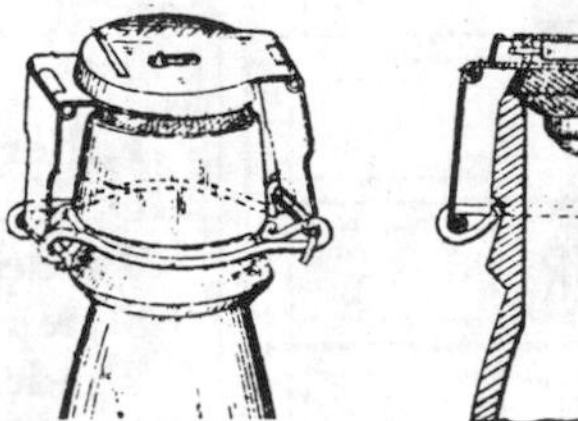

3789/1902
Alexander David Katz *Medicine and Sauce Manufacturer*
16 Castle Street, Birmingham, England

(Charles Bosworth Ketley)

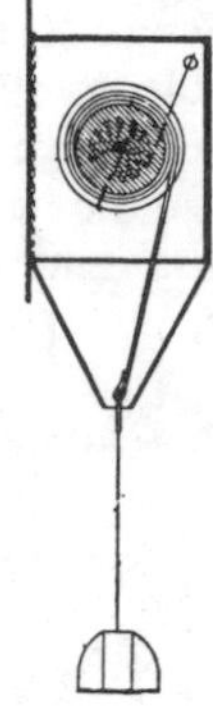

Dean and Farrar's Improved Soap Tablet or Block has as its primary object to effect economy.

For this purpose the soap has moulded into it means for attachment to a tape, band or chain wound on a drum connected with a spring of sufficient strength to wind the tape on the drum when the soap is released by the user and so raise it out of the wash basin into a dry position but in convenient reach for use.

4412/1902
John Dean *Manufacturers' Salesman*
15 Springwood Avenue, Bradford, Yorkshire, England

Edmund Farrar *Worsted Spinner*
Woodlands, Lightcliffe, Yorkshire, England

(Dracup & Nowell)

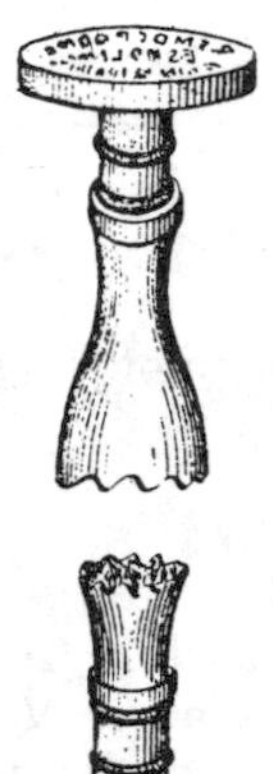

Carter's Improved Means for Sealing Bottles and the like is designed to obviate the difficulties imposed upon licensed victuallers and others in complying with the conditions of "The Intoxicating Liquors (Sale to Children) Act of 1901", namely the sealing of bottles and jars filled with liquor in an expeditious and efficient manner before handing them over to children to carry home. It consists of the combination of a disc of sealing wax moulded or otherwise connected to a cork adapted to close the mouth of the vessel, and a disc or seal for distributing the said wax over the mouth of the vessel so that the seal cannot be broken without being known to the sender or the purchaser, so providing an efficient preventative against tampering with the liquor.

4656/1902
William Alfred Carter *Commercial Traveller*
59 Albion Road, Stoke Newington, London N, England

(Newton & Son)

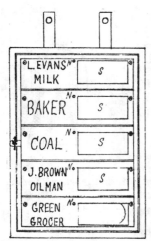

Fuller's Means of Assisting Housewives and Others to Attract the Attention of any Passing Tradesman, and if Necessary, Show a Statement of their Requirements to that Tradesman affords the saving of much time and trouble to all parties concerned.

5230/1902
Frank Fuller *Insurance Agent*
7 Mountford Road, Dalston, London, England *and later* 72 Bouverie Road, Stoke Newington, London N, England

Addison's Improved Foot Stand for use in Viewing Processions, Races and the like enables those people who are at the rear of, or short people who are in the midst of, a large gathering or crowd to easily and comfortably see over the heads of the people in front without any inconvenience or crushing. These elevators will be found of particular service on such occasions as when witnessing a procession, football match, race meeting or sport or game of any kind where a crowd of spectators are assembled.

3814/1902
Henry Addison *Church and School Furniture Manufacturer*
Waterloo Works, Wellington, Shropshire, England

(George Barker)

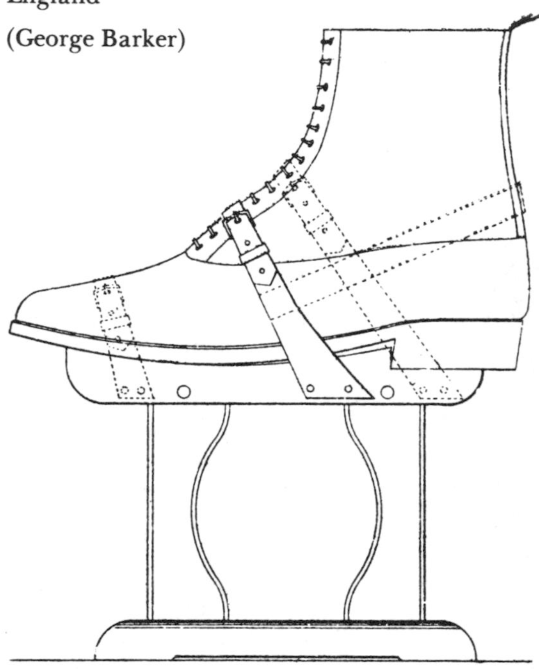

Butterworth's Improved Butter Knife is for cutting off and working butter to its proper consistency for use without warming it before the fire by which means it runs to oil and loses its flavour. This is achieved by making the blade hollow and providing it with an opening furnished with a screw cap through which boiling water can be poured into the blade.

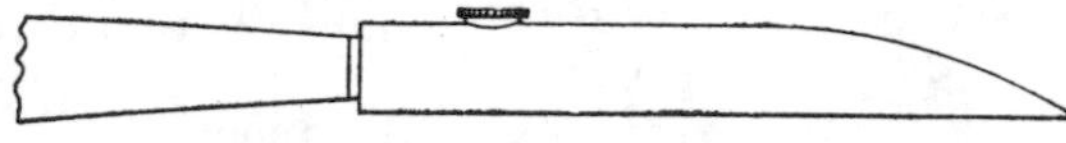

5471/1902
Thomas Butterworth *Life Assurance Agent*
33 King Street, Royton, nr Oldham, Lancashire, England

(Hughes & Young)

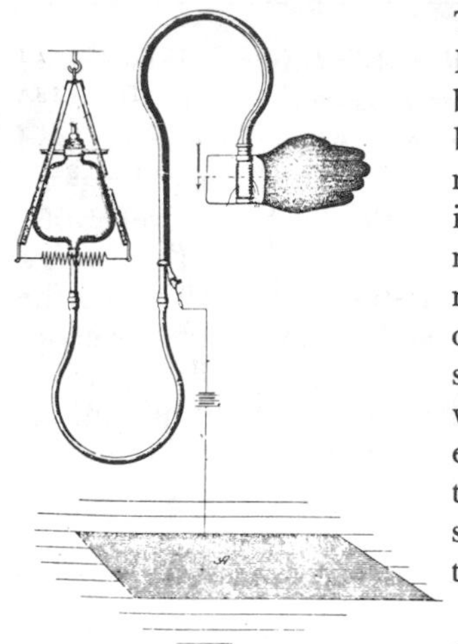

The Personal Hygiene Company's Improved Hydro-Electric Massage Apparatus enables the body to be treated simultaneously with a shower bath, electrical current and frictional manipulation and will prove of inestimable value in care of patients of low vitality. There is a metallic foot-plate or body-contact, a glove or mitten which may be worn either by the bather or by an attendant, and a flexible water-bag suitably connected to the glove. A quantity of water is supplied to the bag, which is then expanded by means of compressed air or gas; this forces the water out of the mitten in minute streams which complete the electrical circuit through the bather's body.

Experiment has shown that with this apparatus the body may be thoroughly bathed and massaged with the use of a comparatively small amount of water. Inasmuch as the surface of the body which is being treated at a given instant is limited, the harmful shock to the system which might follow from the use of a larger quantity of water is avoided. Moreover the massaging operation tends to relieve internal congestion and to increase the exhilarating effect, whilst the application of electric streams stimulates the nerves and greatly increases the local circulation.

6205/1902
Personal Hygiene Company
Chicago, Illinois, USA

(Boult, Wade & Kilburn)

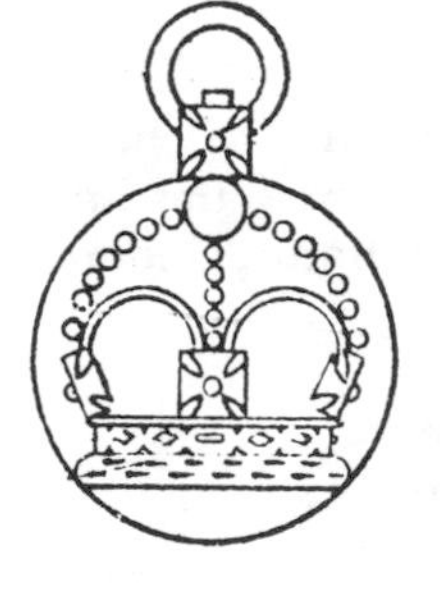

Detmold and Cross's Improvement in Bells, Goblets, Inkstands, Lamp Globes, Lockets, Matchstands, Portrait-holders, Paper Weights, Spoons and Sovereign Purses consists in making all these objects in the formation of a crown. They may be made from metal, wood, glass, celluloid, pulp, sawdust or any other plastic mass and provided with means of affixing thereon portraits of the King and Queen.

The inkstand crown, for example, is made in two pieces partly hollowed out for the reception of one or two coloured inks or fluids, and the bell is made with the top Maltese cross with a push to form the striker in connection with a weight inside. The spoon handle is formed with a crown in relief at its end.

5956/1902
Edward Detmold *Inventor*
9 Crewsdon Road, Brixton, Surrey, England

Rowland Cross *Inventor*
136 Waterloo Road, London, England

(J P Bayly)

Nuttall's Improved Artificial Ear-drum is a disc of india rubber of such size and shape as to fit closely to the natural drum of the ear. A cotton cord is attached to the centre of the disc by being threaded through holes in the centre, the ends of the cord being bound or waxed together, not knotted as this somewhat impairs the transmission of sound.

In use, the disc is held close to the natural ear drum by a suitable adhesive coating such as glycerine or a mixture of glycerine, alcohol and ether, and is fitted using the small rod-shaped instrument shown. This implement with the drum on the end of it is passed into the ear and withdrawn once the drum is adjusted. The projecting cotton cord from the drum is now cut short with a pair of scissors down to the lobe of the ear.

The invention is not restricted to cotton cord for the transmission filament, but this has been found to give the best results and is also the least liable to fray or produce fluffy matter liable to get to the natural drum of the ear and irritate it, which would be most injurious to ears of the kind to which the invention relates.

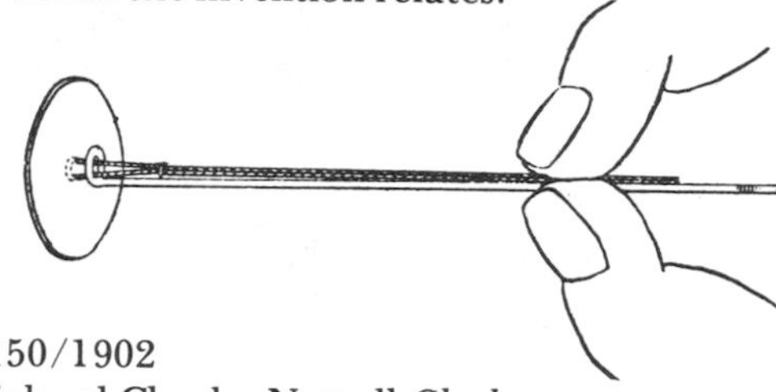

3150/1902
Richard Charles Nuttall *Clerk*
103 Portland Street, Southport, Lancashire, England

(W P Thompson & Co)

Heine's Garment for Protecting the Kidneys or for Healing any Kidney Disorder is worn next to the skin under the shirt. It is shaped to fit the body and provided with pockets destined for the reception of suitable material for keeping the kidneys warm — wadding, wool, flannel or anything similar.

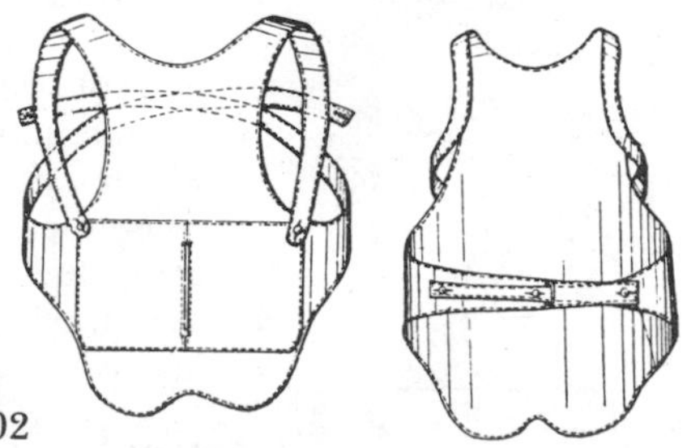

6432/1902
Wilhelm Heine *Gentleman*
Bad Wildungen, Germany

(Ferdinand Nusch *t/a* F G Harrington & Co)

1902

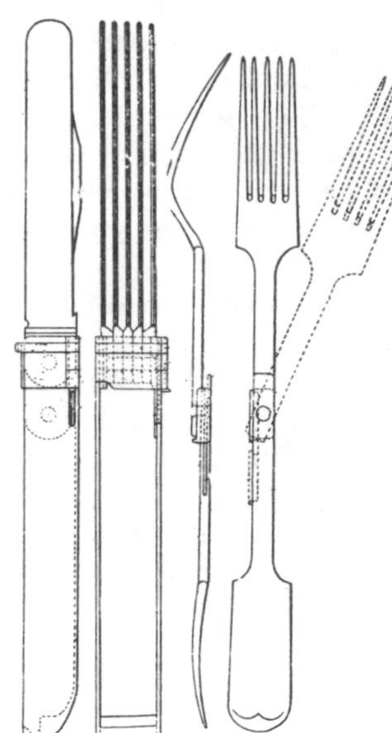

Garda's Portable Combined Fork and many Bladed Table Knife is so constructed that the blades of the knife fit in the spaces between the prongs of the fork, and is to be used for cutting meat and other food into narrow slices. The fork is held in one hand to keep the food in position while the knife handle is held in the other hand and its blades are drawn through the meat between the prongs of the fork. When folded, the knife and fork can be conveniently carried in the pocket.

6284/1902
Giovanni Garda
9 Rue Beccaria, Turin, Italy

(Abel & Imray)

Elmer's Attachment for Safes and Similar Receptacles to Trap and Secure Burglars is constructed so that while a person knowing of the existence of the trap can open the safe without being caught any other person will spring the trap and be caught in it in his attempts to remove the money box from the safe.

Any attempts at grasping the money box cause a toothed blade to move forwards and catch the fingers. The teeth at the extremity of the blade are so shaped as not to cause pain, or the jaws are faced with rubber. A bolt is fitted such that when the trap is sprung and a burglar is caught he is held securely owing to the weight of the safe and cannot escape.

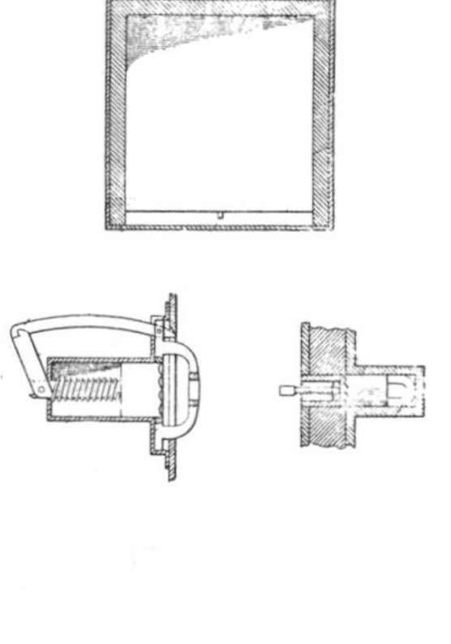

5161/1902
Martin Elmer *Inventor*
Vandyne, Fond-du-Lace County, Wisconsin, USA

(Cruikshank & Fairweather)

Eltz's Cover for Receptacles of Excrements or other Decomposing Matter is designed to prevent the odours thereof emanating in the surrounding air. The cover has one or more rims which fit into corresponding grooves; in consequence the malodorous gases are compelled to mount and descend and are therefore prevented from finding their way out of the vessel.

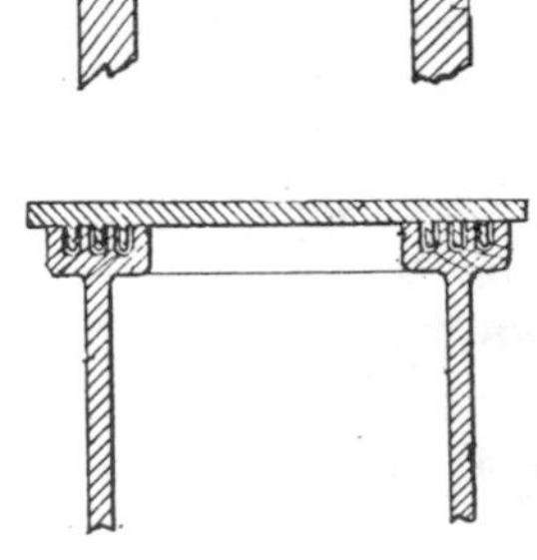

6706/1902
Dr Victor Eltz
Abbazia, Austria

(Max Menzel)

Haynes's Improved Table Game employs a net with two or three circular openings (port holes) near its upper edge. Above the net and suspended from a cross-rod is a hoop. The game is for two players, each of whom has (preferably) five shuttlecocks, rings and tailed balls; a bat; a bat-bag; and a spring shooter for projecting the shuttlecocks etc. through the port holes or hoop, whereupon the opponent endeavours to catch them in the bat-bag or return them through the holes, each time they go through doubling the points.

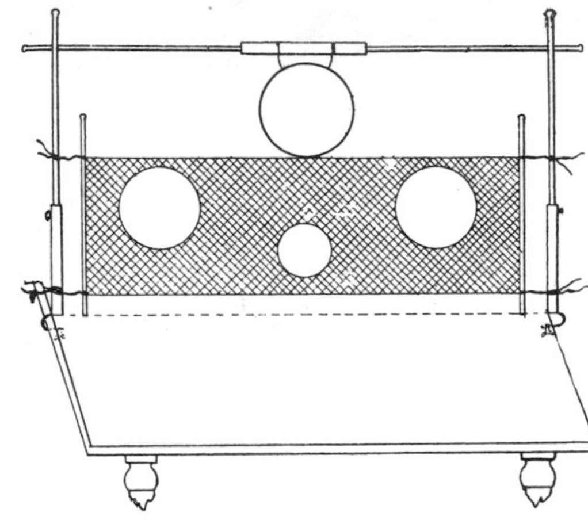

4982/1902
Frederick Haynes, *Retired Master Mariner*
Westfield Bungalow, South Hayling, Hampshire, England

(Hughes & Young)

Bull's Improved Mechanical Toy employs a mechanism as shown to operate the limbs of animals so as to imitate their natural motions.

4239/1902
Einar Bull *Bachelor of Science*
Cross House, Millport, Bute, Scotland

(E Eaton)

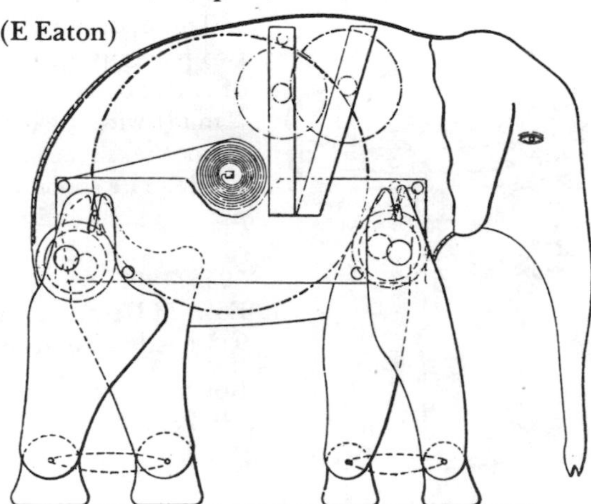

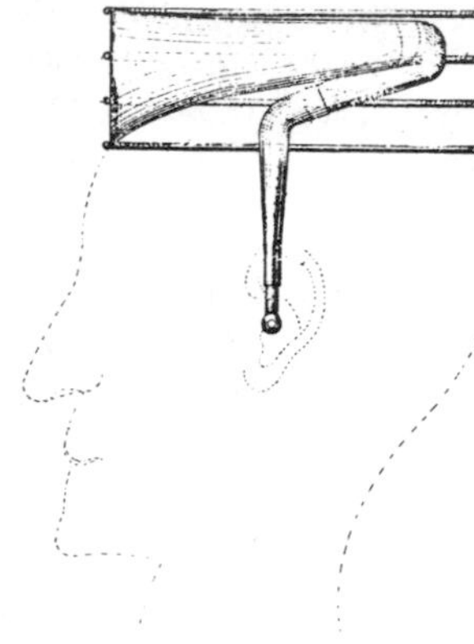

Messenger's Improved Ear Trumpets for the use of those afflicted with deafness provide an effective instrument which can be used with very little inconvenience and will not be unsightly or conspicuous.

This object is accomplished by constructing an instrument with one or more ear trumpets having downwardly projecting ear pieces arranged within a framework which is adapted to form the crown of a cap, hat, or other head covering. The trumpets are so arranged within the framework that their open mouths extend towards the front and are adapted to bear upon the forehead. In this way use is made of the well-known fact that the transmission of sound waves is considerably assisted by a portion of the instrument bearing upon the bones of the head.

9771/1902
Thomas William Messenger *Engineer*
Quorn, South Australia

(Philip M Justice)

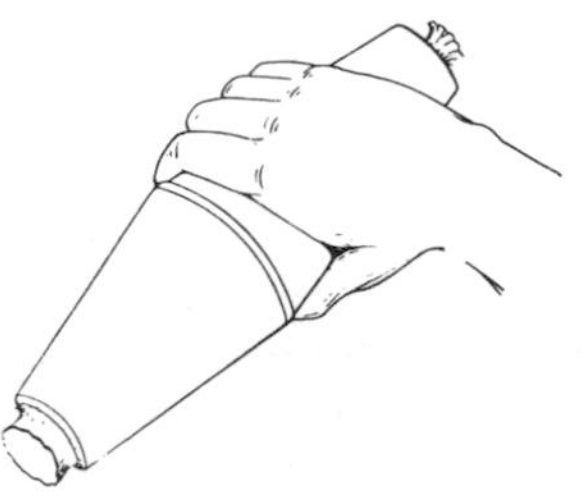

Wheaton's Improved Confetti Distributer has for its object to provide a convenient, cheap and salable [sic] article in the form of a container or case readily adapted to securely hold the confetti without fear of the latter being unintentionally lost or discharged from such container while yet permitting the uniform distribution of the confetti without difficulty and in a speedy manner. It consists of a bi-conical cardboard case with twistable ends of thin paper to retain the confetti. When it is required to distribute the confetti, one end of the case is opened and the contents are shaken out.

10850/1902
Charles Robert Wheaton *Agent*
2 Leswin Road (*formerly* 45 Church Street), Stoke Newington, Middlesex, England

(Herschell & Co)

Helbing's Improved Shirt is made so as to prevent it from riding up in the middle of the back, and is particularly useful for riding or any other kind of exercise. The upper part of the shirt is made on the principle of a coat without a yoke, and the lower part on the principle of a pair of trousers arranged to button in front of fork on right leg.

5040/1902
Thelo Helbing *Tailor*
49 Cherry Street, Birmingham, England

(William Henry Baraclough)

Davies's Game for Bands of Hope and Children in General comprises figures of various sizes and shapes made of wood or other material with a barrel fixed on the top of each figure and a small bell attached thereto. The following and other names of beverages are written or printed on each barrel: ale, porter, rum, gin, whisky, brandy, Old Tom. Balls made of wood on which shall be written, engraved, or stamped the names of Wilfred Lawson; W S Caine; A F Hills; W B Richardson; Canon Wilberforce; Tennyson Smith; Lady Henry Somerset and others are to be used for throwing at and for knocking down the figures.

The whole thing will be a novelty which will amuse and instruct the rising generation.

4931/1902
Joseph Davies *Congregational Minister*
Daylight Villas, Buckley, Chester, England

Friedinger's Improved Train for Ladies' Dresses or Skirts overcomes the great disadvantage of dresses with trains, that the material or border of the train is dragged on the roadway or path so that in a short time the material is worn through and destroyed. Further, the material picks up and absorbs moisture, mud and the like, which dries on and into it and renders it stiff and liable to crack so that its durability is also lessened in this manner. The so-called "brush edging" which has latterly been used on the edges of women's dresses is no protection against this, since it is only adapted to prevent the fraying and tearing of such parts of the edge or seam as may come into contact with the ground.

The improved system provides a border to be arranged inside the train and hold it clear of the ground, and comprises a number of brush trimmings placed side by side in rows; the position of the brushes is such that only a narrow line of bristles touches the ground and the dust on the latter is therefore only lightly touched. In prolonged use, only the ends of the bristles will be worn away.

6875/1902
Ignatz Friedinger *Smith*
Oberthal, nr Hinz, Austria

(Herbert Haddan & Co)

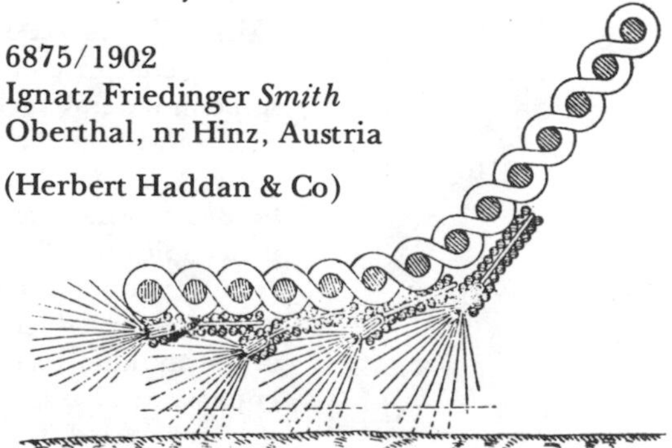

1902

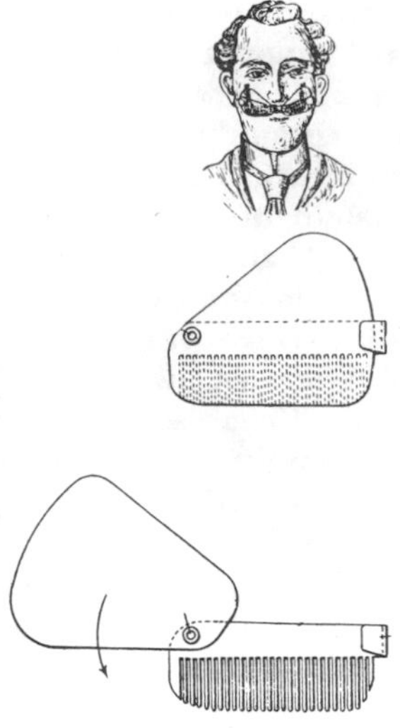

Stephens's Improved Moustache Trainer guides the ends of the moustache so that they assume and maintain certain desired positions or directions in order to conform to a particular fashion or fancy of the wearer.

It consists in a comb with a plate or blade hinged to it, adapted to confine and hold the hairs of the moustache between itself and the comb in such a manner that the said hairs shall have imparted to them a tendency to take and maintain the required shape. It may be made of any suitable material, such as celluloid, tortoise-shell, horn, vulcanite or metal but preferably the blade should be of transparent material to enable the correct position of the hair of the moustache to be verified through it when the device is in place.

7761/1902
James George Stephens *Contractors' Manager*
33 Lansdowne Gardens, South Lambeth, London, England

(W H Beck)

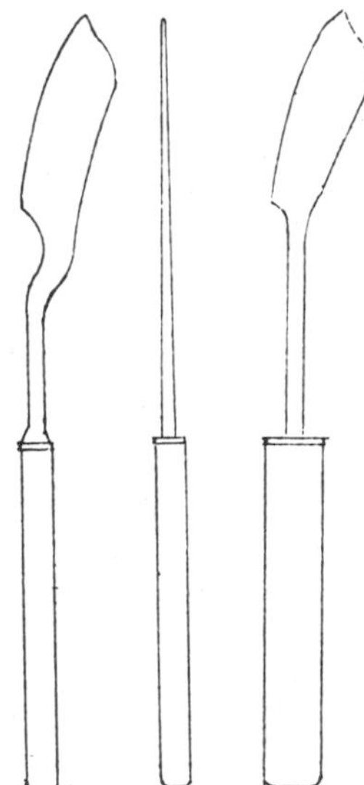

Reid's Improved Table Knife is designed because it is in the nature of things that the old-style table knife cuts only upon its extreme blade point in the plate. The handle, blade and shaft being in a straight line it necessarily follows that, when the handle is raised to cut, practically the whole blade is cutting air and absolutely useless. A blade, say, six inches long employs less than half an inch in the plate. Five and a half inches of blade are, therefore, waste metal.

The improved table knife is shaped with an inclined blade, which, when the handle of the knife is raised, brings the edge of the blade evenly, entirely and effectively upon the surface of the plate.

9738/1902
Andrew Reid *Author*
Lambourne, Romford, Essex, England

Clifford's New Device for Teaching the Golf Swing comprises two independent uprights on stands made of metal or any other material. It teaches what is technically termed, in the art of golfing, as the "golf swing". It can be used in houses or wherever a golf club can be swung, as it occupies but little space. It is, so to speak, a "home trainer". On the stand is a clip which carries a curved length of rubber, or a spring, or other suitable resilient material which it is the endeavour of the player to strike in the upward and downward swing of his golf club.

These arms may, if desired, be provided with a bell or other phonetic device, whereby an audible signal will be given each time they are touched by the player's club.

Care should be taken that the devices be placed so far apart that when a player is effecting the swing, it will be possible for him only to touch the rubber arms. Good players can place the devices themselves; moderate players or beginners should seek the assistance of a professional or good golfer.

7726/1902
Stanley Clifford
Neasden Golf Club, Neasden, London NW, England

(C: Ernest de Pass)

Taylor's Improved Driving Mechanism for Cycles and other Wheeled Vehicles provides means whereby the weight of the vehicle and rider is utilized to store up energy which can be employed to assist in propelling the vehicle.

The rear wheel carries a number of radial cylinders, containing pistons which cause air to be compressed and stored in a vessel under the cross bar. The compressed air drives a small air engine which, via suitable linkages, assists the turning of the pedals.

6702/1902
William Rutherford Taylor *Cycle Fitter & Agent*
Maitland Place, Thirlstane, Bo'ness, Linlithgowshire, Scotland

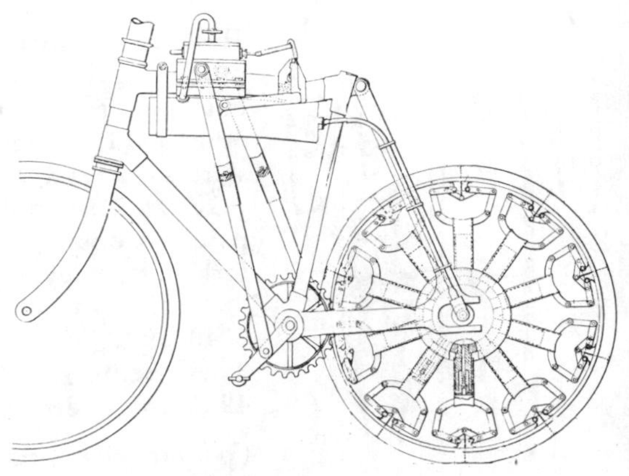

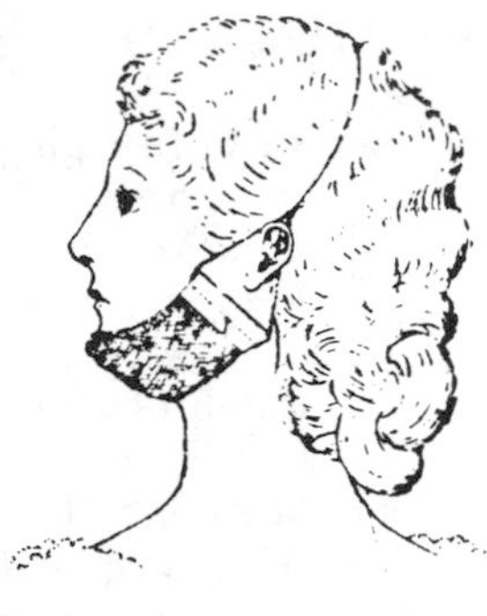

Nicholson's Suspensory Chin Strap is an appliance adapted to prevent, or assist in preventing the formation, which is more especially liable to occur in persons of full habit, of what is commonly known as a "double chin", and which, in ladies at all events, is from an aesthetic point of view a disfigurement.

The appliance consists essentially of a length of light elastic fabric, with a sufficiently open mesh to give adequate ventilation, with its ends stiffened transversely by whalebone so as to maintain the elastic band distended in the transverse direction. This ensures an evenly distributed and comfortable application of pressure to afford support for the relaxed muscles and flaccid tissue beneath the chin during sleep, so helping preserve the original natural contour.

9783/1902
Eleanor Huntley Nicholson *Face Hygienist*
t/a Mrs Adair, 90 New Bond Street, London W, England

(A M & WM Clark)

Withrow's Improved Perambulator may be transformed into a suspensory or basket when it is desired to transfer the child from one place to another where it would be impossible to wheel the carriage, and thus avoids the necessity of unfastening and lifting the child out of the carriage and replacing and securing it again after reaching a place where the perambulator may be wheeled.

9894/1902
Samuel Pottenger Withrow *Manufacturer*
632 West Fourth Street, Cincinnatti, Ohio, USA

(Herbert Haddan & Co)

Lehmstedt's Improved Massage Apparatus comprises a set of instruments for massaging the face, one of which can be used after the other, the form of the instruments being determined according to the shape and form of the part of the face for which they are employed and the particular purpose for which they are intended. Balls and rollers, smooth and ribbed, and a smoothing rod are provided, and are contained in a box which may also act as a receptacle for creams, lotions and other substances for use in massage.

6627/1902
Paul Lehmstedt *Merchant*
1A Potsdamerstrasse, Berlin West, Germany

(Edward Evans & Co)

Newman's Improved Skewer has greatly increased holding and penetrative power whilst being constructed so as to avoid as far as possible injury to the hands.

With the wooden skewers at present employed considerable pressure is needed to insert them and this, where a quantity of skewers are to be inserted, causes the hands primarily to become exceedingly sore and results eventually in the formation of corns upon them. A further objection to wooden skewers is that when the meat is cooked the projecting ends become charred and their removal only with difficulty accomplished.

Accordingly, this invention is effected by constructing a skewer having a body of any suitable length and crinkled or waved either throughout its entire length or for a certain portion thereof. The skewer is provided with a suitable head formed by bending the metal.

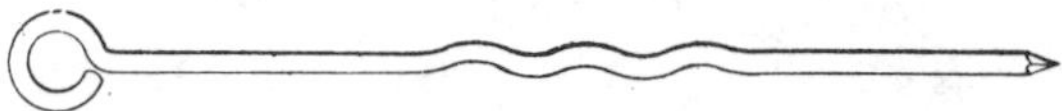

8799/1902
John Herbert Newman *Butcher*
196 Tulse Hill, London SW, England

(J E Evans-Jackson & Co)

Seeger's Collapsible Sun or Rain Shade is adapted to be carried upon the shoulders so that the hands are left free for other purposes, which is of especial importance for cyclists, surveyors and others working in the open air, cripples, tourists, horsemen and others. The awning forms a protection against obliquely falling rain or rays of the sun, either in front of behind, without, however, opposing the passage of air. It may also be employed, in case of need, as a tent (as a single tent for soldiers for example).

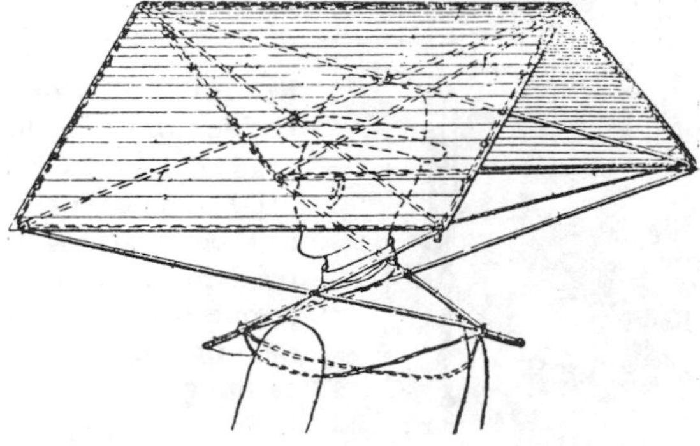

9640/1902
Ludwig Seeger *Merchant*
Feldkirch, Vorarlberg, Austria

(Haseltine, Lake & Co)

1902

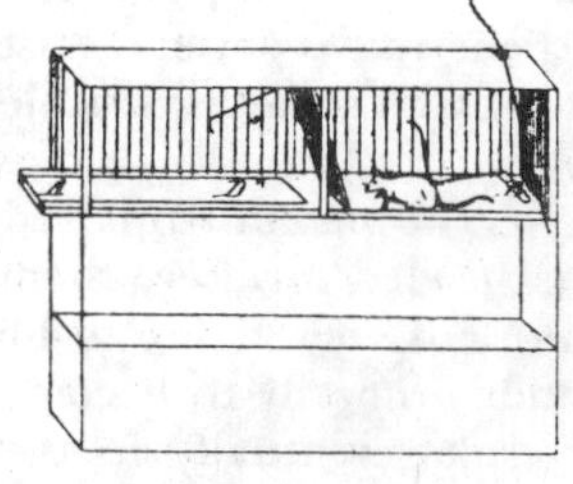

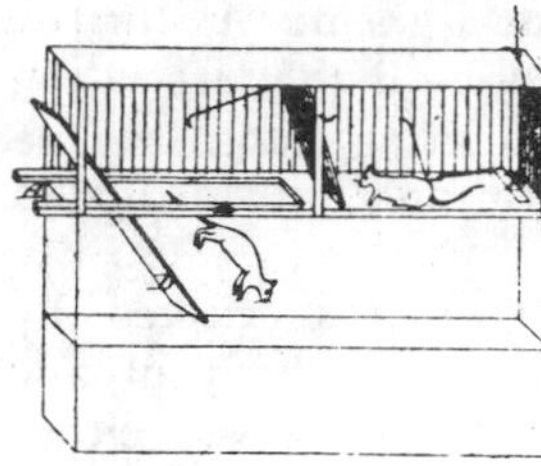

Alston's Improved Trap for Mice, Rats and Similar Vermin consists in a cage, with a tank beneath it, to catch and imprison a mouse which serves as a decoy to other mice which, when entering at the other end of said cage, will be precipitated into the tank beneath by means of a pivoted platform.

When the apparatus is used for field mice, the tank should be sunk into the ground so that the mice can enter the trap at ground level. When it is desired to set the trap for house mice, on the floor of a room for example, the tank may be disconnected, but as by this means only one mouse could be caught at a time it is preferable to use the tank attached and place books, or anything else, to the level of the entrance.

10159/1902
Harry Alexander Alston *Farmer*
4 Barkley Terrace, Cape Town, Cape Colony, South Africa

(Haseltine, Lake & Co)

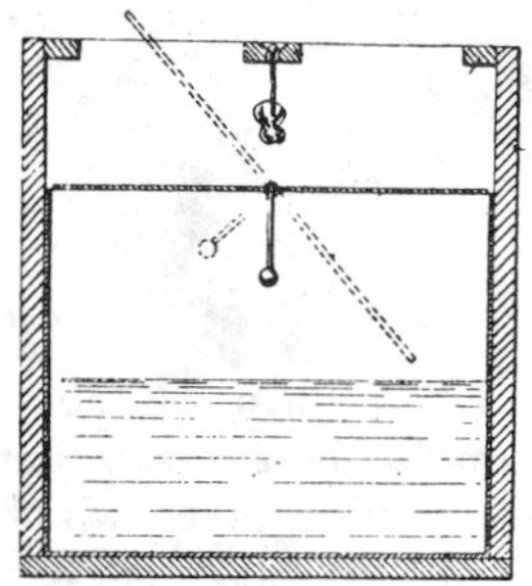

Wolfer's Automatic Trap for Animals has the bait arranged in such a manner that the animal in its attempts to get at it is obliged to leap onto a lid situated over a receiver filled with a poisonous substance. The lid gives way under the weight of the animal, which is thereby deposited into the receiver of poison. Once the animal has fallen in, the counterpoised lid returns to its original position ready to receive another imprudent mouse. The outside of the case is made rough so that a mouse or other animal on the look-out for food will lose no time when attracted by the odour of the bait in climbing up the core and trying to get at it.

11863/1902
Heinrich Wolfer *Manufacturer*
7 Fuhrmanstrasse, Darmstadt, Germany

(George Barker)

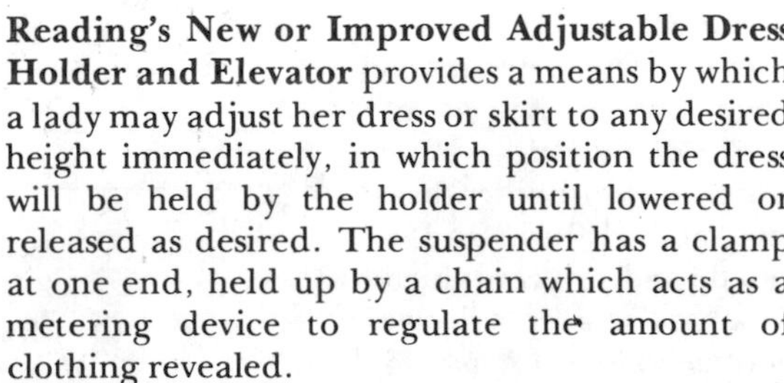

Reading's New or Improved Adjustable Dress Holder and Elevator provides a means by which a lady may adjust her dress or skirt to any desired height immediately, in which position the dress will be held by the holder until lowered or released as desired. The suspender has a clamp at one end, held up by a chain which acts as a metering device to regulate the amount of clothing revealed.

11267/1902
Nathaniel Cracknell Reading *Manufacturing Jeweller*
33 Hall Street, Birmingham, England

(George Barker)

Kreussler's Improved Combination Folding Bath and Bedstead provides an inexpensive and efficient means whereby the necessity of keeping a separate room specially for holding a bath, and the discomforts attendant to a person requiring to use same, consequent upon the difference of temperature usually existent in and between that of the bed and bath rooms, are entirely obviated and a valuable economy of house room thereby effected, the said invention enabling the bath to be situated in the bedroom itself and appear when folded as, and occupying no more space than, an ordinary wardrobe, while, when required for use as a bath, or to serve as a temporary bedstead, the manipulation requisite for either purpose, and for folding up after use, is extremely simple and the labour involved practically nil.

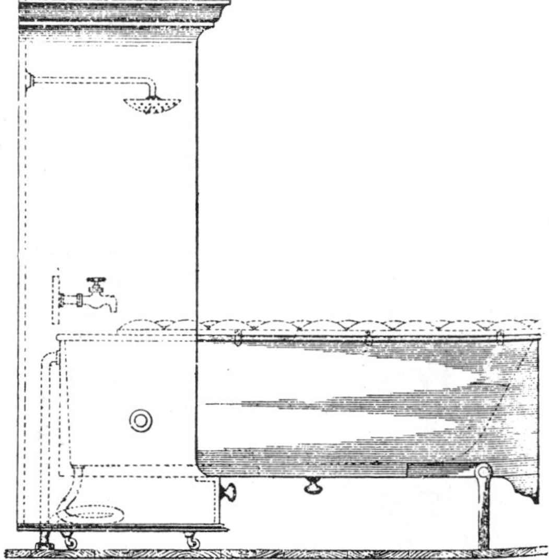

9932/1902
Christian William Kreussler *Hotel Proprietor*
3 Guildford Place, Russell Square, London WC England

(Geo Thos Hyde)

Anderson's Improved Method of Treating Starch and Starch-containing Materials consists in heating the materials, dry, to 125°C – 300°C for from ten to forty-five minutes under pressure. After heating, the pressure is released, whereupon the material is rendered porous by expansion. The expanded material may be used for paste, size etc, or for food preparations.

13353/1902
Alexander Pierce Anderson
629 198th Street, Bedford Park, Bronx, New York, USA

(W P Thompson & Co)

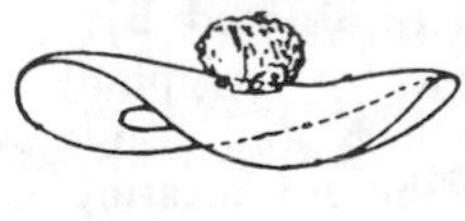

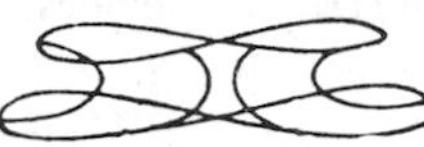

Buckler's Improved Head Cover overcomes the disadvantage of the conventional Tam O'Shanter cap, which, while particularly soft to the head and easy to the wearer, invariably loses its original shape when pulled on to the head and always presents a more or less negligé appearance. Also, the crown of the cap overhangs either side or the front of the wearer's head and in windy weather is lifted by the wind and caused to shift its location, this not only altering the appearance but also being somewhat uncomfortable.

The improved head cover is built on a resilient and preferably endless wire framework to impart a permanent shape to the cap which it will maintain notwithstanding that it may be subject to pressure, squeezing or other conditions. The shaping wire may be bent so as to correspondingly alter the shape of the cap to imitate any of the prevailing fashions of ladies' head gear.

10682/1902
W Buckler & Co and
William Buckler *Hosiery and Tam O'Shanter Manufacturers*
Lansdowne Road, Aylestone Park, Leicester, England

(Boult, Wade & Kilburn)

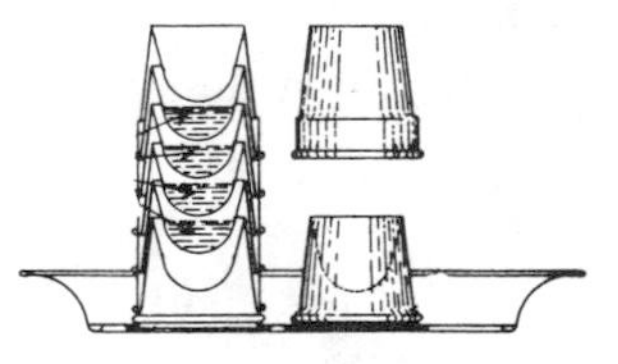

Knowles's Communion Service Tray and Stacking Cups are designed to obviate the risk of contagion by disease germs incurred by taking communion by the usual method. Each communicant uses a separate cup from the stack, rather than taking wine from a common chalice as is the customary procedure. To ensure that no liability to upset a pile of cups and so detract from the decorum of the service is incurred, the cups are of novel form to fit in piles or series to secure compactness and stability in carrying. It is suggested that cups are sterilised, filled and piled in the vestry prior to distribution to communicants. The number of cups in each pile is preferably the same as the number of persons in the pew or table to be served, and the top cup in the pile is fitted with a dome-shaped cover to render it inaccessible to germs. The communicants pass the piles of cups along the pew, and then after having taken communion each replaces his cup, piling them in the reverse order.

11331/1902
Alexander Knowles *Sanitary Inspector*
Burgh Sanitary Office, Inverness, Scotland

(P: Cruikshank & Fairweather; C: Self)

Cayley's Improvements in and relating to Crackers and in the Ornamentation or Display thereof, said Improvements being applicable also to Holders for use with or without Soufflets Suitable for Containing Sweetmeats, Ices or the like and for Table Decoration purposes have attached to them figures or designs of paper, cotton, wool, or any other material, so made as to represent mushrooms, toadstools, lichens, ferns, grass, flowers, plants, fungi, elves, fairies, "brownies", imps or similar forest or emblematical representations, the whole being supported on wires or springs and arranged so that the desired scenic effect or representation may be given thereto.

10186/1902
Frederick William Cayley and
A J Cayley & Son Ltd *Cracker Manufacturers*
Fleur-de-Lys Works, Norwich, Norfolk, England

(T B Browne Ltd)

Krüger's Improved Device for Combing and Smoothing the Moustache consists of a plate with projecting teeth and side guards carried by an arm to which is pivoted another arm carrying a comb with side guards. The illustration shows the device in use by a Gentleman.

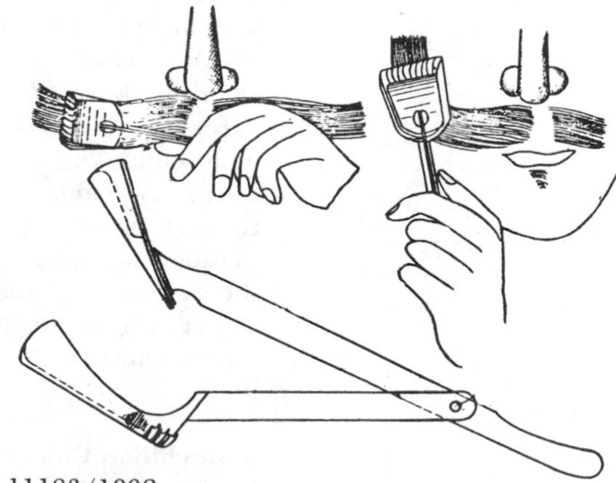

11183/1902
Friedrick Krüger *Gentleman*
26 Westendstrasse, Karlsruhe, Baden, Germany

(Dewitz Krebs & Co)

1902

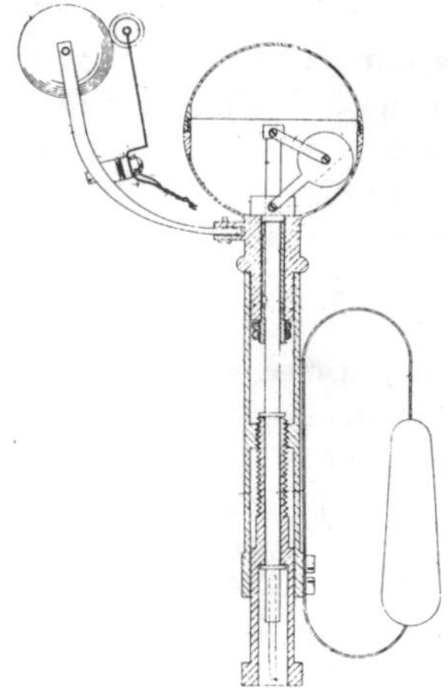

Weinert's Improved Massage Apparatus is the type of device in which, by means of electricity, a shaft with a swinging body or weight attached thereto is rapidly rotated inside a casing and the shocks produced thereby are transmitted to the part of the patient's body to be treated.

Previously, the shocks produced by such apparatus were also transmitted to the doctor or masseur administering the treatment; in the improved apparatus a spring is provided in the handle which effectually intercepts such shocks. The apparatus is so constructed that electricity may be passed to the patient from it and, the part of the body being subjected to massage being electrified at the same time, the intended effect is thereby considerably increased.

10161/1902
Karl Weinert *Arc-lamp Manufacturer*
32 Muskauerstrasse, Berlin, Germany

(Wheatley & Mackenzie)

Sir Charles Cookson's Rack for Holding Toasted Bread is made hollow, and furnished with a drawer to take a heated iron, or hot charcoal; alternatively, it may be filled with hot water.

10623/1902
Sir Charles Alfred Cookson CB, KCMG
96 Cheyne Walk, Chelsea, London SW, England

(P: Jno H Raynor; C: Self)

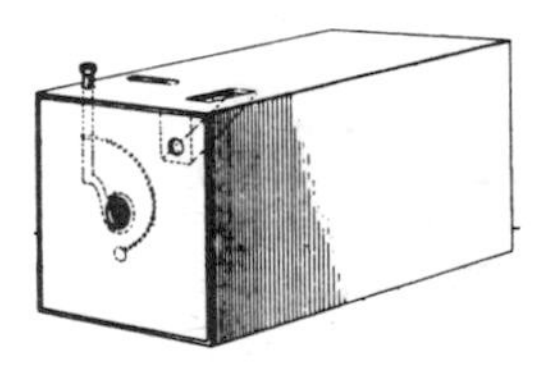

Church and Reynolds's Improved Advertising Device is a novel construction whereby advertisements may be delivered apparently as photographs.

Designed for use by children as a toy, the device resembles a hand camera and is loaded with objects designed to maintain the illusion of the delivery of a picture supposedly made in the camera. Mirrors, and printed or embossed cards, bearing advertisements, are suggested. The child or subject whose photograph is ostensibly being taken deposits a coin through the slot, and the device is operated as if a real photograph were being taken. The operator then opens the holder and delivers one of the objects contained therein, thus distributing the advertisements.

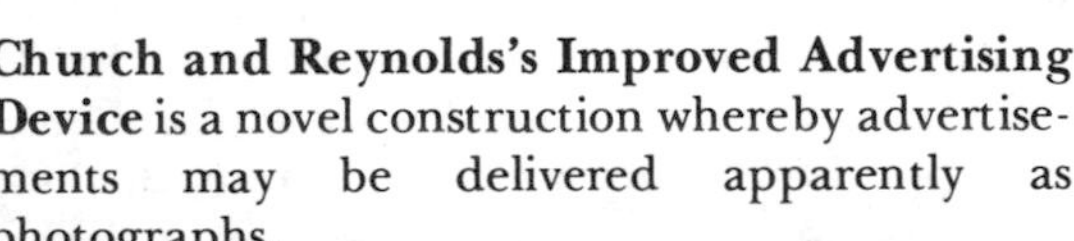

11289/1902
James Elliott Church &
George William Reynolds *Manufacturers*
23 Essex Street, Cambridgeport, Massachussets, USA

(Haseltine, Lake & Co)

Proveanie's Improved Illuminated Bicycle is designed to supplement or supplant the ornamentation of the wheels and frame with glow lamps, by adding thereto rotating cross arms carrying other electrical devices. By causing said arms to rotate in an opposite direction to the wheels a striking and attractive result is obtained. Any suitable number of cross arms may be used and these are fitted with either or both electric glow lamps and vacuum tubes. Showers or flashes of sparks may also be produced by mounting electrical brushes on the bicycle frame and studs or lugs upon the wheels, such that when the wheels are revolved the brushes make frequent intermittent contacts, producing the effect of a circle of sparks to the observer. Electricity may be obtained from a battery carried on the machine or from other sources, such as an installation in the building or place in which the bicycle is used.

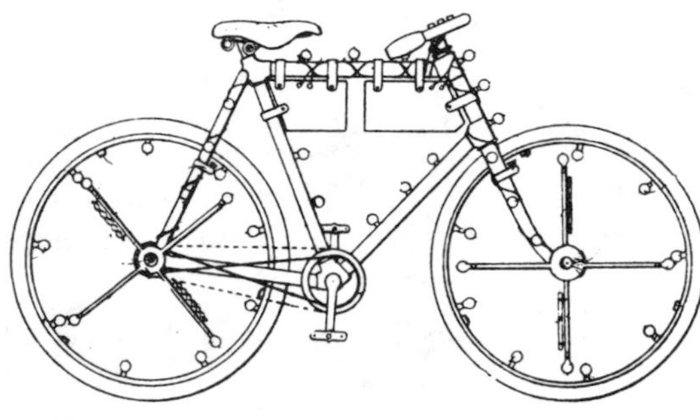

11384/1902
Arthur Proveanie *Cyclist-Expert*
30 Wolfington Road, West Norwood, London, England

(S S Bromhead)

Cole's Improved Lawn Tennis Bat, Golf Club or Croquet Mallet has a handle which is divided and sprung with a piece of whalebone, cane, steel, or other material and an indiarubber buffer. The object is to prevent shocks or jarring being transmitted to the hands of the user, to increase the flexibility of the handle and to improve the driving power of the bat, croquet mallet or the like.

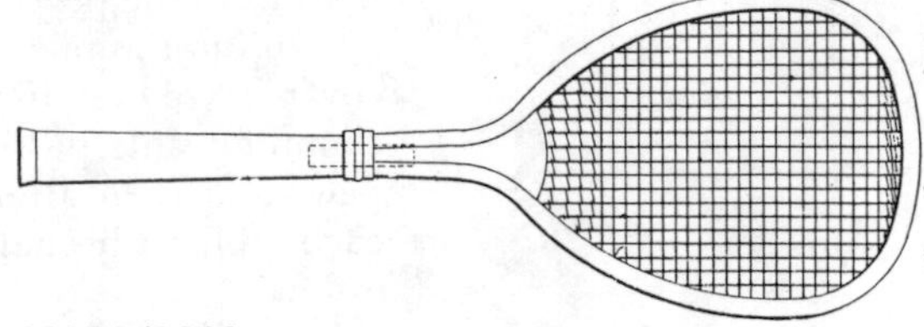

11893/1902
Howard Speare Cole *Clerk in Holy Orders*
The Vicarage, South Brent, Devonshire, England

(Vaughan & Son)

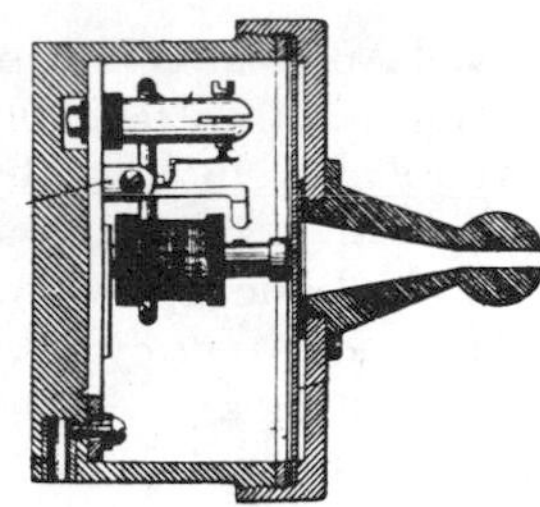

The Hutchison Acoustic Company's Improved Apparatus for Massage of the Ear by Percussive Sounds subjects the ears of deaf persons to a series of rapidly recurring, sharp sounds, thus exercising the ear drum and auricular bones, promoting the circulation and improving the general physical condition. The instrument consists of a casing having an opening on one side for the application of the ear, and a diaphragm vibrated by an electromagnet so as to produce the sharp, clear and well-defined individual sounds necessary satisfactorily to massage the parts of the ear.

11341/1902
Hutchison Acoustic Company
71 Broadway, New York, USA

(W E Heys)

Reagan's Improved Souvenir in the form of a Walking-stick provides an article which can be carried on a festive occasion and thereafter retained as an ornamental commemoration of that occasion. The stick is adorned with suitable favours or portraits and may be hollow for the carriage of chocolate or other articles. The illustration shows a stick designed for use during the coming Coronation Season.

11740/1902
May Roger Reagan *Married Woman*
The Cottage, Llanrhaiadr, Denbigh, North Wales

(Hughes & Young)

Zimmermann's Improved Appliance for Preventing Wetting of the Bed differs from like apparatus hitherto employed by the omission of all screws. It also possesses great simplicity, consisting merely in a springy bent wire, on the arms of which two rubber pads are suitably mounted.

The appliance is placed on the upper part of the male member quite at the rear so that the knobs formed on the free end of the bent arm (which have for their object the avoidance of any wounding) lie uppermost, and the member lies between the pads.

Should a desire to urinate come in the night, it cannot escape, and consequently the patient awakes, displaces the upper pad, and satisfies his requirements.

6865/1902
Carl Zimmermann *Manufacturer*
Heidelberg, (Baden) Germany

(W P Thompson & Co)

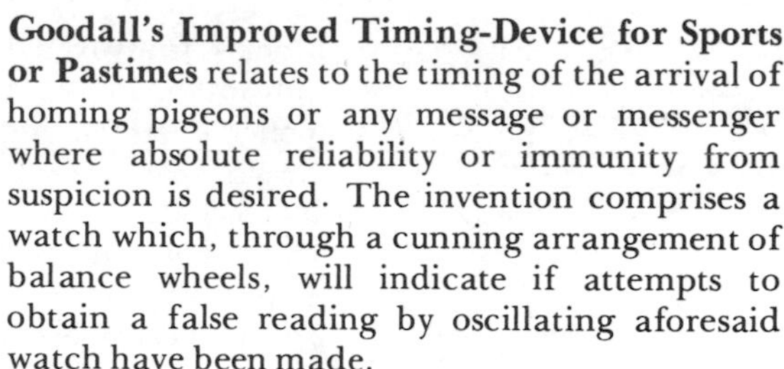

Goodall's Improved Timing-Device for Sports or Pastimes relates to the timing of the arrival of homing pigeons or any message or messenger where absolute reliability or immunity from suspicion is desired. The invention comprises a watch which, through a cunning arrangement of balance wheels, will indicate if attempts to obtain a false reading by oscillating aforesaid watch have been made.

10741/1902
William Goodall *Furnishing Manager*
99 Abbey Street, Derby, England

(W Swindell)

de Lipkowski's Improved Aërial Machine consists essentially in two contrarotating helical lifting planes of special construction which take the machine *up,* and a propeller which pulls it *forward.* Pneumatic buffers deaden the shock of landing, and the resistance which the rudder offers to the wind compensates for the action of the wind upon the lifting screws so that perfect equilibrium may be obtained; the main axis of the apparatus may so be rendered vertical whatever be the velocity of its translation.

11616/1902
Joseph de Lipkowski *Engineer*
104 Boulevard de Courcelles, Paris, France

(Haseltine, Lake & Co)

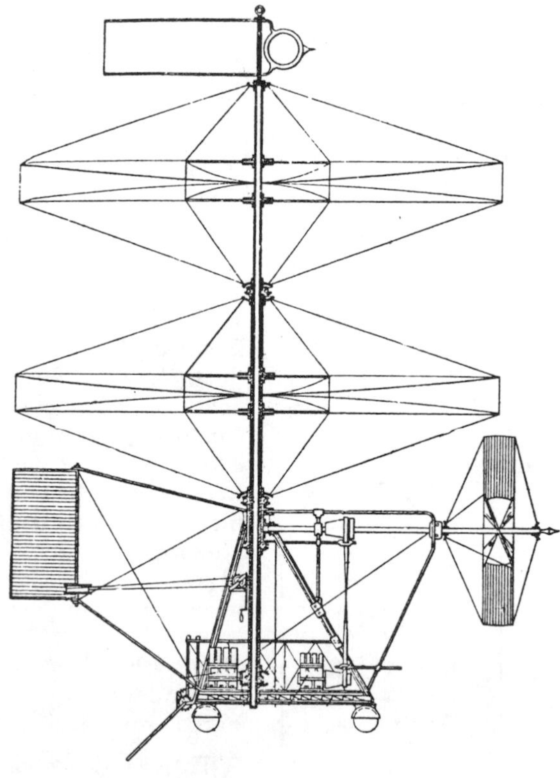

1902

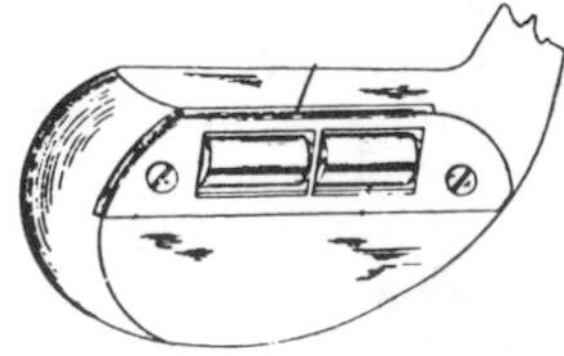

Black's Improved Golf Club has a sole fitted with rollers in order to prevent the stroke being spoiled by its striking upon the ground.

11463/1902
Thomas Hutcheson Bonthron Black *Life Assurance Company's Superintendent*
17 Cluny Gardens, Edinburgh, Scotland

(Johnsons)

Love's Improved Knitted Drawers for Women are so constructed as to dispense with the number of objectionable seams necessitated by the present method of manufacture, thereby rendering the garment more comfortable and hence more generally acceptable than usual. The legs are produced from diagonally overlapping seamless tubular webs, of a proper length for the intended drawers, united at a waistband and finished in any desirable manner.

11798/1902
Henry Milligan Love *Manufacturer of Knitted Goods*
405 E13th Street, Wilmington, Delaware, USA

(Geo Davies & Son)

Ehrsam's Folding Card Serving as Envelope for Pressed Flowers or Plants provides an attractive souvenir or memento, combining a convenient holder for the flora with a view of the place where they were gathered and a space for messages and an address.

12784/1902
Jakob Ehrsam *Merchant and Manufacturer*
Suffenheim, Alsace, Germany

(Abel & Imray)

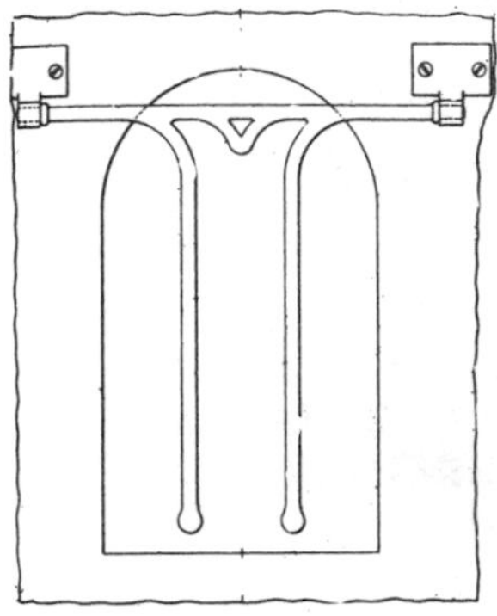

Iles's Device for Preventing Foxes from Entering Fowl Houses, is based on the proved theory that a fox will not place his nostrils against cold metal. The device accordingly comprises a set of metal bars suspended over the entrance hole of the hen-house. The bars swing freely within the entrance hole, permitting free ingress and egress of the fowls; but, if a fox attempts to enter, its nostrils will come into contact with the cold metal bars and it will cease to try to pass through.

12242/1902
Alexander Iles
Park Farm, Fairford, Gloucestershire, England

(E P Alexander & Son)

von Trützschler's Improved Apparatus for use in Teaching Bicycle Riding is easily attachable to any bicycle for the purpose of facilitating the learning to ride. It comprises a light frame serving as a carriage within which the hind wheel of the bicycle is placed, connected to the cycle by rods. The cycle is thus assured against falling over and the cyclist can mount and ride without other assistance. As the rider progresses in his ability the spring connection between support and cycle can be loosened to give greater independence.

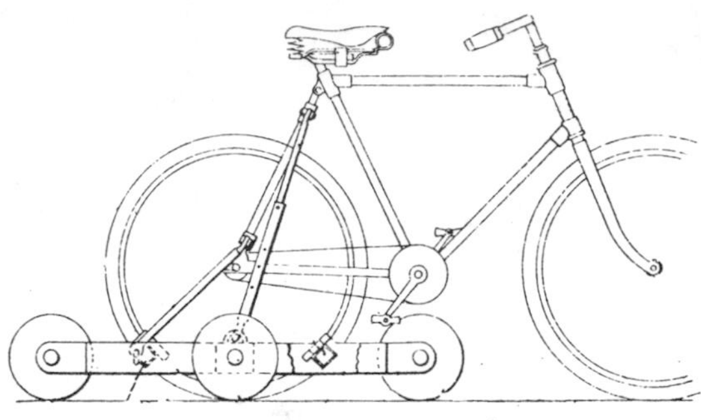

11694/1902
Fritz von Trützschler *Lieutenant and Adjutant in the German Army*
26 Linden Allee, Wismar, Germany

(W P Thompson & Co)

Linnekogel's Improved Method of Manufacture of Cigars and Cigarettes has as its object to render the smoking of cigars and cigarettes less injurious to health by partially or entirely keeping back the poisonous matter contained in the tobacco leaf which, as is known, dissolves when the tobacco becomes heated and passes in liquid form or with the smoke into the mouth. To this end, cigars or cigarettes are provided with absorbent plugs made of ramie fibre, wool, jute, cotton, moss, etc, which has been saturated with a solution of maleic acid and pure vaseline, treated with a solution of potassium platinocyanide or silver cyanide, and finally pressed and dried. Plugs thus prepared have the special effect of absorbing the brown, ethereal poisonous and bitter brenzoel, so that the latter is entirely prevented from mixing with the secretions of the salivary glands and thus entering the body. Cigars and cigarettes provided with plugs prepared in this manner can be smoked without any injury to health so that even invalids can indulge in their use.

11419/1902
Heinrich Linnekogel *Physician*
Silberburgstrasse 150/II Stuttgart, Germany

(Herbert Haddan & Co)

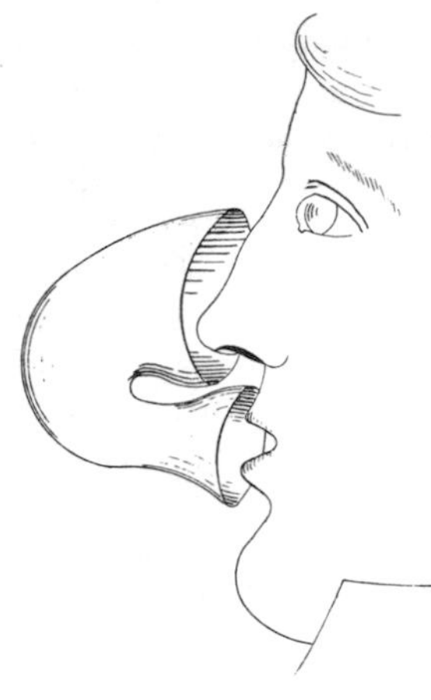

Leder's Foul Breath Indicator is a toilet article to enable persons to test their breath, and comprises an appliance in the shape of a curved tube made of any non-absorbent material such as celluloid, silver or glass, preferably with flared ends. By breathing from the mouth through the tube any foulness or unpleasant state of the breath may be readily detected by the sense of smell.

The drawing is a view showing the invention in use.

16011/1902
Xavier Henry Leder *Seaman*
The Seaman's Institute, Well Street,
London, England

(C: Chas Coventry)

Barlow's Process for the Preservation of New-Laid Eggs hermetically seals the egg shells by a layer of varnish, which is evenly deposited owing to the evaporation of the spirit holding it in solution while the eggs are placed upon the points of sharp steel nails not less than one inch in length and three quarters of an inch apart from each other to dry.

11054/1902
Edward Hovenden Barlow *Medical Student*
21 Great College Street, Westminster,
London SW, England

Hitchings's Improved Milk Receiving Vessel meets the problem that such vessels are often stolen or their contents spilt. The receptacle hangs inside the door, and the milkman carries a suitably formed funnel. The aperture in the door is closed by pivoted covers when not in use.

14125/1902
Annie Elizabeth Hitchings *Teacher*
1 Market Street, Haverfordwest,
Pembrokeshire, Wales

(E Eaton)

Bode's Telescopic or Expanding Whip is intended especially for the use of wheelmen — cyclists and the like — as a protection against dogs. When the handle is swung violently, the thong is jerked out ready for use, and when closed, the whip can easily be carried about in the pocket and can also be fastened to any vehicle.

12763/1902
Heinrich Bode *Merchant*
6 Obstmarkt, Nuremberg, Bavaria, Germany

(S S Bromhead)

The Heards, Nicholson and Neel's Improved Husk Hat aims to utilise maize corn shucks in the manufacture of hats, because of the lightness, toughness and cheapness of the material and the ease with which it can be shaped. This invention is a new use for a material generally wasted and sun hats are admirably adapted to be made from corn shucks because large-brimmed hats are exceedingly light and cool. Corn shucks are cheap and tough and will not chip or crack when bent; they taper in width from butt to tip, and can assume any shape especially when dampened. The hats comprise a plurality of courses of shucks, the outer ends of one course overlapping the inner ends of another and the whole being stitched into place.

11860/1902
Sallie Clara Heard *Lady*, &
Lee Ralph Heard, Clyde Nicholson & George Rowntree Neel, *Merchants*
Thomasville, Georgia, USA

(Boult, Wade & Kilburn)

Scuri's Improved Device for Shaping Moustaches enables the moustache to be shaped without employing either hot irons or cosmetics. A comb, bent to suit the shape of the upper lip to which it is to be applied, is passed through the previously dampened moustaches which are pressed, with more or less considerable force, on to the metal plates attached to the comb. Providing the device is maintained in position for a sufficient length of time, its effect is that the moustache will be formed into the desired shape and retain same for a more or less lengthy period.

15503/1902
Alessandro Giovanni Battista Scuri
Manufacturer
38 Rue Lambert, Lebegue, Liège, Belgium

(Charles Bauer, Imrie & Co)

Henn's Improved Brush for Cleaning Bicycles and the Like is an implement by means whereof a bicycle may be thoroughly, quickly and easily cleansed. The tyre, rim, spokes and hub of a bicycle or like wheel are cleaned simultaneously by a forked brush. Using this single implement a bicycle is completely cleansed most expeditiously and conveniently, the small expenditure of time required being not worth mention.

12238/1902
Wilhelm Henn *Merchant*
Bretten, Baden, Germany

(C Kluger)

Jakobi's Improved Cuff Holder is an arrangement which is for holding back cuffs in coat sleeves, thereby preventing the unpleasant displacing of cuffs which, in the present way of wearing them, at times go much below the sleeves and at other times go up in such a way that they cannot be seen. For uniforms, the holder is placed in preference in such a way that same is at two centimetres from the under seam of the sleeve, its upper end being at one centimetre from the end of the sleeve. A guide-way is fixed in the sleeve of the coat, and carries a block and stud which engages in the button-hole of the cuff. A scale is marked on the guide-way so as to fix exactly the place where the stud shoul be clamped.

13525/1902
Peter Jakobi *Tailor*
Bad-Nauheim, Allemagne, Holland

(E Eaton)

Walther's Improved Head Covering or Screen for Protecting Horses against Heat Stroke or Sunstroke overcomes the disadvantages of the straw hats more commonly used in hot weather, which not only have no cooling effect but, besides, appear ugly and ridiculous. Walther's head covering is made of absorbent material, dipped into cold water before application to the horse's head. The cooling effect is produced by evaporation of the water, too rapid evaporation being prevented by the provision of an air space beneath. This also ensures adequate ventilation.

The screen may be coloured to match the horse; brown, white or black as appropriate.

14048/1902
Franz Cuno Walther *Veterinary Surgeon*
84 Markgrafenstrasse, Berlin, Germany

(Wheatley & Mackenzie)

Baker's Improved Advertising Display Card provides for the ready and convenient exhibition of the effect of variously coloured or tinted trimmings or fittings for the figures or articles represented thereon. A card to advertise goods — for example, prams — has a rotating coloured backing disc so that the effects of different trims can be demonstrated. An almanac or other printed or pictorial matter may be applied to the vacant space on the card. The illustration shows the application of the invention in connection with a show card for the exhibition of perambulators or baby carriages.

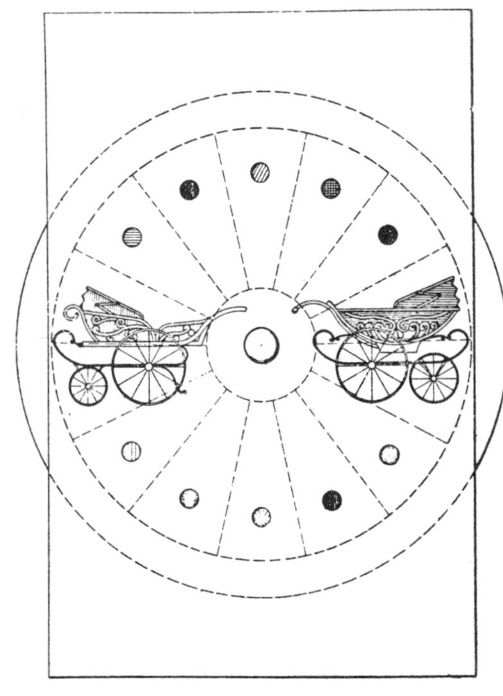

13313/1902
Henry Vincent Baker *Manufacturer*
70 Summer Row, Birmingham, England

(Marks & Clerk)

Bourne's Improved Device for Smoothing the Rough Edges of Starched Linen Articles of Wearing Apparel consists of a small oblong piece of glass, china, or other hard substance having at one extremity two prongs. The rough edge of the apparel to be smoothed is placed between the prongs and rubbed backwards and forwards.

14159/1902
William Bourne *Author*
2 Jackson's Grove, Southport, Lancashire, England

Burr's Improved Fire Escape is a simple and efficient apparatus by means of which people may be lowered from a burning building and the last person may lower himself. A hook provided with a spring safety device is attached to one end of a rope and to the other end is attached an eye, through which part of the rope is passed thereby forming a sliding loop which may be placed under the arms and around the body of the person to be lowered. The rope, which is preferably treated to render it fireproof, is provided at suitable intervals with knobs or "Turks' Heads" to afford a better grip and to prevent the rope slipping too quickly over the window sill. The occupants of the building are lowered out of a window and finally the last man in the building slips down the rope hand over hand.

12575/1902
Daniel Francis Burr *Engineer*
Westview, Eastcourt Road, Worthing, Sussex, England

(Harris & Mills)

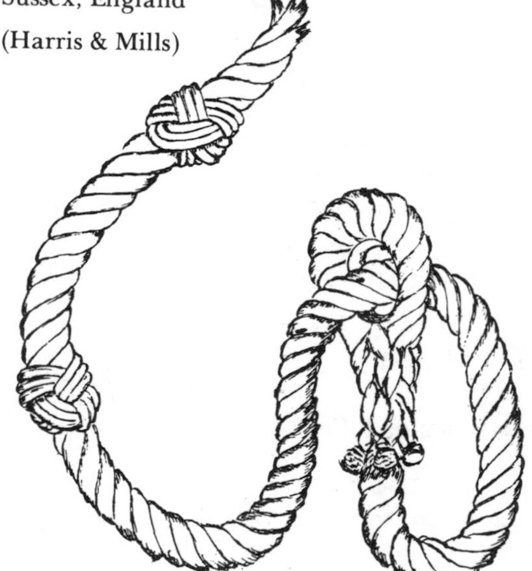

Kleinberg and Fraenkel's Improved Shower Bath provides means whereby travellers and others may have a shower bath without the use of any cumbersome appliances.

The invention comprises an elliptical perforated ring to go over the user's head and rest on his shoulders, the said ring being attached to the water supply via a cock.

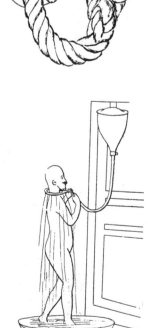

14756/1902
Jacob Kleinberg *Merchant*
3 Lambs Conduit Street, London WC, England

Bruno Fraenkel
30 Guilford Street, London WC, England

(Hughes & Young)

Ellis's Improved Trap or Catcher for Flies and like Insects is easily portable and can readily be set up without soiling the hands. It consists of a piece of paper coated with fly gum which on being manipulated folds into a pyramid or cone. Ungummed edges are provided to prevent adhesion to the surface on which the trap is to be stood or to the fingers etc when the trap is being set up or dismantled. A string can be provided to pull the trap into shape and advertisements and instructions may be printed on it.

15258/1902
James Albert Ellis *Manufacturer*
12 Ruskin Street, Hull, Yorkshire, England

(Phillipss)

The Thornhills' Steam Fire Blower serves a dual purpose: it is either a domestic appliance for exciting sluggish fires or a kind of scientific toy or paradox.

A hollow vessel, constructed preferably of copper or bronze, is made in some fanciful animal shape and filled with water. By heating the body, and allowing it to cool with the snout under water, the water becomes forced into the body of the vessel and partially fills it.

The generator is then placed on the fire (more particularly, a charcoal fire) to be urged with the outlet tube directly directed towards or into the heated embers, so that when the water boils the steam generated emerges in a jet so directed amongst the embers as to cause them to be roused into activity.

14696/1902
Bensley Thornhill *Civil Engineer*
Central Hotel, Simla, India

Hilda Thornhill *Spinster*
Hautboy Hotel, Ockham, Woking, Surrey, England

(A M & WM Clark)

1902

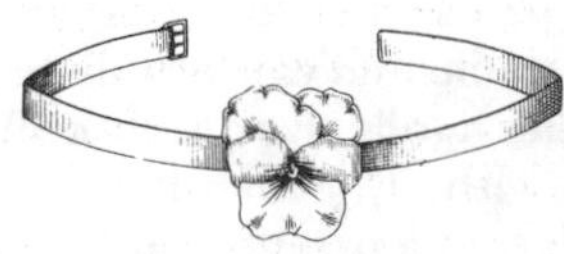

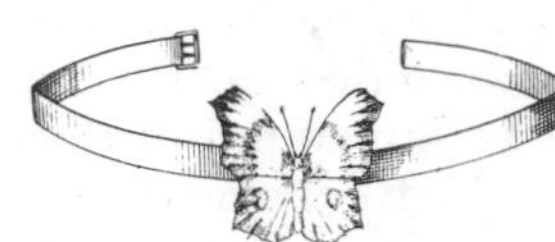

Blanck's Improved Neck-tie has as its purpose the substitution of the customary bow, scarf, cravat etc. by an imitation of a suitable figure, animal, flower, badge or other tasteful or decorative design. These designs and imitations can be made of any desired or suitable material and provided in the customary manner with a ribbon or band fitted with a buckle or other well-known closing means for attaching the novel necktie around the collar or bare neck. Neckties of this kind distinguish favourably from the customary cravats, scarfs, bows, etc. For instance, same can be chosen to be in harmony with the dress worn, and especially ladies will find them a suitable ornamentation in connection with all sorts of dresses.

13603/1902
Arno Blanck
13 Schlorumpfweg, Ricklingen-Hannover, Germany

(Edward Buttner)

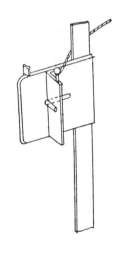

Schmidt's Device for Stopping Runaway Horses comprises a pin which can be pulled from the device attached to the blinkers so that, if a carriage horse runs away, its eyes become blindfolded by a cover piece without hurting the horse in the least.

17294/1902
Ernst Franz Schmidt *Harness Maker*
18 Zschoscherstrasse, Leipzig-Plagwitz, Germany

(Ferdinand Nusch *t/a* F G Harrington & Co)

Squires and Morehen's Improved Advertising Hat consists essentially in having a hinged crown which can be caused to move at regular or irregular intervals, as desired, in order to attract attention and at the same time display an advertisement.

The inventors use a hat of the kind known as a "top hat" and advantageously employ a pneumatic mechanism fixed inside it, coupled with an air ball carried in the pocket and connected to the mechanism by a flexible tube.

An electric lamp or lamps may be arranged so as to light up when the crown is lifted.

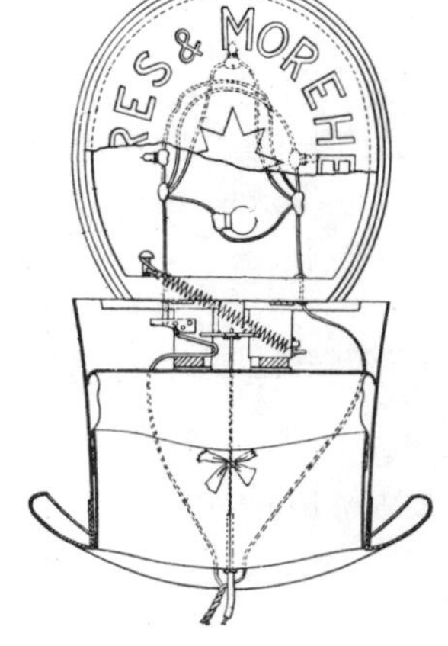

7607/1902
Sidney Robert Squires *Theatrical Manager*
Egyptian Hall, Piccadilly, London, England

Edward Morehen *Electrical Engineer*
32 Herbert Road, Wimbledon, Surrey, England

(G F Redfern & Co)

Oleszkiewicz's Improved Driving Mechanism for Cycles is designed to accelerate the motion of cycles of any of the existing systems. The rider wears a harness such that when he throws back his shoulders with a great effort the action transmits an impulse to the free wheel, thus giving the rear wheel of the cycle to which it is fitted an energetic forward impulsion. As the rider again bends his body forward a spring returns the actuating lever and all the elements or parts back to their initial positions. When it is not desired to use the device the rider must neither bend forward nor throw his shoulders back as otherwise he will impart motion to the lever operating the device.

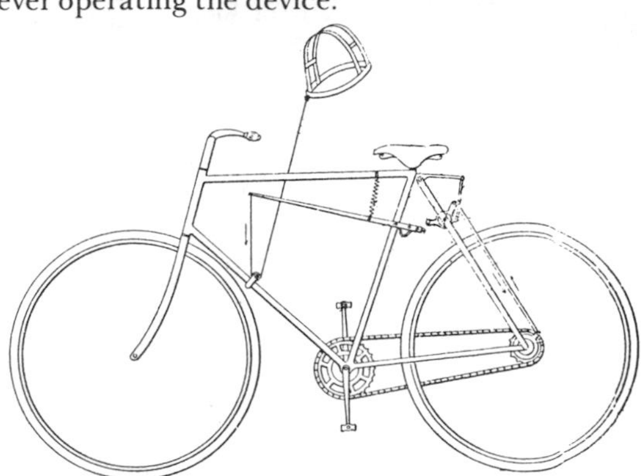

25417/1902
Anton Oleszkiewicz
Bielaia, Tzerkov, Kiev, Russia

(Boult, Wade & Kilburn)

Bürel's Apparatus for Cramming Poultry to Cause Enlargement of their Livers is provided with an extension to force solid food, such as grain, down a bird's throat and is an improvement on existing machines of this kind which are only able to deliver ground meal.

The feed-screw is operated by a hand-crank or a pedal.

14207/1902
Xavier Bürel
8 Bei den Gedecten, Brücken, Strassburg, Alsace, Germany

(W P Thompson & Co)

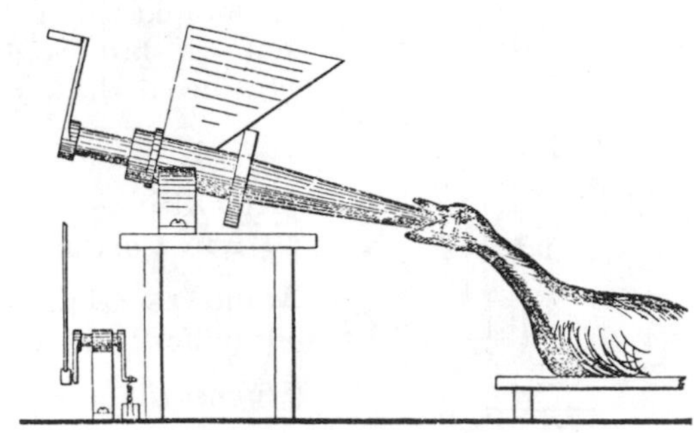

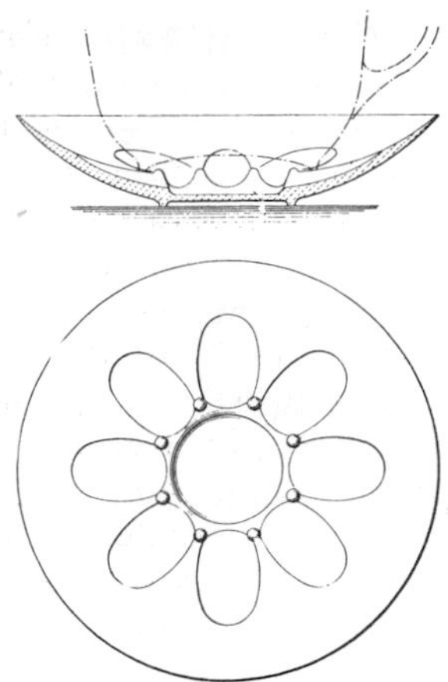

Pryor's Improved Cup and Saucer obviates the inconvenience which arises when the contents of the cup have been spilt into the saucer. When this happens in the ordinary way, the bottom of the cup is wetted and drops are formed thereon, which are liable to fall off when the cup is raised. In Pryor's device the bottom of the cup is prevented from becoming wetted by spillage by providing a well and drainage channels in the saucer. Beads between the channels steady the cup and the capacity of the well is sufficient to accommodate any reasonable quantity of liquid likely to be spilt.

15342/1902
William Alfred Pryor *Gentleman*
12 Mowbray Road, Brondesbury, London, England

(Newton & Son)

Heidelberger, Foettinger and Hagel's Protective Paper Cover for Water-closet Seats is designed to provide a changeable or renewable paper cover of convenient form, whereby it is believed that all danger of the communication of contagious disease may be averted. It consists in a strip of paper with perforated sections corresponding to the hole in the pan which may be pulled from a roll. The invention is further distinguished by the projecting flap at the front of the cover which is turned down inside the opening in the seat for the protection of the sexual parts.

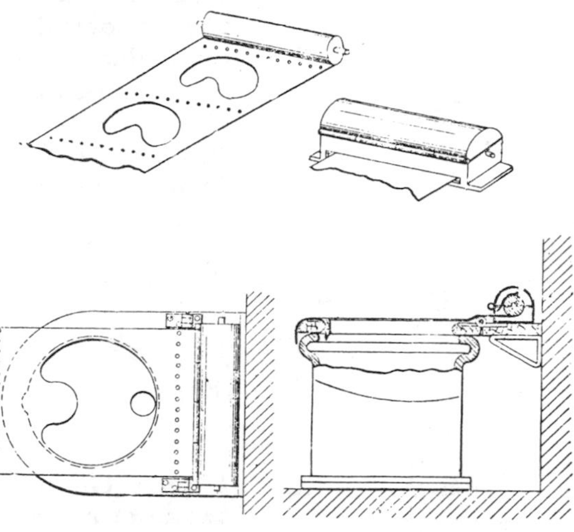

16029/1902
Joseph Heidelberger *Manufacturer*
Johann Foettinger *Hotel Keeper*
R Hagel *No Occupation*
all of 4 Jasmirgottstrasse, Vienna, Austria

(S S Bromhead)

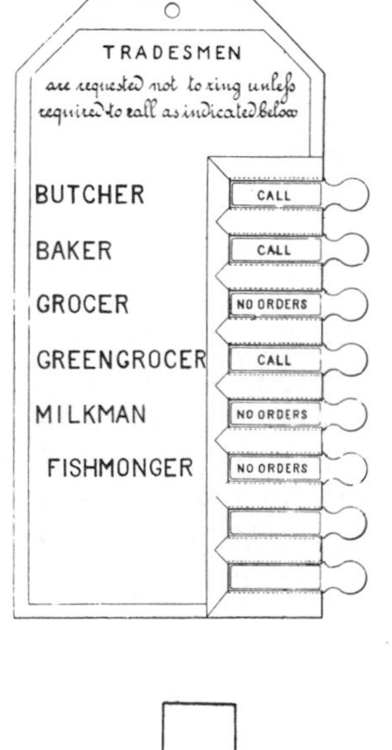

Mummery's Improved Indicator for Tradesmen or Other Callers at Houses is a device which can be so set or exposed at the entrance of a house that tradesmen or others can ascertain at once, without troubling the occupants of the house, whether their services are required. A tablet is provided with removable strips having 'call' on one side and 'no orders' on the other, recessed to prevent the inscriptions from being rubbed off. The tablets may be enclosed in a glass case, and may be locked or secured against interference; the board may also be charged with a number of cards on which orders may be written.

11894/1902
Ernest Stephen Mummery *Envelope and Account Book Manufacturer*
77 St John Street, Clerkenwell, London, England

(Mewburn, Ellis & Pryor)

Dunne's New or Improved Cigarette or Cigar is self-lighting and has on one end a composition which is capable of being ignited by rubbing it on a suitable surface, so kindling the tobacco. The end of the cigar is, in fact, furnished with an "ordinary" or "safety" match, and suitable surfaces recommended for the match end include charcoal, saltpetre, chlorate of potash, phosphorus or sulphate of antimony.

18380/1902
John Francis Dunne *Stenographer and Typist*
78 Westbourne Street, Liverpool, England

(Cheesebrough & Royston)

Dando's Improved Chamber Utensil provides a means by which a chamber pot may be used noiselessly. Special interior surfaces are provided to receive the impact of fluid and so avoid the striking of fluid upon fluid which is what causes the objectionable sound.

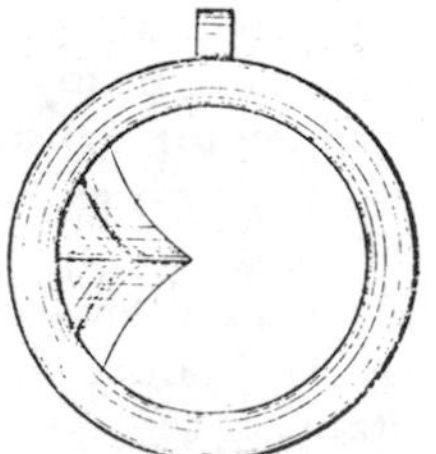

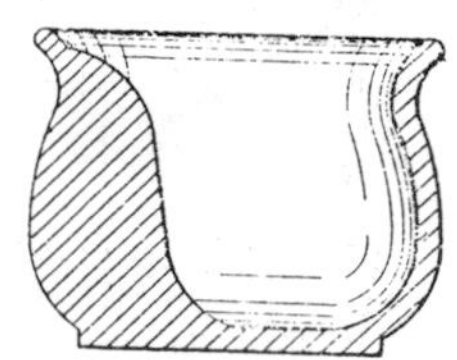

18141/1902
Percy Herbert Dando *Cabinet maker*
Whangarei, New Zealand

(Herbert Haddan & Co)

1902

Biucchi's Improved Sun or Weather Bonnet or Shield for Horses can be easily and firmly secured in position and has the further advantage of not disfiguring the animal as is the case with the more or less absurd-looking bonnets now in use which, owing to their instability and insecurity of fastening, are in almost constant motion and thereby frequently irritate nervous animals and often cause serious accidents.

The bonnet is made of leather as protection against inclement weather, rain or snow, and textile or plaited straw may serve as protection against the sun. It is secured by straps, and worn under or over the bridle. Coverings of the same or suitable material are provided as protection for the ears, as long as such coverings are not made too stiff so as to obstruct free play of the ears and thereby inconvenience or irritate the animal. A frame provides free passage of air.

15608/1902
Basil Biucchi *Manufacturer*
13 Cremer Street, Gray's Inn Road, London WC, England

(Edward Evans & Co)

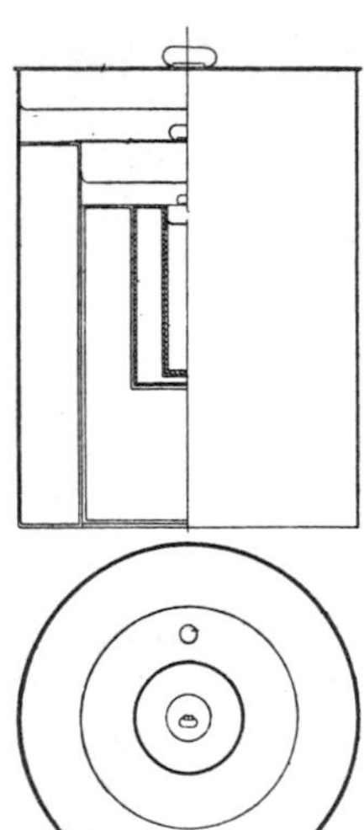

Huish's Thermal Chamber Container for the Semen of Animals is designed to preserve alive and vigorous the zoosperms in animal semen for a sufficient number of hours for it to be transmitted from one end of the country to another and to the continent and to be subsequently used for the production of progeny from mares, cows, bitches and other animals and so avoid the expense and risk of sending females to the distant stud. The semen is obtained by placing upon the penis an envelope made of fine silk coated with rubber and held in place by means of rubber bands. The male thus equipped copulates with a female in the usual way, but the semen is deposited in the envelope, which upon dismounting is removed and the semen transferred to a bottle in a felt-lined vessel, in a chamber filled with boiling water, in a wooden box. Upon arrival of the semen, the female in oestrus will be injected with my Improved Patent Inseminator (British Patent 15362/1898) in which a syringe bulb is connected to a tube of celluloid as opposed to the more usual glass, which may break and cause wounds. This has the additional advantage of husbanding the vitality of the sire in a large degree.

16903/1902
Charles Henry Huish *Surgical Instrument Maker*
12 Red Lion Square, London, England

(J P Bayly)

Hickman and Heard's Improved Telescopic Life Preserver for Defence Against Sudden Assault is a weapon constructed in such a way that it can be rendered so compact as to be carried in the pocket without discomfort and yet be immediately lengthened into a formidable bludgeon and thus provide "a very present help in time of trouble". The weapon is made of a number of pieces which are adapted to telescope within one another when assuming the compact condition and to freely emerge when the elongated condition is desired. An advantageous form of construction consists of telescopic tapered tubes, weighted at one end. Helical springs cause the device to erect itself when a catch is released. Screws, rivets, or flutings may be employed to limit the movement of the sections and provide additional rigidity.

16565/1902
George Charles Hickman *Cycle Works Manager*
74 Old Bedford Road, Luton, Bedfordshire, England

Herbert Cyril Heard *Manager*
Mohawk Motor and Cycle Works, Chalk Farm Road, London NW, England

(P: Phillips & Leigh; C: Edward Evans & Co)

Brösel's Improved Toupée Comb consists of a curved toothed portion, and a number of separable or combined bars or strips of different curvature which entirely takes the place of wool or hair toupées as generally used and which are unhealthy and by no means cleanly. The new comb admits of the free circulation of air between and under the bars, thus greatly promoting the hygiene of the hair.

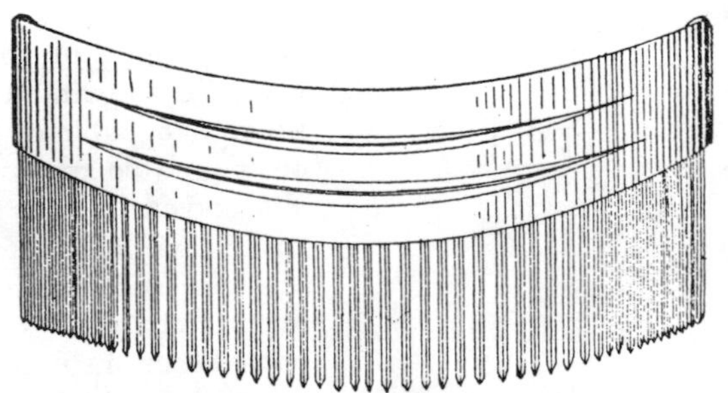

20611/1902
Erhard Brösel *Clerk*
32 Teltower Strasse, Berlin, Germany

(Walter Reichau)

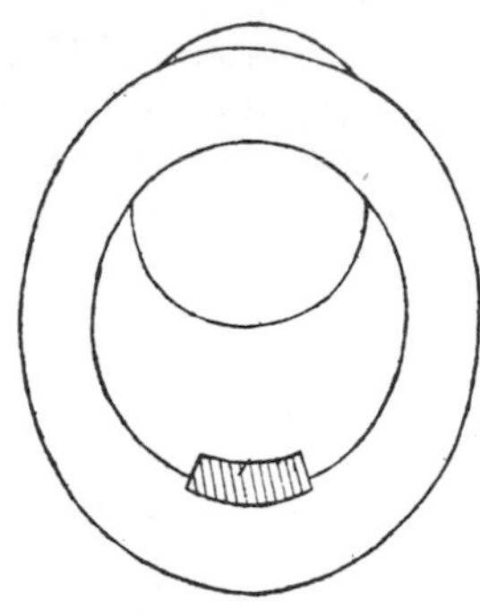

Ogden's Improved Means of Ventilating Silk or Gossamer-Bodied Hats aims at providing cheap and efficient ventilation by the simple means of a hole or holes in the hat, camouflaged with a strip of material sewn on top.

21860/1902
Thomas Ogden *Silk hat manufacturer*
Bradshaw Street, Shudehill, Manchester, England

Hugle's Improved Advertisement Board consists of a frame in modern style and in the most elegant finish provided with a clock and a barometer to attract the attention of the public to the advertisements in the frame, which may be fixed, or change automatically.

20635/1902
Otto Hugle *Engineer*
L12, 8, Mannheim, Germany

(Ferdinand Nusch *t/a* F G Harrington & Co)

Bradbury's Improved Collapsible Mail Cart for Children has a frame made in two parts which may be folded by withdrawing a locking pin, so constructing a toy that will occupy only a small space when out of use, but at the same time be perfectly safe and rigid and not liable to collapse when arranged for use.

22209/1902
Bradbury & Co *Engineers*
William Henry Phillips *General Manager*
William Ernest Ashton *Manager of Basinette Department*
Wellington Works, Oldham, Lancashire, England

(George Davis & Son)

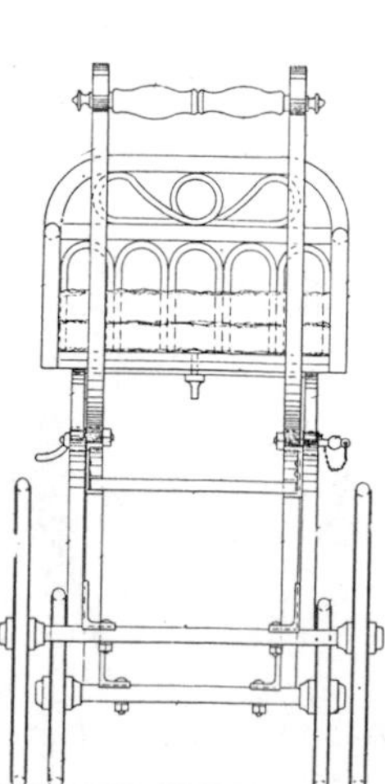

Tibbits's Improved Marine Life-saving Device and Swimming Appliance has as its object to provide a device which is simple of construction, comparatively inexpensive of production, light in weight and of so little bulk about the body, when inflated ready for use, that it can be conveniently worn at all times while on the water to prevent a person from drowning if thrown suddenly and without warning into the water and may also be used to good advantage by persons learning to swim, it being absolutely impossible for a person to sink.

16587/1902
Wiley Preston Tibbits *Book-keeper*
1447 Lincoln Avenue, Toledo, Ohio, USA

(Wheatley & Mackenzie)

Hooley's Improved Means for and the Method of "Tarring" Broken Slag, Macadam, and similar Materials makes use of a steam-coil to melt tar, to which is then added pitch, Portland cement and resin. The mixture passes to a heated reservoir and then passes to a trough wherein it is mixed with stone broken in a crusher, or with slag broken while hot from the blast furnace.

7796/1902
Edgar Purnell Hooley *Civil Engineer*
The Shire Hall, Nottingham, England

(W H Potter)

1902

Schillberg's Scalp Massager for Self Treatment comprises a pair of handles bearing hollow conical rubbers, ribbed to give adherence to the scalp by suction. Any desired friction movement can be given between the scalp and the skull to increase the circulation and promote the growth of hair.

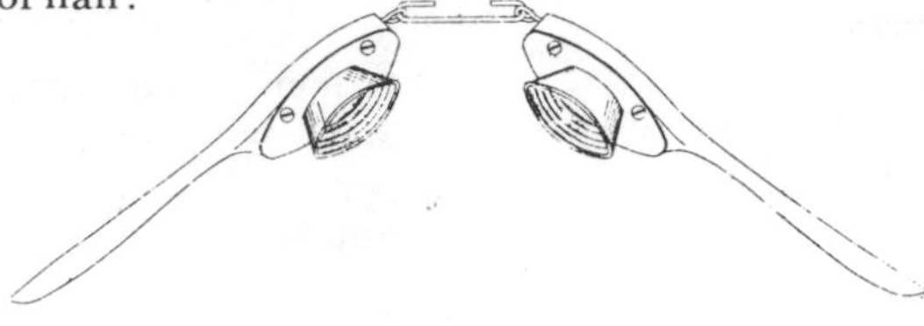

14381/1902
Torsten Schillberg *Masseur*
353 Bath Stret, Glasgow, Scotland

(Johnsons)

Baranouits and Marguth's Improved Ladies' Fan provides a means for producing a gentle fanning motion by pressure of the thumb or finger on the handle instead of as hitherto by a movement of the arm or wrist. The invention further comprises means for automatically unfolding the fan on the release of a catch to make it erect.

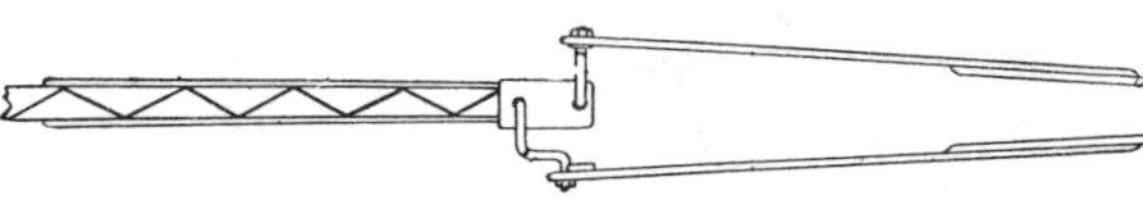

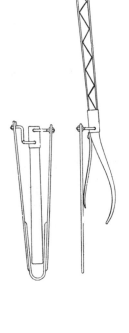

24826/1902
Géza Baranouits *Apothecary*
Monor, Hungary

Ilona Marguth *Gentleman*
Gaderos, Hungary

(Cruikshank & Fairweather)

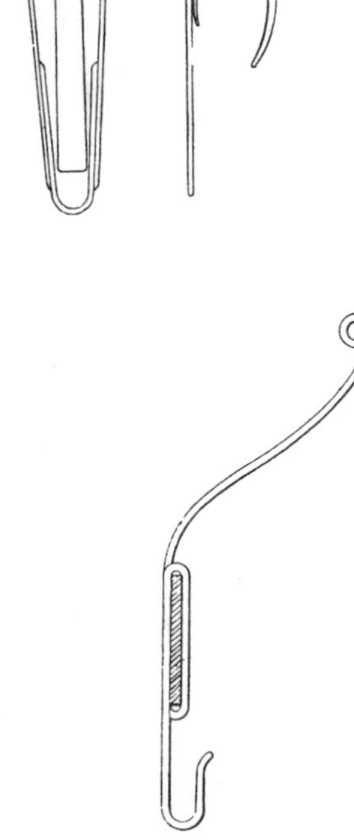

Rumpfkeil's Improved Means for Fixing Clothes Hanging Hooks may be employed at any desired place where there is no opportunity for suspending wardrobe from fixed hooks and provides a convenient portable means which will be found of great utility on excursions, picnics and parties, and will also be found very useful to field-labourers, geometers and like persons, making it possible to hang clothes from trees where it is inconvenient to knock in nails because no tools are to hand or for fear of injuring the tree.

19547/1902
Adolf Rumpfkeil *a Subject of the King of Prussia, Merchant*
Gross Berkel, near Hameln-on-the-Weser, Germany

(Edward Buttner)

Prince Syud Hozoor Meerza's Improved Hat is ventilated with a fan in the top driven by electricity, clockwork, pneumatic action or suitably operated mechanical or other means. When driven by electricity, this may be obtained from an accumulator contained in the helmet or in the wearer's pocket.

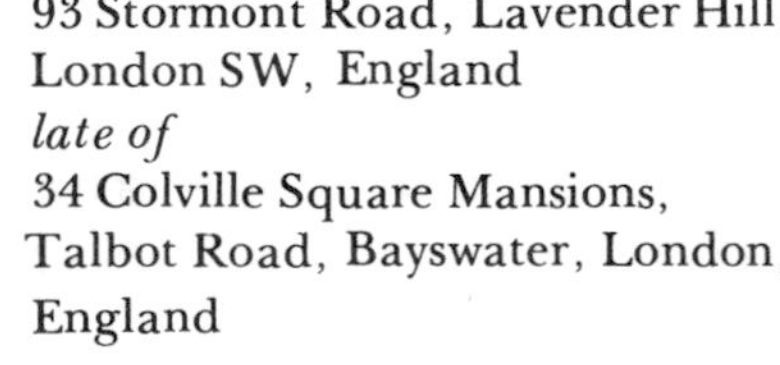

19015/1902
Prince Syud Hozoor Meerza
93 Stormont Road, Lavender Hill, London SW, England
late of
34 Colville Square Mansions, Talbot Road, Bayswater, London, England

(Hughes & Young)

Franken's Collapsible Protecting Device for Closet Seats meets a hygienic want especially for those persons who, through travelling or other courses, are compelled to make frequent use of public closets. The protecting device is simply laid on the seat, and consists of a number of celluloid segments flanged on their inner and outer edges so that, when not in use, they may be telescoped together and conveniently carried in a small space.

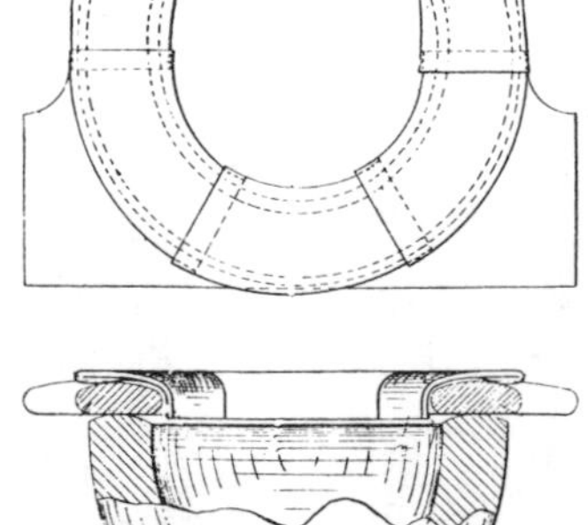

18897/1902
Isak Franken *Merchant*
45 Neue Zeile, Frankfort-on-the-Main, Germany

(H D Fitzpatrick)

Selkirk's Improved Cup is made with a projecting lip which allows the free flow of liquor but prevents sediment, tea leaves or coffee grounds from leaving the cup while drinking, assuming that the handle is held by the right hand of the drinker.

21343/1902
William Robert Selkirk *Hardware Agent*
58H Hatton Garden, Middlesex, England

(Jensen & Son)

Manchester's Improved Ladies' Veil is provided with elastic, mounted advantageously in two or more parts meeting at or about the centre of the veil such that the veil can be drawn or gathered to produce fullness wherever desired and permanently secured in that position.

21361/1902
Clara Manchester *Married Woman*
Sherwood Firs, Nottingham, England

(Tongue & Birkbeck)

Stanger's Improved Apparatus for Curative Treatment by Aid of Electricity and Oxygen produces pure, nascent oxygen, of full concentration, purity and potency, by means of electricity, through the action of electrolysis, so that it is conveyed into the body at the moment of its production, the electric current being suitably passed through a beverage and through the body. Not only is the pure, beneficial oxygen produced when the electric current passes through the liquid, there is also an action in the liquid known as *katophoresis* whereby particles from the fluid are transferred with the electric current in its passage from one pole to another and deposited on the tongue, mouth, throat, lungs and so forth. There is also a general internal electrification.

The treatment is specially adapted for diseases of the throat, for tuberculosis and the like, but may be further used with good results for curing toothache. It is also useful for healthy persons because all kinds of drinks, such as wine and beer, can be taken in this way, and the apparatus may be further modified for application to the arm, foot, urethra etc as desired.

17902/1902
Johann Jakob Stanger *Tanner*
15 Ehinger Strasse, Ulm, Germany

(W P Thompson & Co)

Smith's Improved Baby-Jumper or Analogous Device is designed to provide a chair or receptacle for the child which may be suspended clear of the floor in such a manner as to be able to swing or travel freely in various directions horizontally so as to readily adapt the same as a creeper or walking-guide for infants. The chair may also be collapsed and also adjusted in various positions ready for use. Improved means for adjusting the length of the suspension cords for the chair are also provided.

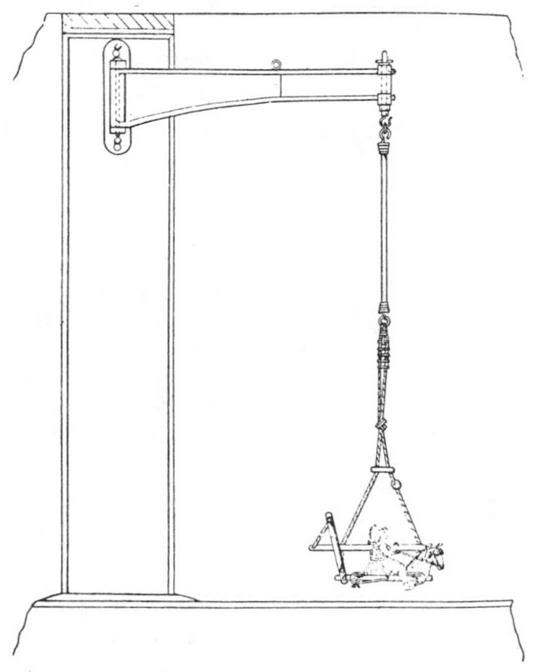

19163/1902
Eugene Charles Smith *Mechanical Engineer*
53 Greenwich Avenue, Manhattan, New York, USA

(S S Bromhead)

Darby's Electrical Heat-indicator and Fire Alarm indicates any change of temperature in the apartment where it is fixed, either in the same building, or at the fire-brigade office, or at the private residence of the tenant of the premises where the indicator is situated. The device operates by closing an electrical circuit to sound an alarm if the temperature rises above the safe limit. The contact is made by bridging a gap with a conductor, or allowing one plate to fall on another; this movement is caused by a block of butter which melts as the temperature rises.

25805/1902
George Andrew Darby *Electrical Engineer*
211 Bloomsbury Street, Birmingham, England

(Alfred William Turner)

1902

Nathan's Improved Method of Manufacture of Clock Cases also applicable to Pendulums overcomes the objection that, hitherto, the fronts of clock casings, and also other parts such as the pendulum, have been made of wood or metal, the decoration of which by carving or similar process is expensive and takes time. This invention pertains to the making of pendulum bobs and clock cases from flexible, elastic material such as lincrusta or linoleum which can have suitable ornaments embossed on the front, back and sides.

25012/1902
Carl Nathan *Stockbroker*
25 Luisenstrasse, Berlin, Germany

(R W James)

Calantarients's Improved Dung Trap for Carriage Horses and the Like prevents the excrement or dung and urine of horses or other animals harnessed to carriages from falling onto the ground and so dirtying and contaminating the roads. To attain this object, a receiver with a removable lining is carried on the vehicle and is fitted with a tube or shoot to catch dung when the horse defaecates. Alternatively, the action of the horse lifting its tail may cause a trough to extend by means of cords suitably attached. A funnel or tube may also be provided to collect urine and convey it by gravity to the receptacle. As a modification, a spring-loaded lid may be caused to open when the tail is lifted, and closed by the momentum of the falling dung or the lowering of the tail. Alternatively, an electric motor may be employed to bring the receiver into position.

26710/1902
Johannes Avetician Calantarients *Doctor of Medicine*
8 Alma Square, Scarborough, Yorkshire, England

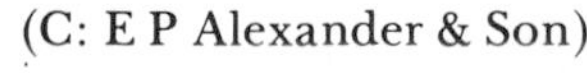

(C: E P Alexander & Son)

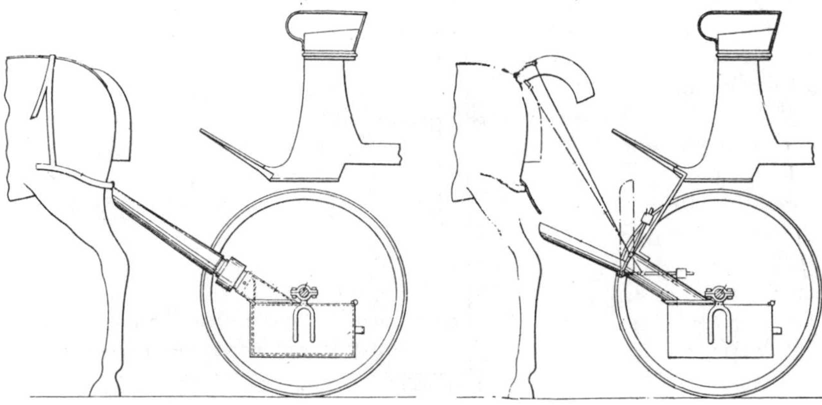

Tooley's Improved Hansom Cab is provided with two hinged arms, which, when horizontal, form arm-rests and so obviate the liability of the occupants being thrown out on to the road or street and sustaining serious injury should the horse slip or fall or the cab otherwise suddenly stop when in motion. The arms can be raised up against the sides of the cab by the hand when not required and when needed can be easily lowered into position, so that they form a protection across the cab and prevent any sudden ejectment of the occupants in case of unexpected stoppage.

If preferred these arms may be worked automatically.

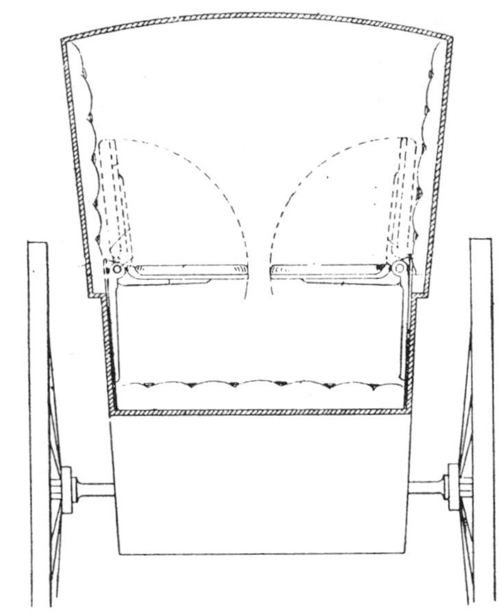

21362/1902
Alfred William Tooley *Forage Contractor*
Ravenstone, Leighton Buzzard, Bedfordshire, England

(C: Reginald W Barker)

Pitney's Improved Machine for Embossing or Stamping Envelopes Labels and Cards and for Sealing and Stacking Envelopes is to speed up the process of mailing by impressing stamps on the material to be posted, thus obviating the use of adhesive stamps. The machine is so designed that it will record by means of a mechanism inaccessible to unauthorised parties the cumulative value of the stamps it has impressed, and automatically lock itself when a certain predetermined number of impressions have been made.

21234/1902
Arthur Hill Pitney *Superintendent*
145 Wabash Avenue, Chicago, Illinois, USA

(D Young & Co)

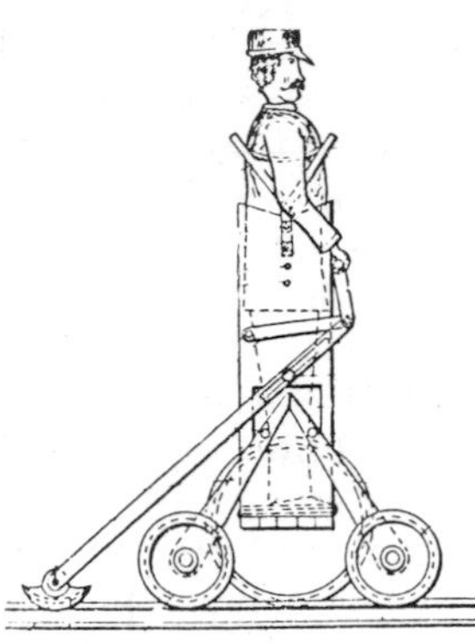

Barry's Improved Railway Velocipede provides a light, strong, durable vehicle to be propelled by manual power which is also capable of being folded into compact form so that it may be conveniently carried upon a railway train and also readily set up on the track for use in emergencies – for instance to run back from or ahead of a wrecked train to secure aid and to signal approaching trains.

19908/1902
Henry Barry *Inventor*
547 Howard Street, San Francisco, California, USA

(P R J Willis)

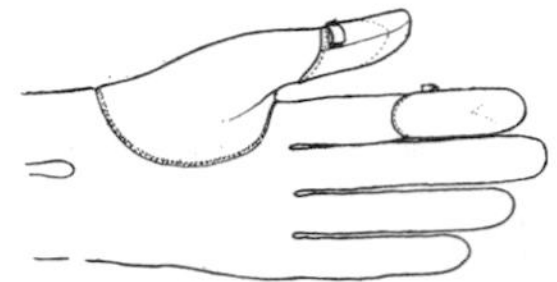

Walley's Glove Protector is designed to prevent the thumb and forefinger of a glove from being soiled by contact, at meals or like occasions, with food containing grease or colour such as cake, chocolate, biscuits, tea, coffee, wine and the like. All such articles soil and often spoil a light-coloured glove, and ladies wearing such gloves will often refuse refreshments of which they would gladly have partaken because they do not wish to soil or injure their gloves. The invention consists of finger slips or stalls for the thumb and forefinger, either separate or connected, and made of paper, cotton, silk, gutta-percha, india rubber, xylonite or other pliant material so as to allow proper use of the finger and thumb.

It is desirable that it should be washable or cleanable unless it can be made cheap enough to be thrown away after use. To make the design pretty and ornamental, while not expensive, it may be held on by rings or ribbon. Those for refreshment rooms may bear advertisements.

22696/1902
Winifred Buckland Walley *Spinster*
170 High Road, Lee, Kent, England

(S S Bromhead)

de Folleville's Improved Head Covering serves as an eye protector and is designed to replace sculptors' or motorists' spectacles or masks and the like. The head covering can be made of any suitable material, such as felt, cloth or linen, and can be made of any convenient shape according to the requirements of fashion. The front part of the covering is provided with openings corresponding with the eyes. These openings contain protecting glasses, which can be replaced by optical glasses according to the sight of the wearer. The back may be prolonged to protect the neck, or bear buttons to take an extension for that purpose, and a bib may be provided to protect the throat against the effects of weather or atmosphere. The whole lower edge of the covering may be bordered with stiffening material so as to prevent deformation by the wind in the event of the covering being intended for motorists or cyclists.

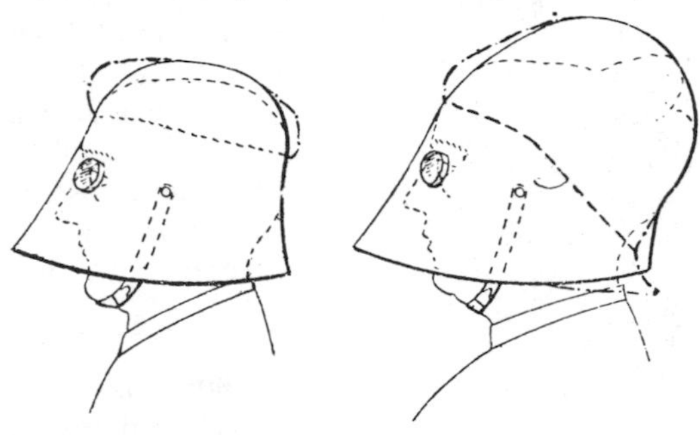

23574/1902
Jules Pierre Despois de Folleville *Sculptor*
St Maurice, Seine, France

(Herbert Haddan & Co)

Röhr's Improved Dog Kennel provides an arrangement by means of which a cover is spread over the animal on entering the kennel and lying down and which when it leaves the kennel is automatically brought back into its original position. The lowering of the cover is brought about by the action of the dog's weight when it enters the kennel; it is raised by counterweights, adjustable according to the size of the canine occupant, when it leaves it. To effect this invention, the entrance of the kennel is provided with a false bottom, and when the dog treads on this platform, its weight causes a blanket to be lowered over it, slowly, as there is a fan to retard the motion. Eyes are formed in the material of the blanket so that it may adapt itself to the recumbent animal's movements.

23483/1902
Johannes Röhr *Hotel Proprietor*
Rullstorf, Thüringen, Germany

(J Owden O'Brien *Successor to and late of* W P Thompson & Co)

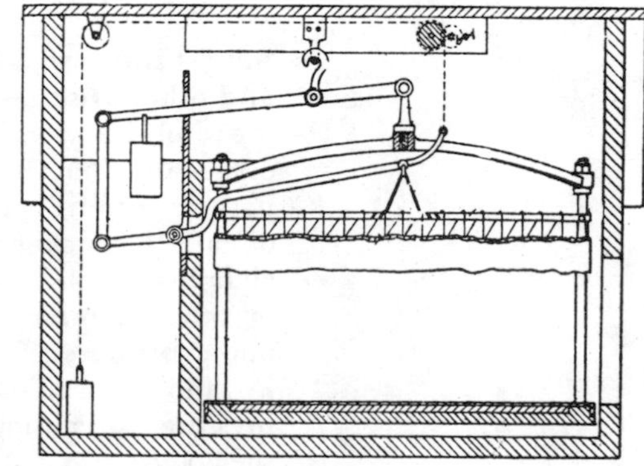

1902

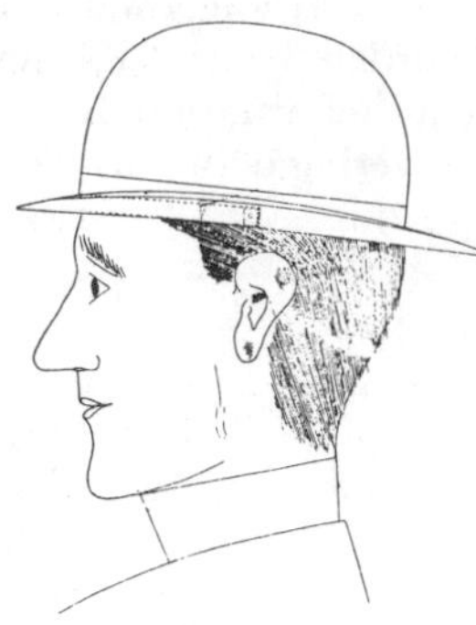

Smith's Improved Hat Guard has as its object to provide more than usually effective means for preventing these articles from being blown off by the wind. The interior of the hat is provided with a spring clip which fits round the front semicircle of the head, from which depend two pieces to rest on top of the ears. The lower ends of these pieces bear an elastic band which encircles the back semicircle of the head. The ear pieces, when not in use, are turned up into the hat. The whole or any portion of the guard may be covered with suitable material of any desired colour.

25300/1902
David Benjamin Smith *Merchant*
28 Wrotham Road, Gravesend, Kent, England

(Hughes & Young)

The Ballard Hygiephone Company's Hygienic Device for Speech Receivers and Transmitters meets the well-known contention with physicians that contagion can be spread through promiscuous use of telephones by diseased people; besides, it is not pleasant to use a public telephone which may have been breathed, smoked or even coughed into, or held against the perspiring ear of a person of doubtful cleanliness. This invention obviates the likelihood of such contagion and promotes a feeling of safety when properly used. It comprises a clip for holding a plurality of protecting-sheets in close proximity to the mouth- or ear-piece of a sound transmitter or receiver, a fresh piece of paper being swung down in front of the instrument before use and torn off after use. This also provides a useful means by which advertisements may be kept in constant view where they would not otherwise be admitted.

25948/1902
Ballard Hygiephone Co.
605 Homer Laughlin Building, 315 South Broadway, Los Angeles, California, USA

(W P Thompson)

Scott's Improved Method of Treating Septic and other Diseases has as its object to disinfect the whole body, or any portion thereof, of micro-organisms that are injurious to health. The patient suffering from an infectious or other disease is enclosed in a hermetically sealed chamber which is supplied with formalin, ozone, or other disinfectant forced into the chamber under pressure. The gas, or liquid, is pumped in, and then air is pumped in to increase the pressure, so causing the molecules to permeate the body.

A specially shaped cover may be used for treating selected areas, such as wounds, or inlet and outlet tubes may be inserted into the body cavity or tissues, finely pointed to penetrate the flesh. In a modification specially adapted to treat the bowels and surroundings, an inlet tube is introduced at a suitable position and an outlet tube is introduced, preferably at the back, and the disinfectant is caused, by pressure, to permeate the bowels. A slight modification of this method is to connect the disinfectant supply with one tube and, by a suction pump applied at the other tube, gradually draw the disinfectant through the bowels. A similar method may also be applied for disinfecting the lungs.

The process may be applied to animals as well as to humans.

23918/1902
Robert Scott *Gentleman*
8 Graingerville North, Newcastle-on-Tyne, England

(A F Spooner)

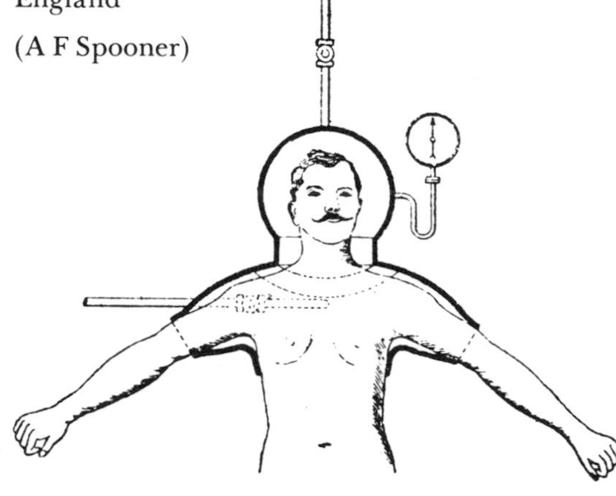

Friday's Improved Reversible Cooking Tongs are a novel and useful adjunct to the kitchen range for effecting with ease and safety the lifting, straining and turning over of fish, steaks, chops, kidneys, sausages, pancakes, eggs, bread, pastry and all kinds of vegetables. Two square, round or oval plates are mounted on a handle of wire, or handles of wood with a spring. The handles are crossed so that the tongs *open* when squeezed. A similar apparatus with flat or oval tongs, or with prongs, may be used for serving sugar, ice, pickles, olives or sandwiches.

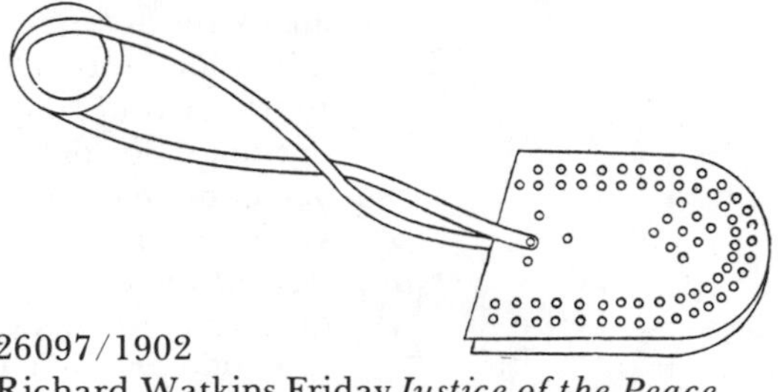

26097/1902
Richard Watkins Friday *Justice of the Peace*
Trigoni Hall, Largs, Ayrshire, Scotland

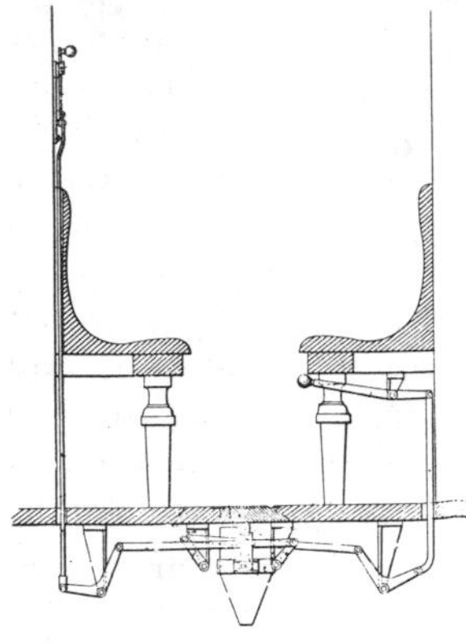

Podwenetz's Improved Spittoon for Railway Carriages is arranged beneath the floor of the same so that it not only occupies no space in the carriage itself but also remains concealed from the eyes of the passengers; in case of necessity it can, however, be made accessible at once and ready for use by means of a suitable device, its lid or cover, which forms part of the floor, being opened by a lever beside the chair or above the head, and closing automatically under the action of a suitable spring.

If desired, the spittoon may also be disinfected, for which purpose the exhaust steam of the locomotive can be used to advantage.

22223/1902
Isidor Podwenetz *Merchant*
2 Kirchen Platz, Lugos, Hungary

(Allison Bros)

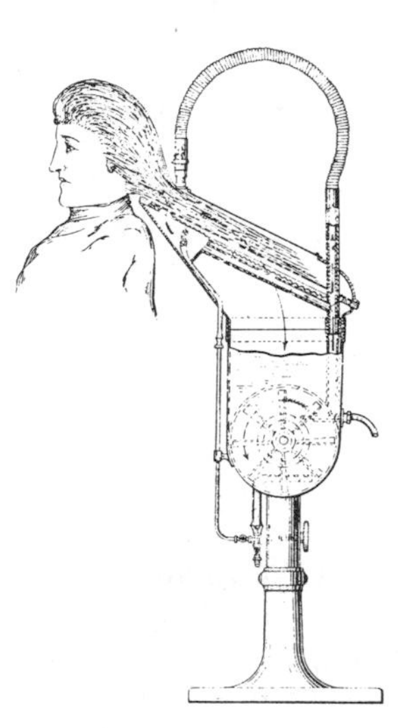

Hornung's Improved Apparatus for Drying the Hair effects that end both by hot and cold air, thus preventing the hair from becoming brittle, which, as experience has shown, is the case when it is dried exclusively with hot air.

The hair is laid on a gauze plate over a Bunsen burner, and the hot air is sent round to the hair by means of a flexible tube. Next, cold air is drawn through the hair by the action of a rapidly-rotating fan driven by a water jet.

22463/1902
Jean Hornung *Hair Dresser*
8 Kronenstrasse, Chemnitz, Germany

(W P Thompson & Co)

Mott's Improvements in Machines for Navigating the Air pertains to that class of machines which are heavier than the air displaced by them and provides means for sustaining them in the air against the force of gravity and coincidentally controlling and directing their movements.

Mott's flying machine has the elemental simplicity of a bicycle, in that it comprises two wheels and a frame. The design is concentric and formed largely of wire-tension construction giving it the important characteristics of low cost, compactness, great strength, simplicity and ample factor of safety. Should the factor of safety fail, the machine would descend slantingly like the apparatus used for gliding flight.

The machine may be dismantled for easy transportation and is therefore specially adapted for army and navy use; for travellers and explorers, for geographical, meteorological and kindred sciences, and the general uses of man.

The size of the machine is six feet high and nearly twelve feet in diameter; the weight is about two hundred and twenty five pounds. It must be understood, however, that it is not limited to any size or weight as these things may be indefinitely varied. The heart of the machine is a tank of gasoline, etc. to supply a motor whose cylinders drive its crankshaft either direct or via gearing. The crankshaft drives two contra-rotating wheels to lift the machine, constructed with wire spokes, rims of wood or light tubing, and blades of celluloid, aluminium or other light material. The direction of motion of the machine is determined by the inclination of its axis based upon and explainable by the law of resultant or compound forces known to physicists as the parallelogram of forces.

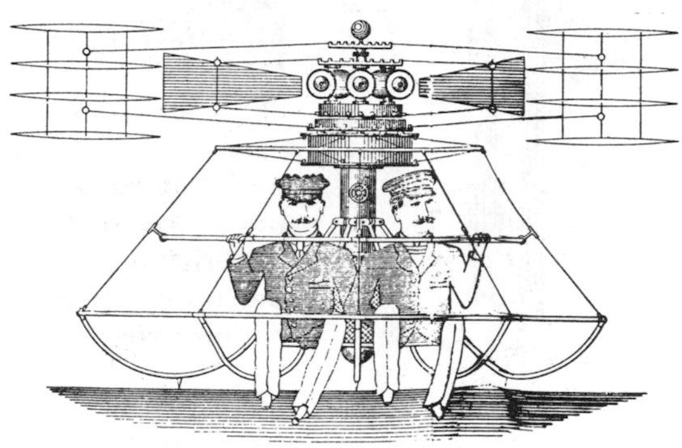

24587/1902
Samuel Dimmick Mott *Engineer*
130 Autumn Street, Passaic, New Jersey, USA

(Allison Bros)

Wheeler's Improved Shield for Frying Pans and analogous Cooking Utensils is a simple and inexpensive means of preventing the grease from spattering out of the pan on to the stove, thereby causing an unpleasant smell besides dirtying the stove. The centre of the shield is open to allow of the steam and fumes to escape and also has an opening in the side towards the front so that articles being cooked can be turned without necessitating removal of the shield.

5075/1902
Mary Jane Wheeler *'Lady' wife of John Wheeler*
Endwood, Wimborne Road, Bournemouth, Hampshire, England

(W Lloyd Wise)

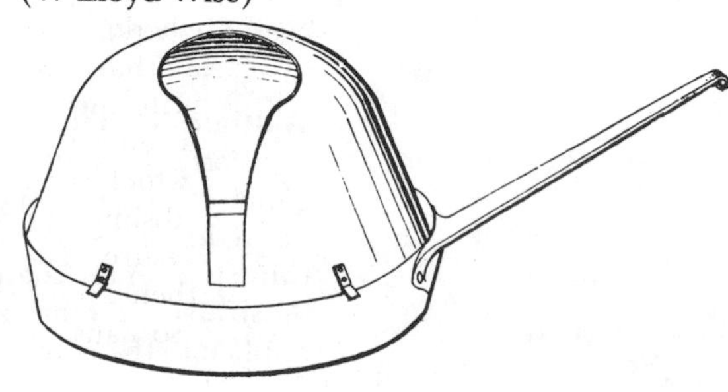

1902

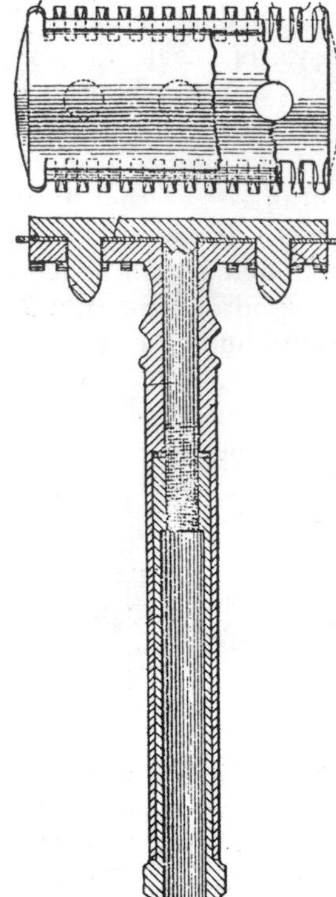

Gillette's Improved Razor employs a thin, flexible blade which is made of so small an amount of material and is capable of being sharpened so quickly and so easily that it may be produced at a very low price, and, when it becomes dull, may be thrown away and replaced by a new one.

28763/1902
King Camp Gillette *Manager*
94 Marion Street, Brookline, Massachusetts, USA

(Boult, Wade & Kilburn)

Ostwald's Improved Process for the Manufacture of Nitric Acid and Nitrogen Oxides makes use of his discovery that a mixture of ammonia and an excess of air is converted catalytically into nitric acid or higher nitrogen oxides by passage over solid or spongy platinum, or a compound of the two, or over other materials such as metallic iridium, rhodium or palladium, oxides of heavy metals, silver, copper, iron, chromium or nickel.

698/1902
Professor Wilhelm Ostwald *Doctor*
2/3 Linnéstrasse, Leipzig, Germany

(Wheatley & Mackenzie)

Barbary's Improved Egg Boiler replaces those appliances hitherto in use such as egg cup, sand-box or saucepan. The egg to be cooked is placed inside the receptacle upon a skeleton support to hold it in a vertical position. About 1½ cubic centimetres of water is sprinkled over the egg and the cover is placed in position. A pad of asbestos is then impregnated with alcohol, placed in the cup under the receptacle, and ignited. The flame lasts nearly 2½ minutes, the pad having absorbed about 1 gram of alcohol. The egg is now sufficiently cooked to be eaten. It only remains to take off the cover and the egg is served.

27651/1902
Antoine Barbary *Civil Engineer*
50 Rue Henri Regnault, Courbevoie, France

(Fell & James)

Williams's Improved Game of Skill is an electrostatic appliance well adapted for the game of croquet or other games in which it is desired to cause the balls to traverse passages or tunnels or enter compartments. The invention is illustrated in the accompanying drawing as applied to the game of French billiards.

The balls are of elderpith and are displaced on an electrostatically charged table by modifying the electric state of the system. The table surface is ebonite and is charged by rubbing with a pad of felt, skin or cork. A block is provided to facilitate electrification by rubbing it on the surface. The small light balls are placed on the electrified surface and thus become charged with electricity. If the pad which has served for electrifying the ebonite table is now brought into proximity with them, displacements of the balls are produced which may be readily determined in advance. In this manner movements of various kinds may be imparted to the balls, which may be caused to cannon for example, these operations constituting the game of skill.

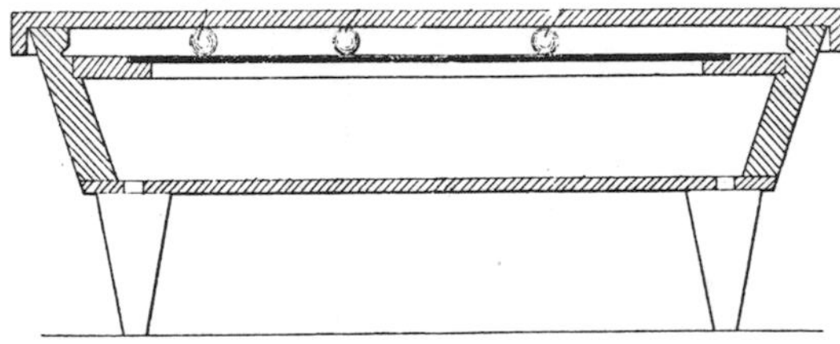

25043/1902
Reginald Stansell Williams *Gentleman*
28 Faubourg Street, Honore, Paris, France

(Haseltine, Lake & Co)

Henderson's Improved Hat Ventilator for Tall or Dress Hats has two distinguishing features: firstly, an additional lining to prevent draughts and, secondly, a combination of discs to prevent ingress of rain and regulate the amount of air allowed access to the wearer's head.

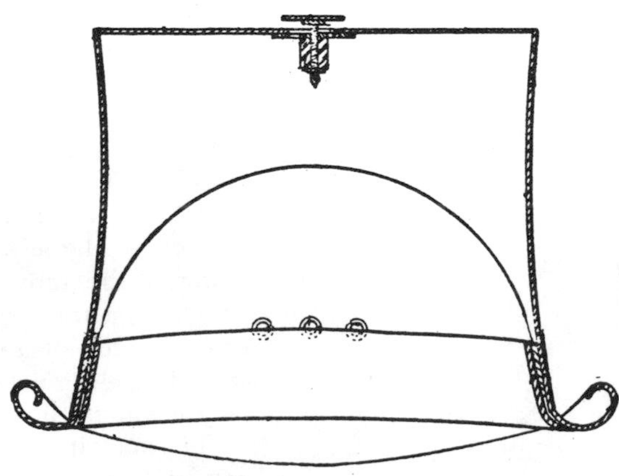

21754/1902
James Henderson *Engineer*
18 Snowdon Place, Stirling, Scotland

(W R M Thomson & Co)

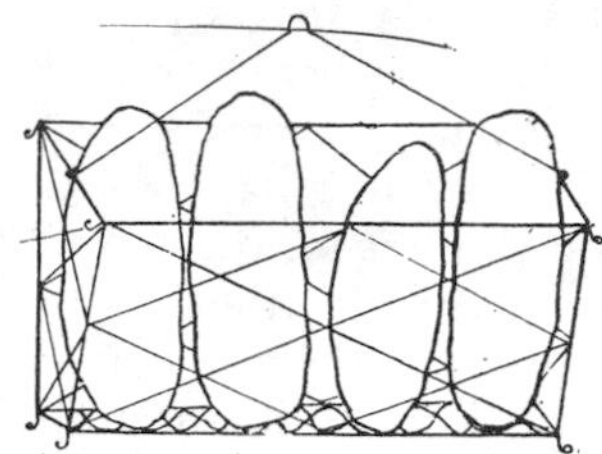

Millard's Multiple Toaster and Roaster for Chops, Potatoes, Chestnuts and the like is intended to facilitate the automatic and simultaneous toasting of several slices of bread at once before a kitchen fire, or before any fireplace used in a breakfast room or any other living room while leaving the attendant free to attend to his business; equally, other viands may be conveniently toasted or roasted on the invention.

When in use the only attention which the invention will require from the attendant will be to turn the slices of toast when it is found that the side facing the fire has been sufficiently done. The invention may be constructed from any suitable material, and has a number of trellised cages each of which takes four slices of bread. Owing to the peculiar shape of the cage, wide at the top and narrow at the bottom, it will be feasible for the attendant to manipulate, remove, replace or handle in any way one slice of toast without interfering with any other slice. Thus it ensues that the invention will automatically toast from four to a dozen or more slices of bread simultaneously as well as automatically in the same brief space of time as is at present required with the toasting fork for only one slice.

7127/1902
Edwin Havelock Millard *Indian Railway Contractor*
63 Croxted Road, Herne Hill, London SE, England

Gordon and Dobson's Improved Method of Constructing Stairs, Staircases and Landings in Mixed Schools provides separate staircases for the passage of boys and girls to the classrooms on the upper floors, separated by fences or screens where they cross or pass one another such that boys and girls can enter their respective classrooms, without coming into contact with each other either in the hall or in the flats above.

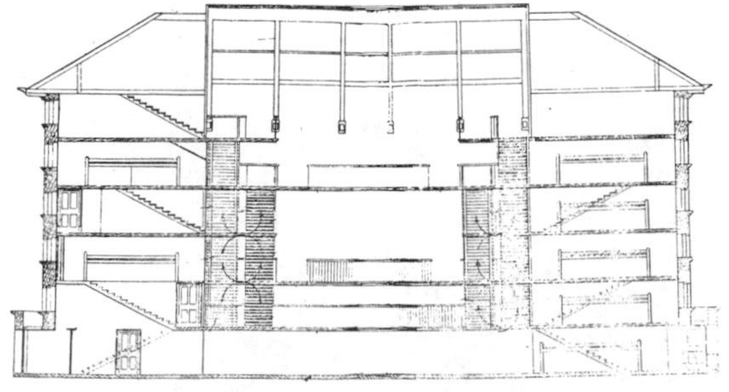

26627/1902
John Gordon & David Bennet Dobson *Architects*
261 West George Street, Glasgow, Scotland

(Bottomley & Liddle)

von Orth and Duisberg's Improved Suspender for Skirts and the like allows the garment to be lifted to any desired height, gripped in place firmly, and then let down again at the wearer's will.

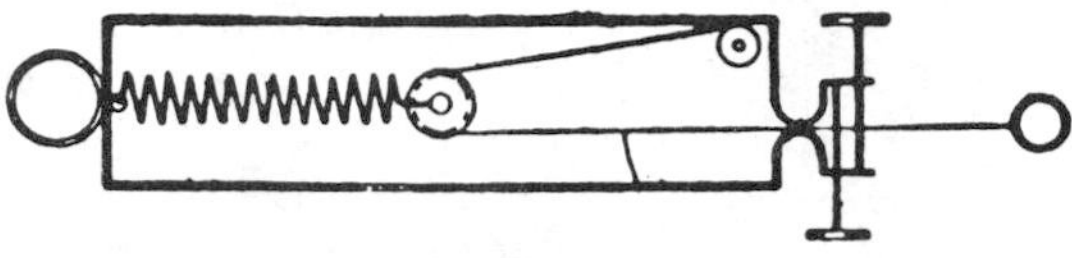

11106/1902
Dr Ludwig von Orth *Engineer*
9 Ludwigkirchstrasse, Berlin, Germany

Martin Duisberg *Manufacturer*
133 Koepenickerstrasse, Berlin, Germany

(Herbert Haddan & Co)

The Mortons' Improved Sewing-machine is disguised by being contained within a figure such, for example, as a lion standing on a rock with its forepaws on a higher level than the back, and capable of containing the working parts of the machine. The head of the lion and part of the rock are so hinged and pivoted that they can be simply swung out of the way to gain access to the movable parts of the machine without having occasion to detach any of them.

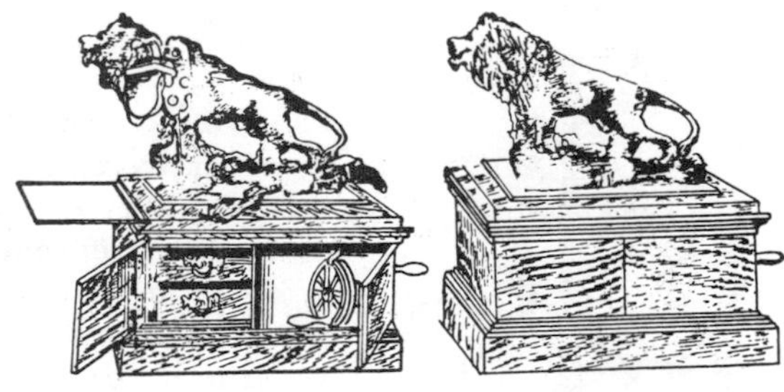

21933/1902
William Urie Morton & James Raeburn Urie Morton *Sewing machine manufacturers*
11 Bothwell Circus, Glasgow, Scotland

(Bottomley & Liddle)

Lanchester's Improved Brake Mechanism for Power-propelled Road Vehicles comprises a metal disc carried by the wheel hub, subjected to the action of a pair of gripping jaws.

26407/1902
Frederick William Lanchester *Engineer*
Armourer Mills, Montgomery Street, Sparkbrook, Birmingham, England

(Marks & Clerk)

1903

Levaillant's Improved Harness enables horses to be quickly released, as in cases where they are running away, by dividing those portions of the harness which prevent the horse from moving out of same and connecting them together at each division by quickly disengaging fastenings. All the fastenings can be operated by pulling on them from a single strap or the like.

1675/1903
Louis Levaillant *Merchant*
Stockerstrasse 58, Zürich, Switzerland

(Philip M Justice)

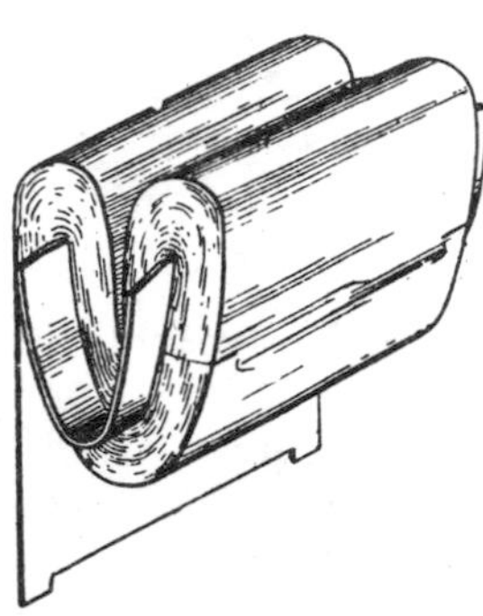

Scott's Improved Fixture for Serving Toilet Paper overcomes the objections to conventional toilet paper dispensers: that they are complicated, noisy, require great care and loss of time in loading and unloading and occupy a great deal of space.

This device provides a new and novel form of paper roll and case or fixture for its reception, being arranged to sag instead of being circular in cross-section, so that when the end is pulled the loop hanging in the roll tightens and tears, and the new end flops back for the next use.

339/1903
Arthur Hoyt Scott *Manufacturer*
7th Street, Philadelphia, Pennsylvania, USA

(Haseltine, Lake & Co)

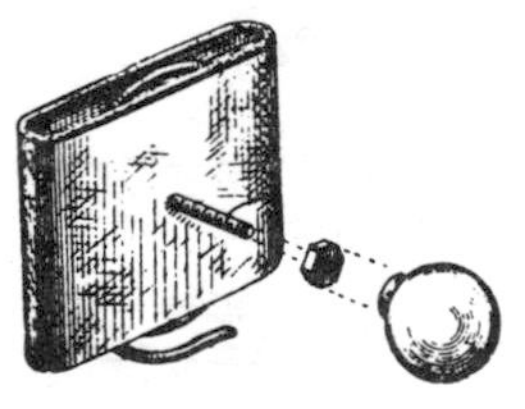

Chevalier and Schultz's Improved Hat Fastener is a spiral of wire inside the hat which is manipulated by a knob on the outside of the hat so that it will engage with the hair. The fastener is also provided with a shield to absorb any moisture that may be deposited on the spiral and thus prevent damage thereof.

360/1903
Joseph Chevalier *Gentleman*
Eugene Daniel Schultz *Letter carrier*
61 Ellicott Avenue, Batavia, New York, USA

(Abel & Imray)

Turner's Improved Ear Cap is primarily intended for keeping children's ears close to the head, and is constructed so as to allow free ventilation of the head while it is worn. A piece of openwork material covers each ear, and the two are kept in place by tapes and ribbons.

944/1903
Adelaide Sophia Turner (*née* Claxton) *Artist*
28 Bath Road, Chiswick, Middlesex, England

(F W Golby)

Prince Hozoor Meerza's New Game is played by two persons with pieces on a board, rather like draughts except that one player has two large pieces (cats) and the other has twenty smaller pieces (rats). The object is for the cats to take the rats as in draughts, or for the rats to corner the cats and prevent their moving, thereby affording much amusement to both players. The pieces can equally well represent other animals, soldiers, or be of any other desired shape and the details of the game may be varied to suit.

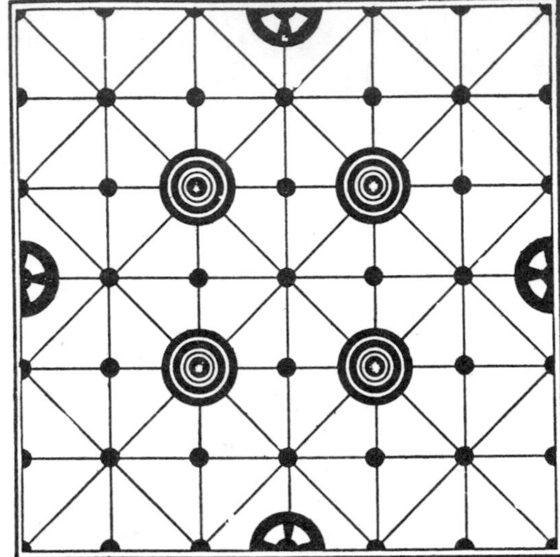

538/1903
Prince Hozoor Meerza *No occupation*
Stormont Road, Lavender Hill, London SW, England

(P: Hy Fairbrother; C: Self)

Clark's Improved Means for Protecting and Securing Burial Cases in Graves comprises a frame to take the coffin, which is placed at the bottom of the grave and fastened in position by lateral "bolts". The coffin is put in and a metal hood secured in position, thereby simply and effectively anchoring the casket in the grave and locking the case hood.

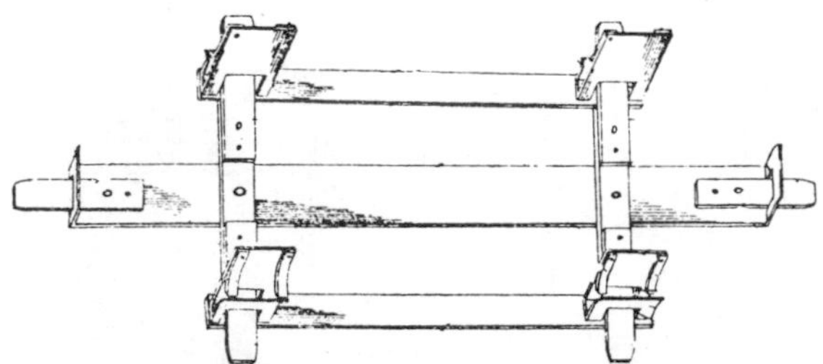

3544/1903
Hugh Dale Clark *Manufacturer*
658 Oak Street, Columbus, Ohio, USA

(Marks & Clerk)

Dichtl's Improved Combined Tent and Cloak provides ample room for two men with their kits in its simplest form, and a larger tent adapted to accommodate several persons can be erected by inserting two covers, for instance camp blankets, between the cloaks.

A separate tent pole is not necessary but is preferably improvised when required, for instance a rifle for military purposes or a climbing or walking stick for tourists.

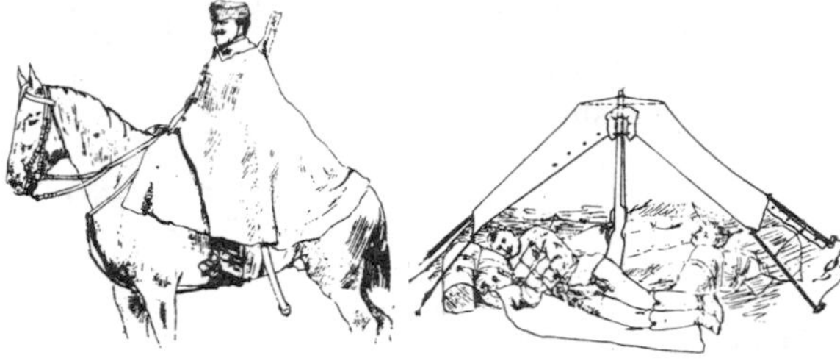

132/1903
Herman Dichtl *Captain on the General Staff*
Lemberg, Austria

(Herbert Haddan & Co)

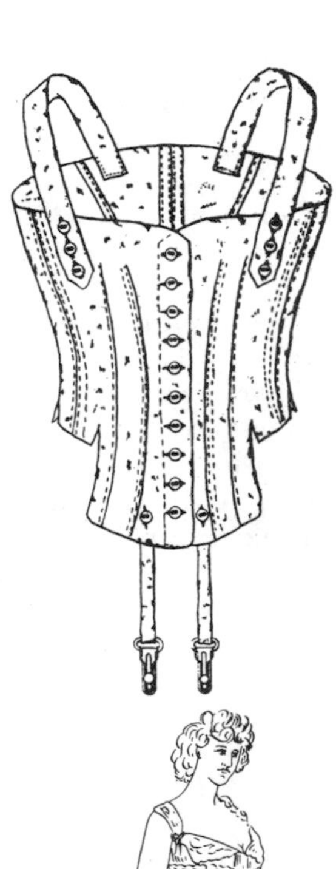

Ehrlich's Improved Undergarment is a bodice serving to replace corsets which, merely by reason of its cut and without the employments of steels or bones, maintains the body in its natural position. Since the waist is not confined by the employment of a string or band there is no prejudicial pressure upon or unnatural confinement of the internal parts, thus providing for the hygienic requirement as regards women's underclothing and making the garment suitable for wear at any time, even at night.

351/1903
Emma Ehrlich *Gentlewoman*
76 Mariahilferstrasse, Vienna VII, Austria

(Haseltine, Lake & Co)

Harman's New and Improved Skirt or Petticoat is designed to enhance the bodily charms of the wearer and may be worn either under or over the corsets. It is designed to prevent sagging and may be so padded or constructed as to impart the necessary fullness at the back and hips where hollows occur, to set off the figure and consequently the dress to the best advantage. The whole garment, in fact, allows a perfectly full and smooth surface to be attained.

1365/1903
Emma Elizabeth Harman *Dressmaker*
89 Palmerston Road, Southsea, Hampshire, England

(Benj T King)

Tkast's Improved Holder for Cravats and the like securely fastens the necktie and prevents it from sliding by means of a shield which can be sewn to the tie, coupled with specially-notched undisplaceable studs.

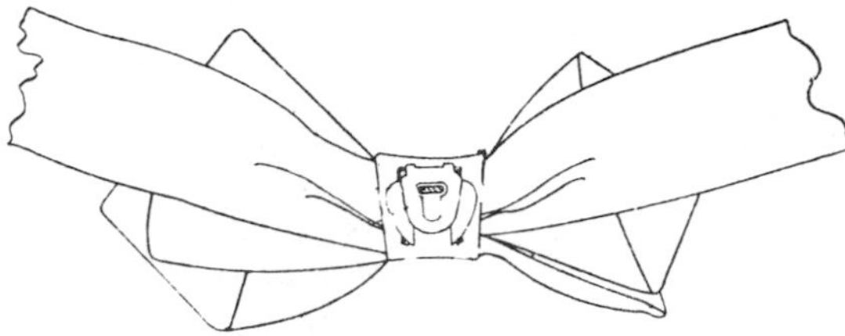

2404/1903
Franz Tkast *Merchant*
10 Hauptstrasse, Mährisch-Ostrau, Austria

(Haseltine, Lake & Co)

Dove's Appliance for Carrying and Supporting Umbrellas, particularly for the use of Ladies enables the umbrella to be attached to the person so that when not in use it can be closed and allowed to hang down.

1668/1903
Mabel Bertha Evelyn Dove *Married Lady*
23 Binswood Avenue, Leamington, Warwickshire, England

(P: Hughes & Young; C: Self)

Billet's Improved Means for the Removal of Dust from the Interior of Buildings seizes the particulate matters suspended in the air and gets rid of them as effectively as possible; having fixed the apparatus on his body, the operator can easily work the bellows with one hand.

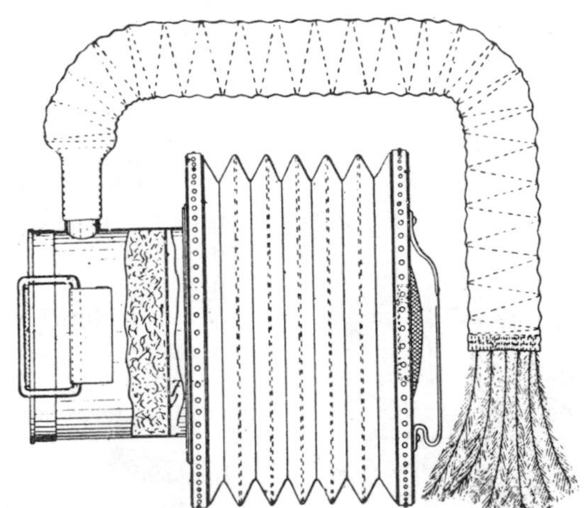

1411/1903
François Billet *Engineer*
15 Rue de Richlieu, Paris, France

(W H Wheatley)

1903

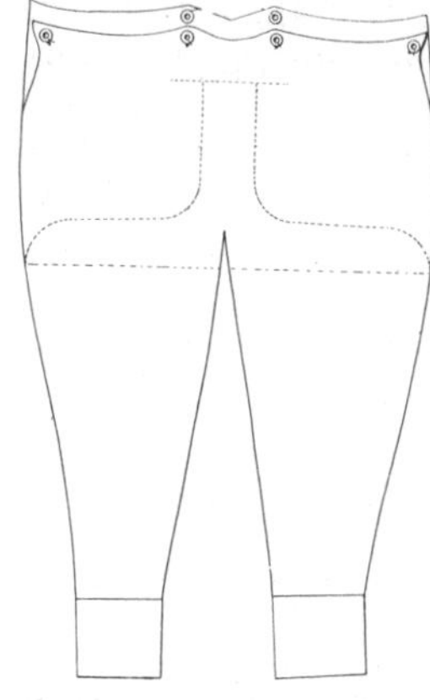

Orchard and Blackmore's Improved Trouser and other Nether Garment for the use of Men is made with an open seat having a combination of horizontal and vertical flaps which can be buttoned over each other, the object being so to make these articles that when under certain circumstances it is necessary to let them down there will be no need to unfasten them from the braces.

3359/1903
Frank William Petherick Orchard
Alfred James Blackmore *Merchants,* both of
1 Exe View Terrace, Exmouth, Devonshire, England

(Hughes & Young)

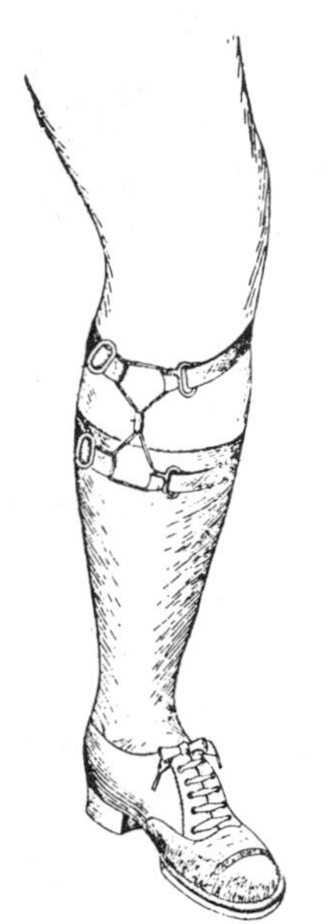

Bauer's Improved Garter is particularly adapted for use by men to hold up half hose, but may also be used by women and children for holding long stockings. The garter has two bands, the upper one supporting the lower, and which may be readily adjusted to the exact size desired. The first band encircles the leg over the stocking, and the second encircles the leg. By engaging the stocking entirely around the leg the first garter holds the hose up evenly so that it will not slip and wrinkle or work down on one side as frequently happens with the ordinary garter now employed, which is attached to the stocking at one point only by a clasp. The improved garter is not attached to the stocking at any point and therefore cannot tear the stocking as clasps frequently do.

5599/1903
Perry Sidney Bauer *Inventor*
283 25th Street, Chicago, Illinois, USA

(Marks & Clerk)

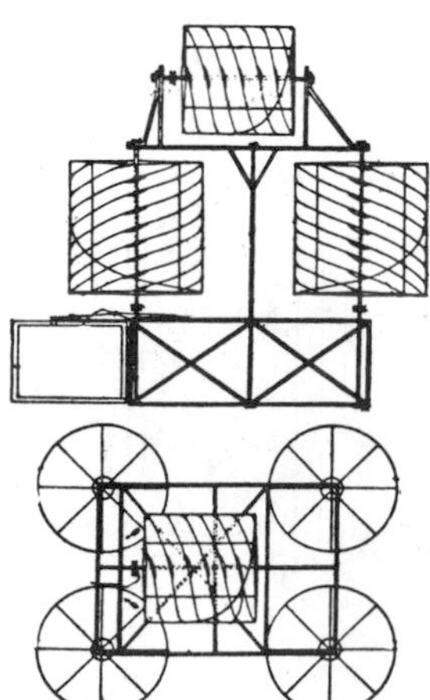

The Haylocks' Self-Lifting Flying Machine consists in a car provided with four or more vertical screws and one or more horizontal screws. These lifting fans are made of aluminium and, when not revolving, act as parachutes.

The machine may be made to any size by increasing the numbers of pairs of screws and any shape may be given to the cars, such as round, square, hexagonal, octagonal or oblong.

7763/1903
David Haylock *Architect and Surveyor*
32 Merton Road, Wandsworth, Surrey, England

Richard Arthur Haylock *Manual Instructor*
41 Bolton Gardens, Chiswick, Middlesex, England

Cressierer's Holding Device for Feeding Bottles is so constructed that the receptacle can be mounted on a perambulator, easily tilted and its flow regulated, whilst it is impossible for the infant within to dislodge the bottle.

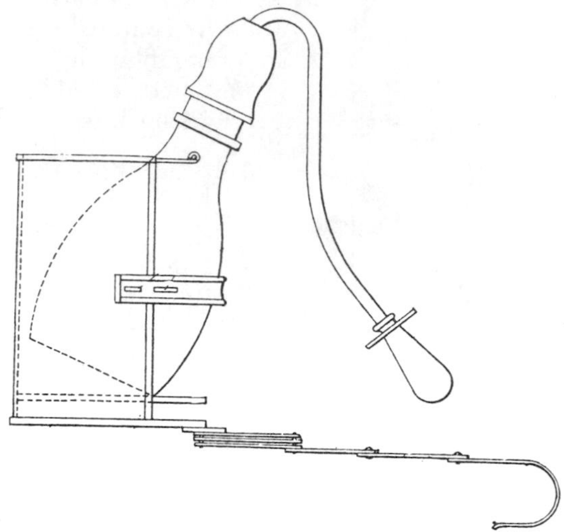

2095/1903
Mrs Albertine Cressierer
8 II Obeliskenplatz, Landshut, Germany

(Ferdinand Nusch *t/a* F G Harrington & Co)

Altenpohl and Wiegand's Improved Rocking Horse is adapted for locomotion as well as for a rocking movement.

The principal feature of the invention consists in an arrangement of rollers adapted to be pressed against the ground by bearing on the stirrups so that the momentum imparted by the rocking movement drives the horse forward. By means of continuous bearing on the stirrups a continuous forward movement on the horse can thus be obtained.

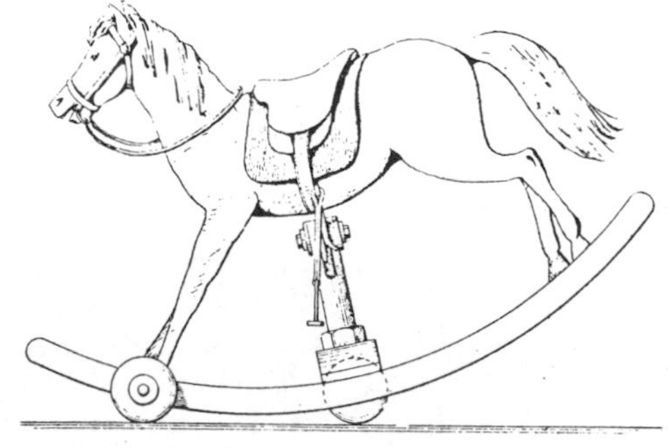

3303/1903
Friedrich August Altenpohl *Merchant*
Vallender-on-the-Rhine, Germany

Ludwig Wiegand *Engine Fitter*
Metternich-on-the-Mosel, Germany

(Herbert Haddan & Co)

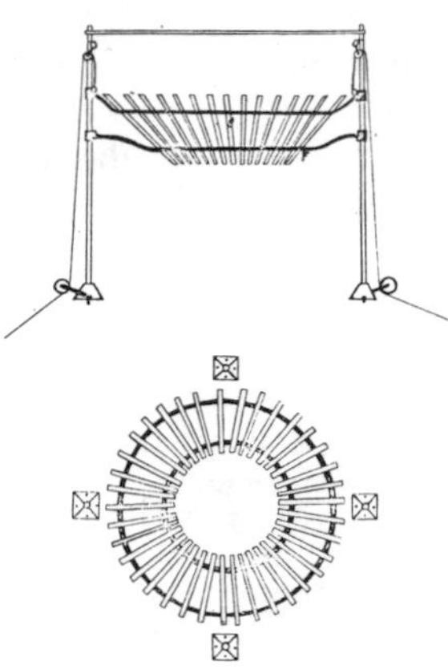

Davies's Improved Cycle Track is an inverted conical ring of slats which may be made to rise and fall on a frame while a cyclist is riding round it, thus enabling the cyclist to perform a new feat termed "whirling the whirl", performed as follows:- the "track" being on the ground, the cyclist commences on the ground inside the track and when he has attained sufficient velocity rises on to the inside of the track. Windlasses are then put into motion and, by means of chains and pulleys, the track is raised from the ground while the athlete continues to cycle round the interior conical surface thereof. When the performance in mid-air is completed the track is lowered to the ground; by increasing his speed the cyclist can then rise up over the outer edge of the pillars and so reach the ground, conveniently, or by slackening his speed can descend to the ground on the inside of the track.

6464/1903
Thomas Davies *Champion Cyclist*
54 Vine Street, Hulme, Manchester, England

White's Apparatus for Facilitating the Warming or Drying of Gloves and other Articles of Wearing Apparel consists of a skeleton frame, the shape of a glove or a sock, on which the article of wearing apparel is placed so as to distend it laterally and facilitate warming, drying, cleaning and stretching of the aforesaid article.

534/1903
Edward Lewellin White *Manager*
121 Cheapside, London, England

(Jensen & Son)

Bagnall's Improved Device for Toasting provides a means whereby bread or the like may be clipped or held in such a manner that the sides may be readily reversed during toasting without removing same from the device. Using this device, the bread or muffin may be reversed by quickly rotating the handle, thus causing the spring-loaded clip holding it to reverse its position.

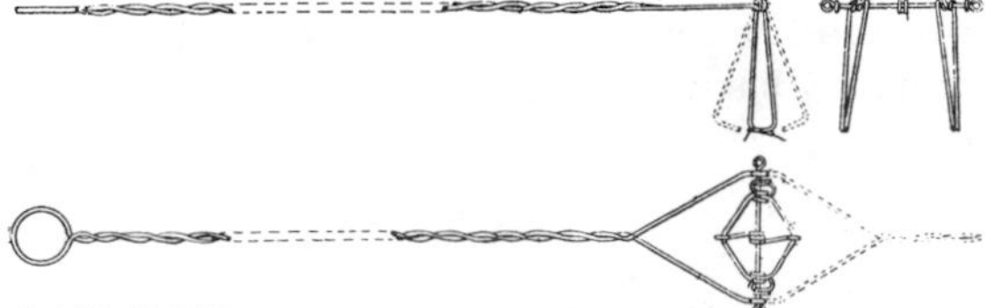

3403/1903
Edwin Bagnall *Engineer*
128 Coventry Road, Birmingham, England

(William Henry Baraclough)

Payne and Broadbent's Improved Instrument for Reproducing Sound improves the quality of sounds produced by appliances which employ a "record" as the primary means of producing vibrations which are given out as musical sounds or articulate speech. It renders these sounds more akin to the original instrument or voice by combining with the record a box-like resonating chamber such as a violin body, mounted so as to allow a stylus attached to it to follow the grooves in the record. A trumpet is not required.

3723/1903
Reginald Herbert Payne *Violin Maker*
43 High Street, Aylesbury, Buckinghamshire, England

Thomas Broadbent *Engineer*
East Lynn, Summer Road, East Molesey, Surrey, England

(Boult, Wade & Kilburn)

Schaeffer's Improved Seat, Carrier and Table Attachment for Baby Carriages and Other Vehicles allows a nurse or other attendant to be seated when desired, and, subsequently, to completely fold the seat for concealment beneath the carriage or partly fold it for the purpose of using it as a carrier. Thus used as a table, the attachment will be found invaluable at picnics and similar outings.

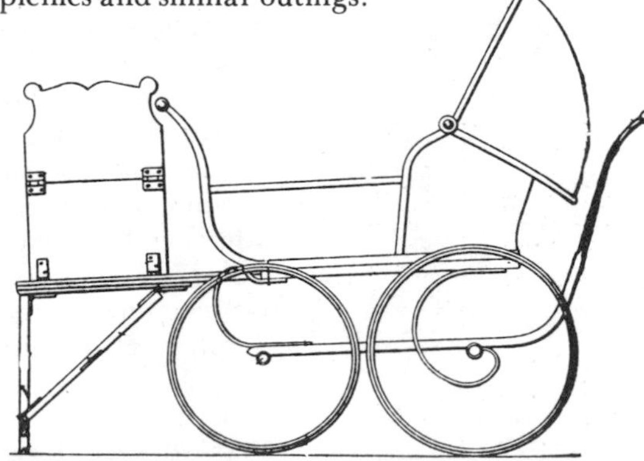

4453/1903
Anna Schaeffer *Gentlewoman*
Independence, Missouri, USA

(P R J Willis)

1903

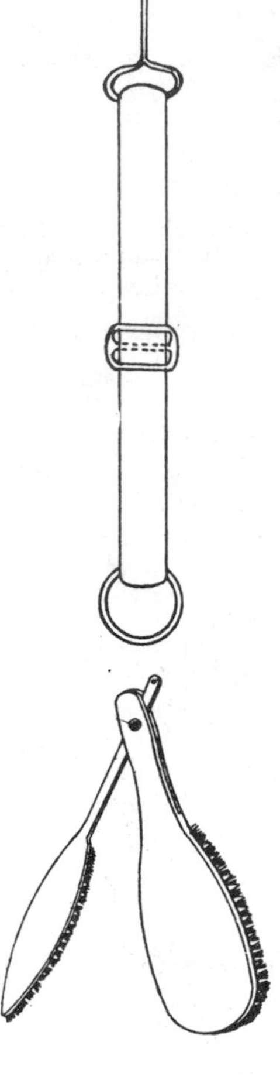

Miller's Appliance for Retaining Neck Ties in Position is an elastic cord or band having at one end a ring and at the other end a hook. The cord may be regulated in length by means of a sliding adjuster and by means of the ring the appliance can be attached to the shirt front. The hook at the other end is affixed into the tie.

The inventor is at pains to point out that any other means for attachment to the tie may also be employed; he, for example, always uses a safety pin.

818/1903
Hugh Miller *Merchant*
2 Fox Street, Greenock, Scotland

(P: Johnsons; C: Self)

Lashlie's Combined Hat and Clothes Brush consists of two parts, one of which, the smaller, can be used for a hat while the other, or the two together, will be utilisable for brushing clothes. Means may be provided for locking the two brushes together, and an arrangement for hanging the combined brush up is furnished so that the whole is always ready for use. The article may be made in any shape, but the hat brush should preferably be slightly curved so that it can be operated in the turned-up brim of a hard felt hat.

892/1903
Cowper Lashlie *Salesman*
Temuka, New Zealand

(Bernhard Dukes)

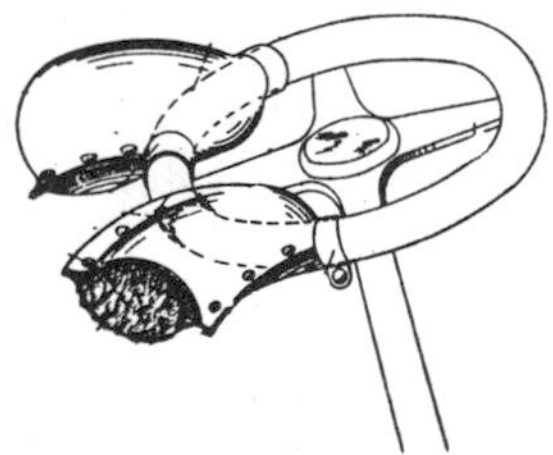

Morris's Improved Hand Shields for Cycles, Motor Cars and like Vehicles consists in muffs, preferably made of a waterproof material, lined with fur, and provided with stiffening strips of metal or whalebone to preserve the shape. They may be fastened to bicycle handlebars, steering wheels, speed and other levers of motor cars, and controllers and brakes of tramcars in such a manner that the hands of the rider or driver are adequately protected from cold, wet, wind, and other inclemencies of the weather and the danger of the rider or driver losing control over his machine owing to his hands having become numb is entirely obviated.

1383/1903
Frank Brooks Morris *Boot Manufacturer's Manager*
121 Harrow Road, Narborough Road, The Fosse, Leicester, England

(Boult, Wade & Kilburn)

Allen's Necktie Retainer comprises a piece of wire so bent as to ensure that it not only holds the necktie firmly in place but also cannot damage the collar, stud or wearer's throat.

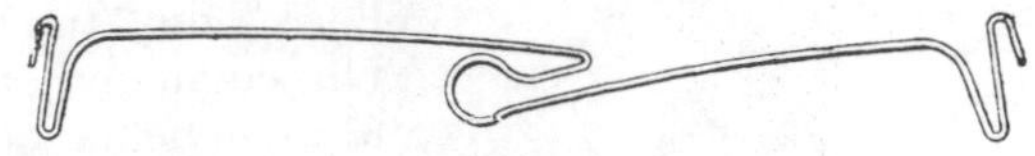

321/1903
Herbert Thomas Allen *Civil Engineer*
4 The Shrubberies, George Lane, South Woodford, Essex, England

(Hughes & Young)

Wickstrom's Improved Nut Cracker is preferably made in the form of a squirrel, but it must, however, be understood that the mechanism employed may be arranged to form a device of any other suitable design, without departing from the spirit of the invention. To operate this improved nutcracker, the lower jaw is first opened by pressing the tail inwards; the nut is placed in the jaws, and the tail pressed outwards.

1213/1903
August Wickstrom *Gentleman*
506 18th Street, Denver, Colorado, USA

Krebs's Improved Hairpin overcomes the well-known disadvantages of ordinary hairpins, namely that they slip too easily from the hair so that a rather considerable number of hairpins are required. Other hairpins also have the disadvantage of almost always pulling the hair out on withdrawal. This hairpin overcomes these problems by fitting the short leg of the straight part into a socket so that the twisted part is free to rotate about it upon insertion and withdrawal from the hair, and cannot adhere to single hairs.

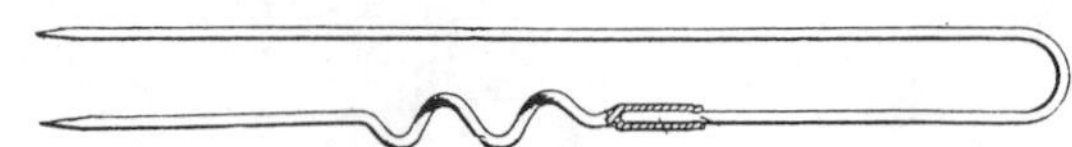

1969/1903
Otto Frederik Krebs *Engineer*
3 Romersgade, Copenhagen, Denmark

(F W Golby)

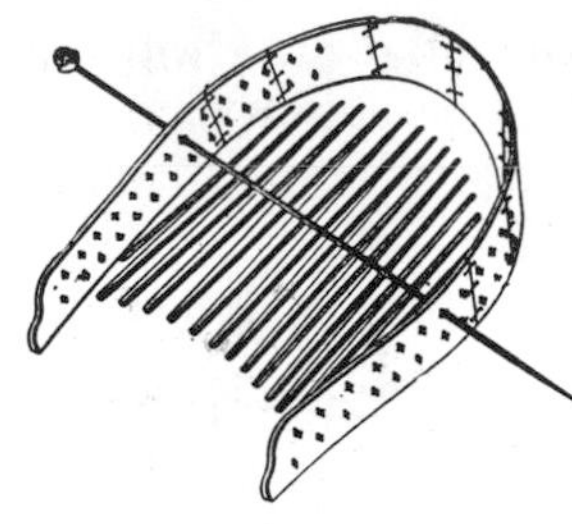

Gordon's Improved Method of Firmly Fixing Ladies' Hats to the Head is a comb with upturned edges roughened to grip the interior of the hat, and liberally pierced with holes through which a hat-pin can be speared. It may also be possible entirely to dispense with hat pins by making the comb, or back thereof, of steel with a jagged surface which should then grip the inside of the hat firmly when compressed into the crown.

1118/1903
Millicent Gordon *Married Woman*
Haddenham Hall, Haddenham,
Buckinghamshire, England

(C: Benj T King)

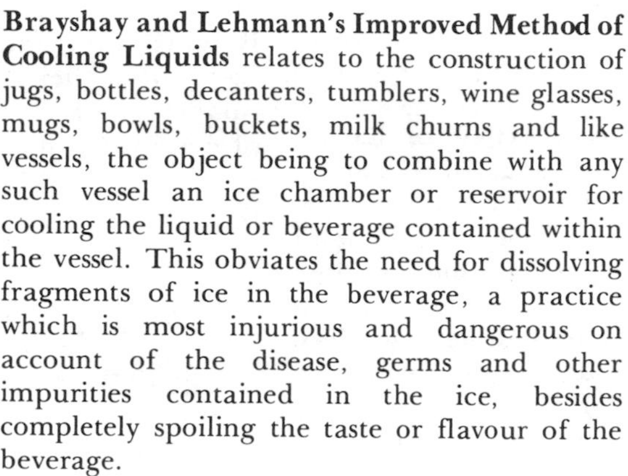

Brayshay and Lehmann's Improved Method of Cooling Liquids relates to the construction of jugs, bottles, decanters, tumblers, wine glasses, mugs, bowls, buckets, milk churns and like vessels, the object being to combine with any such vessel an ice chamber or reservoir for cooling the liquid or beverage contained within the vessel. This obviates the need for dissolving fragments of ice in the beverage, a practice which is most injurious and dangerous on account of the disease, germs and other impurities contained in the ice, besides completely spoiling the taste or flavour of the beverage.

1375/1903
Arthur Edwin Brayshay *Refreshment Contractor*
Hyde Park Terrace, Leeds, England

Frederick William Bruno Lehmann *Artist*
91 Meadow Lane, Leeds, England

(John E Walsh)

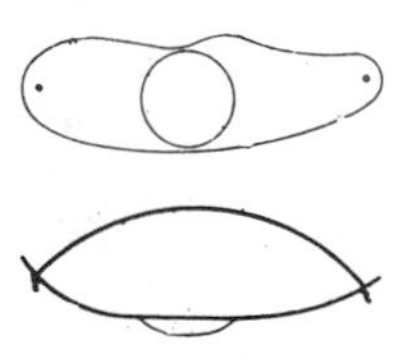

Martin's New or Improved Writer's Appliance has for its object and effect to enable writing to be done at a greatly enhanced speed, obviate writer's cramp, improve writing and lessen the labour. It consists of a shaped arrangement of Xylonite, metal, ivory, bone, or other material which fastens to the hand with a rubber band and may be provided with rollers, balls or other devices to facilitate the hand sliding on the paper by rendering the friction extremely small, so increasing speed in writing from 25 to 50 per cent.

645/1903
Harold Sheen Martin *Electrical Engineer*
22 Sir Thomas Street, Liverpool, England

(Cheesebrough & Royston)

Foerster & Schulze's Improved Perambulator can be completely disinfected with great facility so that it can be employed as an ambulance carriage for children suffering from infectious diseases.

This object is attained by making the entire carriage, including the body or basket in which the child rests, of metal, and only providing it with a readily removable upholstered lining. The object of making the body of metal is not to increase its durability but to render an effective disinfection possible, for it is only necessary to push the carriage after removal of the lining into a baker's oven or disinfecting stove, in order to kill or destroy absolutely by means of heat all germs of disease, while the upholstered lining and the removable cover or hood are disinfected in any other suitable way.

155/1903
Foerster & Schulze *Manufacturers*
80 Dresdenerstrasse, Berlin, Germany

(Carl Bollé)

Lacy's Improved Hat Fastener facilitates the operation of securely fastening a lady's hat or bonnet upon the head. It comprises a piece of resilient material, spring steel is particularly suggested, attached to the hat. In putting on the hat, the free ends are forced apart and are laid one on each side of the head.

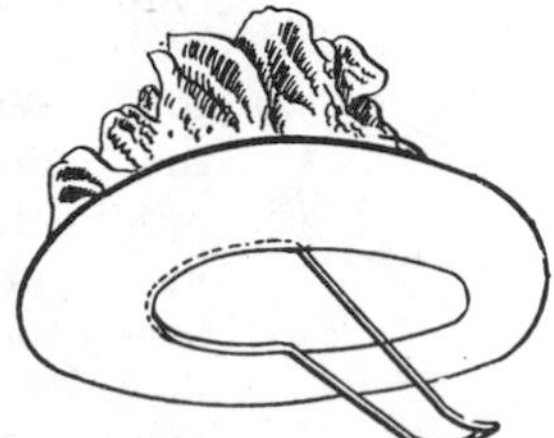

1818/1903
George Edgar Lacy *Inventor*
17 West Park Street, Newark, New Jersey, USA

(P R J Willis)

1903

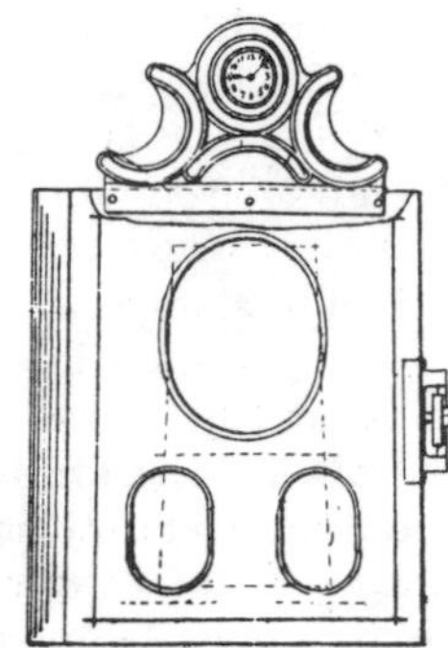

Thompson's Improved Album serves other useful and ornamental purposes besides being a receptacle for photographs, cards, stamps or the like. It is provided with a strut so that it may be propped in a reclining position, ensuring that the clock, barometer etc in the top projection is clearly visible, the shape, configuration and ornamentation of this top projection being variable as desired. One — or more — photographs may be mounted on the front cover and when the clasp is opened, it allows the clockwork musical instrument cunningly concealed in the back to entertain the assembled company.

1942/1903
Denis Richard Thompson *Merchant*
27 High Street, Sheffield, Yorkshire, England

(Boult, Wade & Kilburn)

Wolstenholme's Improved Article of Ladies' Underwearing Apparel is adapted to support the bust to retain it in form and preserve the contour of the figure. It is designed to be worn beneath an evening or other dress and the upper portion may be folded down, say for one inch and a half, for wear with a décolletage. The seamless bust support is designed to afford ease and comfort to the wearer and prevent injury from the stiffening portions of the corset. It consists in a seamless tapered band shaped by pleats and secured by an elastic band, the whole shaped or bowed to the contour or desired contour of the bust of the wearer.

3161/1903
Kate Morgan *t/a* Kate Wolstenholme *Corset and Surgical Belt Maker*
21 Wells Street, Oxford Street, London, England

(Wm Brookes & Son)

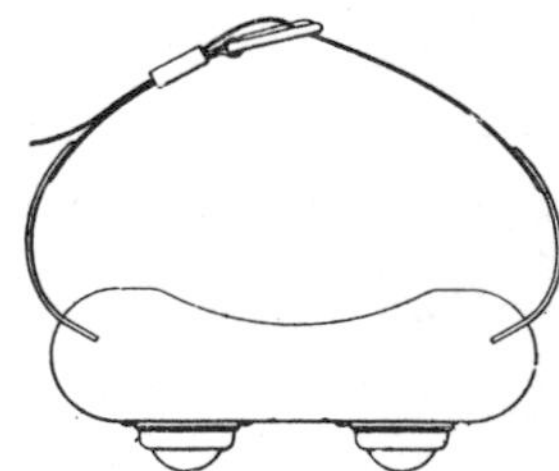

Rockwell's Improved Handwriting Developer consists of a pad of cork, or other suitable lightweight material, which straps to the wrist and runs on balls for the purpose of facilitating the movements of the muscles used in writing, so that children learning to write acquire a free movement of the arm and correct position of the hand and fingers to insure a rapid and legible formation of letters.

1677/1903
Frederick Clark Rockwell *Manufacturer*
West Hartford, Connecticut, USA

(D Young & Co)

de Koningh's Improved Collar Stud will also act as a tie-clip, and has the appearance of a jewelled scarf-pin.

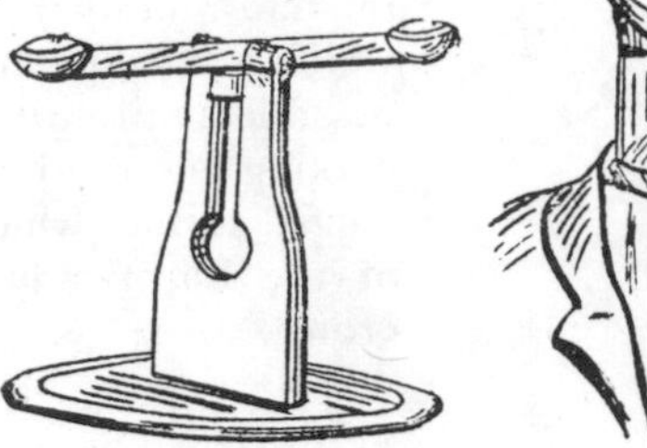

3267/1903
James de Koningh *Working Jeweller*
120 Liverpool Road, Islington, London, England

(Browne & Co)

Wilkins's Cutter for Removing the Tops from Eggs in a Clean and Effective Manner is guarded so that it is impossible to cut the fingers when using the device.

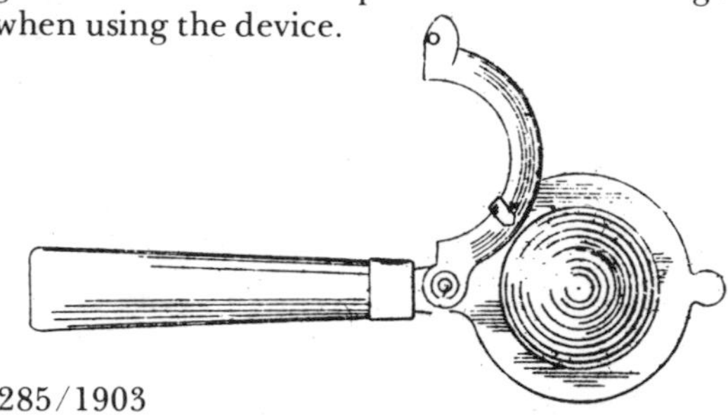

285/1903
Frederick Charles Wilkins *Watchmaker*
Malvern, Worcestershire, England

(Lewis W Goold)

Fischer and Thieme's Improved Machine for Loading Dung, Manure, Field Produce and Other Materials consists of two endless belts on a stand, each bearing a series of forks pointing towards the other. The bands are caused to separate as they approach the dung, and bite it up by closing as they pass through the heap. If beet-root is to be dealt with, the point of each fork is fitted with a small ball to prevent the roots from adhering to it.

2405/1903
Wilhelm Fischer
25 Hartz, Halle, Germany

Max Thieme
8 Uhland Strasse, Halle, Germany

(Jensen & Son)

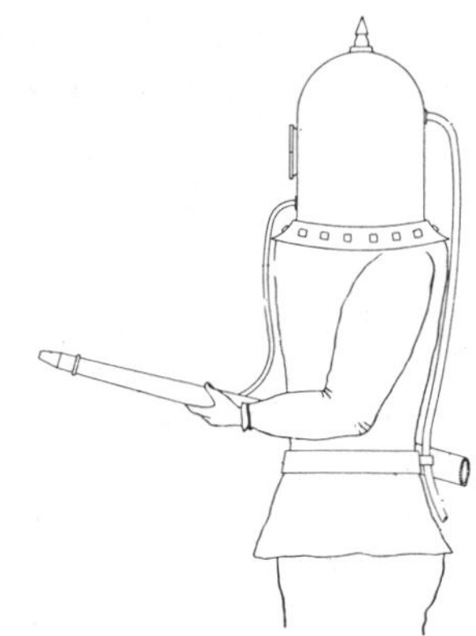

Alonso's Improved Means for Supplying Air for Respiration to Firemen's and Like Helmets enables firemen to remain in places filled with smoke or charged with some other asphyxiating gases or vapours by simply using the extinguishing stream of water to induce a continuous flow of air into the helmet.

It is simple in construction, continuous in action, and overcomes the defects of all other smoke helmets, which are complicated and difficult to manipulate and only imperfectly meet the purpose for which they are intended.

Furthermore, it fulfills completely all the tests to which it may be subjected and has the advantage that, since the portion of the pipe between the helmet and nozzle is flexible, it performs the dual office of also serving as a speaking tube; it can be flattened by the fingers so as to stop at any desired moment the noise caused by the outward rush of water and to admit of speaking and hearing being carried on with clearness. A movable crystal window placed in the helmet also enables the fireman to see clearly.

3600/1903
Jose Maria Gorde Alonso *Manufacturer*
13 Rue Chapitela, Pamplona, Spain

(Cheesebrough & Royston)

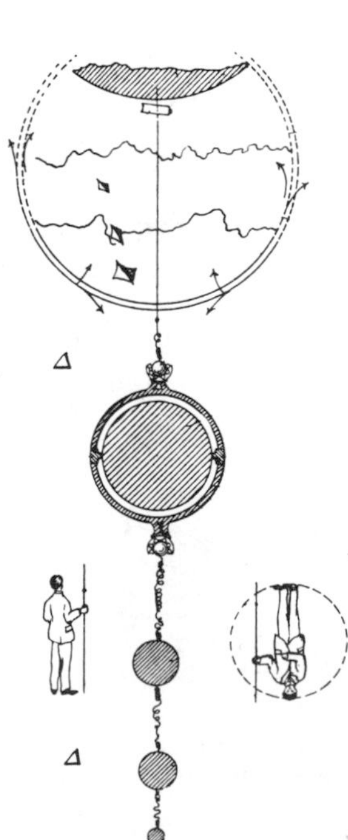

Whitney's Method for Collecting and Putting to Practical Use the Electricity from the Interplanetary Ether for power, lighting, heating and other purposes employs a specially-constructed cable some 150 miles long which is projected through the earth's atmosphere by any suitable form of terrestrial energy, such as a small powerful cannon or an airship. It should be noted, however, that it is only necessary to furnish energy to lift the furthest end of the cable the first seventeen miles above the earth's surface; beyond this, the electric force in the ether itself will raise the furthest end of the cable through the miles remaining without the necessity of employing any extra force.

Furthermore, it should not be forgotten that even though the cable is a total of 150 miles long, constructed in five sections of wires of diminishing strength according to the decreasing strain to which they are subjected in passing through the atmosphere to the ether, and that the first section alone, the seventeen miles needed to reach to the edge of the earth's atmosphere above Chicago, would weigh approximately 152,592 pounds flat on the earth, its weight in an upright position stretching towards space is considerably less than this because the particles of the wire will not push towards the earth but towards each other, towards the centre of the wire. Thus, once it is in space the self-attraction of the particles of the wire and the electric force of the ether itself will serve to keep the cable taut while it conducts electricity from the ether to charge secondary batteries.

The receiver is mounted upon a tower, the foundation of which carries a circular rail track on which moves a truck carrying a winding spool so that the length of the cable can be adjusted according to the currents of air bearing upon it in the atmosphere.

4413/1903
Albert Gallatin Whitney *Lawyer and Scientist*
1322 Wolfram Street, Chicago, Illinois, USA

Welsh's Improved Device for use in Connection with Railways for Recreation is designed for use upon what are known as roller coasters or scenic railways, now so frequently used in amusement parks, whereon a light passenger car travels by gravity. Upon such railways covered passageways are sometimes built to represent tunnels and this device may very well be used in such tunnels. It consists in the arrangement of mirrors whereby objects properly located with relation thereto may be reflected many times and so heighten the attractiveness of the railway.

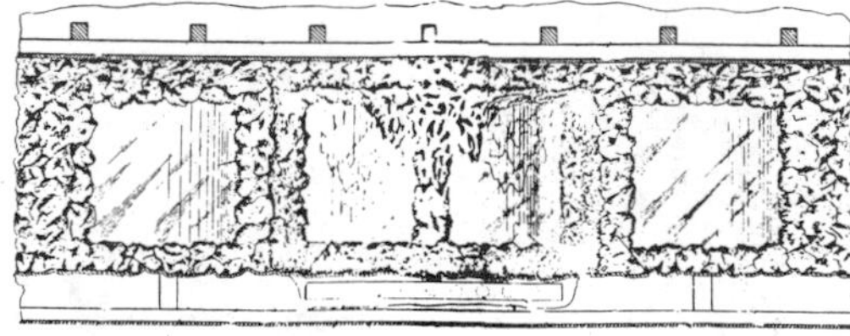

3288/1903
Adam A Welsh *Inventor*
105 Summer Street, Boston, Massachusetts, USA

(Haseltine, Lake & Co)

Leach and Smith's Improved Pickle Spoon is for taking olives and the like from bottles, jars and dishes. It is provided with an open draining bowl and upturned prongs on the edge of the bowl to catch the pickle and hold it.

2479/1903
Arthur Henry Leach *Pattern Maker*
Henry Dexter Smith *Druggist*
Middleboro, Massachusetts, USA

(Marks & Clerk)

1903

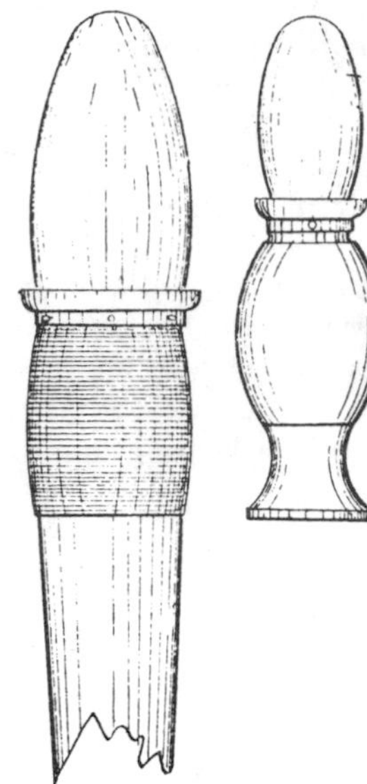

Jewell's Improved Paint, Shaving, Whitewash or Other Brush is provided with a ledge to collect drips and so protect the user's hand when the brush is used overhead or with the hair uppermost.

7764/1903
Frank Jewell *Sergeant, Permanent Staff, 3rd Royal Munster Fusiliers*
Charles Fort, Kinsale, Co Cork, Ireland

(Hughes & Young)

Schmid's Improved Tooth Brush provides a means of performing buccal injections during the operation of cleaning the teeth, so ejecting the alimentary detritus which is impossible to remove with an ordinary brush. It has a bulb on the handle, communicating with the bristles by means of suitable passages. The bristles may be of badger-hair, india-rubber etc. The bulb is filled by squeezing it and allowing it to regain its shape whilst immersed in a suitable liquid, which is injected on to the teeth during use.

28391/1903
Alfred Schmid *Surgeon-Dentist*
111 Boulevard de Magenta, Paris, France

(Phillips & Leigh)

Bywater's Rational or Divided Skirt for Cycling consists of knickers or bloomers and an overskirt. The legs of the knickers fit tightly round the ankles, and are surrounded by kilted outer legs. A pocket is provided. The overskirt consists of two short side pieces to which the back and front panels are attached. The panels are gored and pleated, the pleats being kept in shape by elastic bands. The panels are provided with buttonhole tabs, which engage with buttons on the front and rear of the legs of the bloomers. It is quite new and nothing like it in the market being made separately, the two garments give the appearance of a very pretty kilted skirt; it is a most comfortable dress if worn in an ordinary way, in cycling there is no danger of exposure; neither is there anything to come in contact with the pedals of the machine; a lady requires no petticoat or bloomer, except the bloomer and overskirt, which is a most comfortable and stylish skirt for cycling, riding and walking.

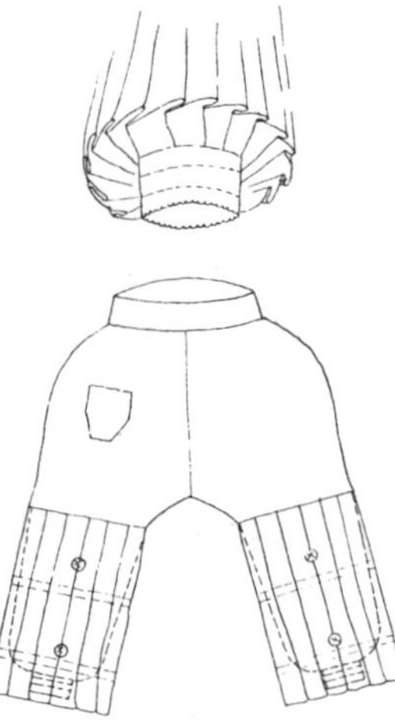

1542/1903
Oretta Bywater *Wardrobe dealer*
3 Charles Street, Briton Ferry, Glamorganshire, Wales

(C: Astley Wm Samuel)

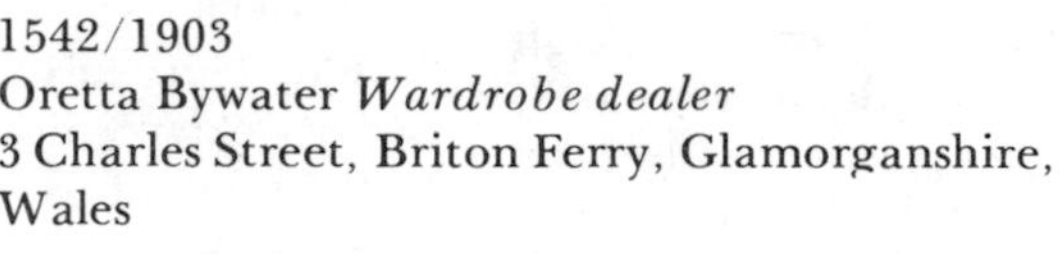

Glade's Improved Foot Cycle or Skate is provided with a suitable mechanism in order that by a slight up-and-down leg movement and transference of weight from one to the other the person wearing them is propelled forward at a much greater rate of travel than the ordinary walking movement of the feet on the ground could produce. The improvements consist in the employment of spring-actuated compass or toggle levers which form a resilient medium to act against the weight of the person wearing the cycles and at the same time store power to allow of its being retrieved in assisting to lift the foot plate with the upstroke of the foot in the walking action and to assist the rider when ascending grades or travelling against wind pressure.

The cycles are also provided with a ball bearing and collar arrangement to prevent the wheels from locking when turning sharp curves.

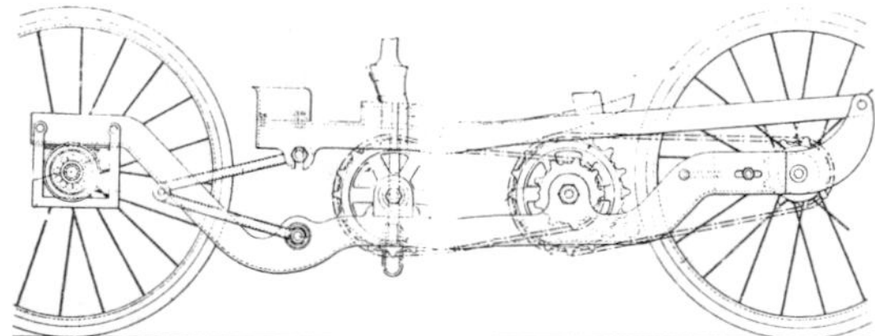

4257/1903
Henry Glade *Mechanical Draughtsman*
4 Delbridge Street, North Fitzroy, Victoria, Australia

(P: A M & WM Clark; C. Self)

Neff's Improved Apparatus for Displaying Advertisements consists of a vehicle with endless curtains for exposing a series of advertising pictures, and is provided with a musical instrument to attract the public's attention to the material displayed.

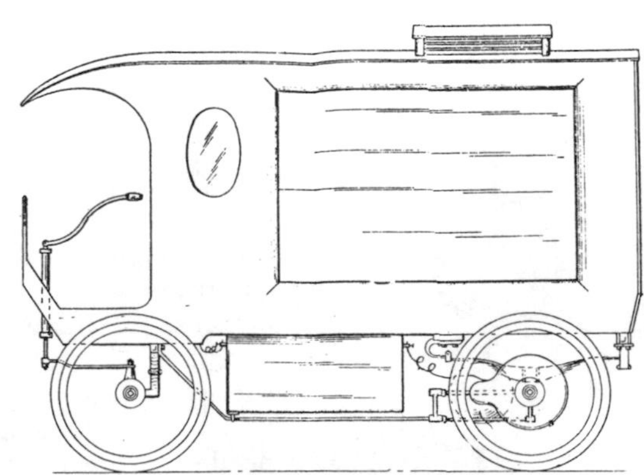

9393/1903
Richard Henry Neff *Gas Inspector*
1403 South Eastern Avenue, Indianapolis, Indiana, USA

(Boult, Wade & Kilburn)

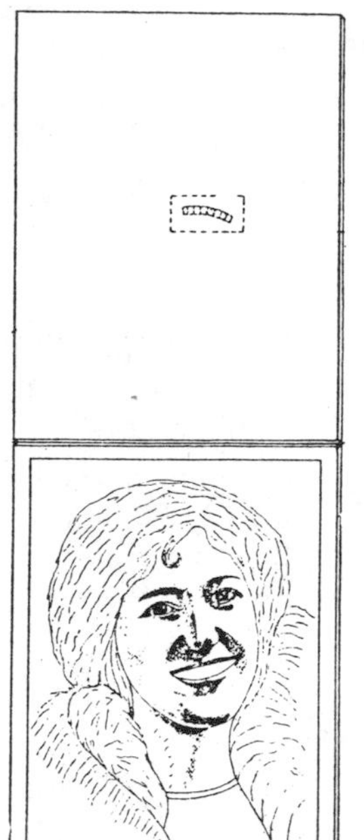

Sim's Improved Dentists' Advertising Device comprises a picture of a face with an aperture where the teeth would be backed with a sheet showing teeth, which can be seen through the hole in the face. By these means the effect and appearance of the teeth, and different kinds of teeth, can be illustrated.

13062/1903
John Sim *Printer*
21A Seel Street, Liverpool, England

(Cheesebrough & Royston)

Tweedale's Improvements in the Preservation of Eggs for Edible Purposes provides a process which enables eggs to be preserved from the day they are laid, and collected and transported or exported in a sound edible condition. These objects are attained by first soaking the eggs in a dilute preservative solution of boracic or salicylic acid which penetrates and fills up the pores and then coating them with syrup or varnish or waterglass to keep them moist. The coated egg is now protected by applying a wrapper of impervious material, preferably waxed tissue paper, on which advertisements or directions may be printed.

These coatings may either be broken when the egg is gashed on the edge of a cup to remove its contents or boiled away, leaving the egg in its normal condition.

4716/1903
Charles Lakeman Tweedale *Clerk in Holy Orders*
The Vicarage, Weston, nr Otley, Yorkshire, England

Fulton's Improvements in Means Acted upon by Changes of Temperature for Indicating such Changes or for Regulating Temperatures or Obtaining Movements for other purposes relates to a double bellows, into each half of which is sealed a gas, the two gases having different coefficients of expansion, but unaffected by changes of atmospheric pressure. The apparatus may be used to indicate changes of temperature, regulate such changes by operating valves, or to set off automatic alarms, fire extinguishers, or other mechanisms.

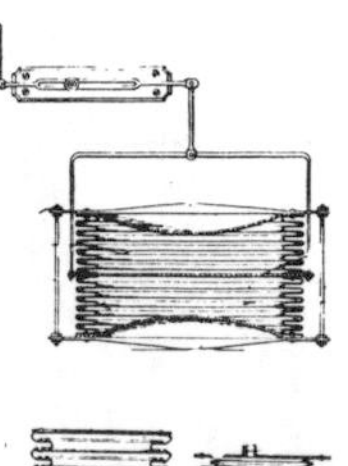

11547/1903
Professor Weston Miller Fulton *Professor of Meteorology in the University of Tennessee*
Knoxville, Tennessee, USA

(Johnsons & Willcox)

Noxon's Improved Cow-tail Holder is a simple and effective device for holding a cow's tail while being milked to prevent the same from being switched into the milker's face or from brushing dirt into the pail. It consists of a clip to fit the cow's leg which holds a spring clamp to hold its tail and may easily and quickly be attached and detached without annoyance or injury to the animal.

4566/1903
William Noxon
Bloomfield, Ontario, Canada

(Marks & Clerk)

Wacker's Improved Device for Preventing Mistakes in Taking Hats and other Head Gear consists of an impediment pivoted in the hat so that it can be turned down, when the hat is not in wear, and turned up, when the hat is in wear.

If anyone attempts to put the hat on, the impediment comes into contact with the head causing the person who has put the hat on by mistake to take it off to ascertain the cause of the obstruction, so catching sight of the name of the owner of the hat placed at a suitable point in its interior, so that the error cannot help being noticed and the changing of hats prevented.

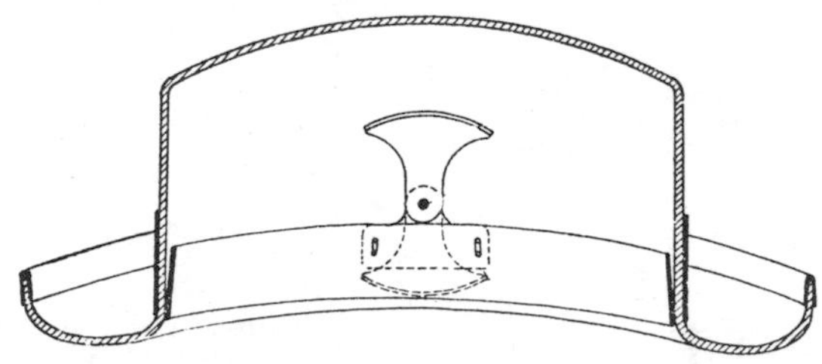

8293/1903
Albert Wacker *Engineer*
Bayerische Celluloidwaarenfabrik vorm.
Akt-Ges, 44 Landgrabenstrasse, Nuremburg, Germany

(W P Thompson)

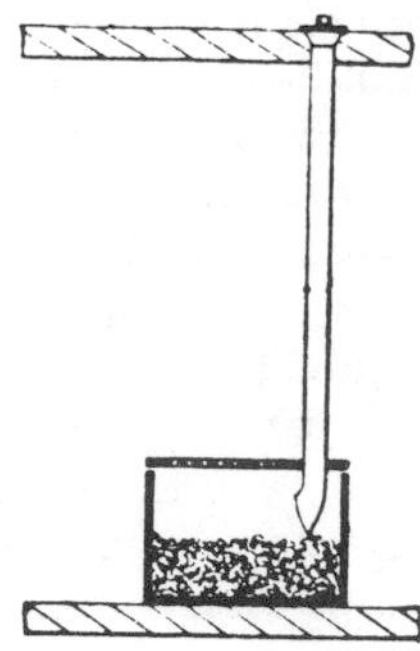

Staff's Apparatus for Extinguishing and Preventing Fire at Sea and for the Destruction of Rats on Board Ship employs the principle of making carbon dioxide gas in ships' holds by pouring sulphuric acid down a system of pipes on to marble chips in specially constructed vessels. The gas thus generated extinguishes any fires — or rats — which happen to be around. When the apparatus is required for use, all hands should be ordered on deck before pouring the acid down the pipes.

10347/1903
George Thomas Albert Staff *Physician and Surgeon*
Carrack Dhu, St Ives, Cornwall, England

Phillips's Improved Animal Trap is an attractive device which may be shaped to represent a steam-roller, automobile or other miniature vehicle. The trap is on wheels, and when the mouse is caught in the cage, it runs in the casing and causes the cage to move along.

Its mode of operation is as follows: attracted by the bait, the animal pulls upon its detaching trigger and thereby closes a door. Caught alive in the cage, in his roaming to find an exit he will cause the device to move forward in the manner illustrated. The cage is made entirely of wire gauze or closely perforated tin, to reduce the weight of the trap and afford a clear view of the movements of the animal. This also renders it comparatively inexpensive to manufature.

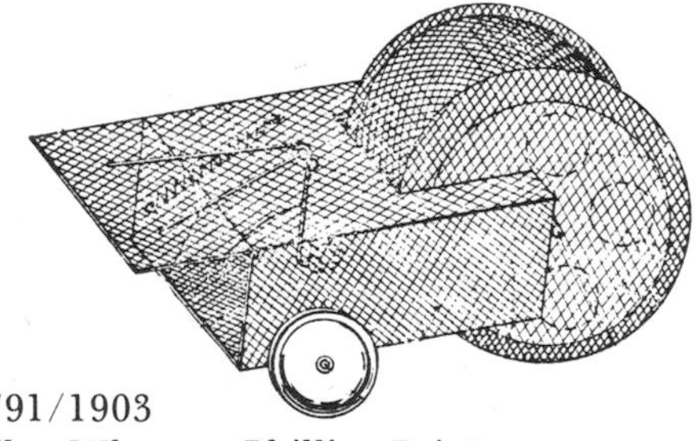

6791/1903
Allen Wheaton Phillips *Printer*
75 Bassett Street, Providence, Rhode Island, USA

(George Barker)

Howells's Improved Swimming Appliance is a contrivance, constructed of aluminium for lightness, which is fastened to a swimmer's legs to facilitate propulsion by the opening and shutting motion of the flaps behind the wearer's feet.

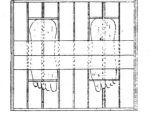

12092/1903
Arthur Morgan Howells *Carpenter*
8 Best Lane, Canterbury, Kent, England

(Hughes & Young)

Harig's Improved Trap for catching Mice, Rats and other Animals is a novel device in which the bait, such as grain, may remain permanently and, while serving as a line to entice the animals to enter the trap, is inaccessible under all conditions, so that the trap requires but one baiting.

The trap consists of a double-walled mesh cube with a covered runway — preferably embodying staircases and landings — extending from top to bottom. The remainder of the space is filled with grain which the animal can see but not reach. Lured by the bait, the animal falls through a hinged lid into the inner trap vessel; by making the inner vessel removable from the outer vessel, the animals may be readily gotten rid of.

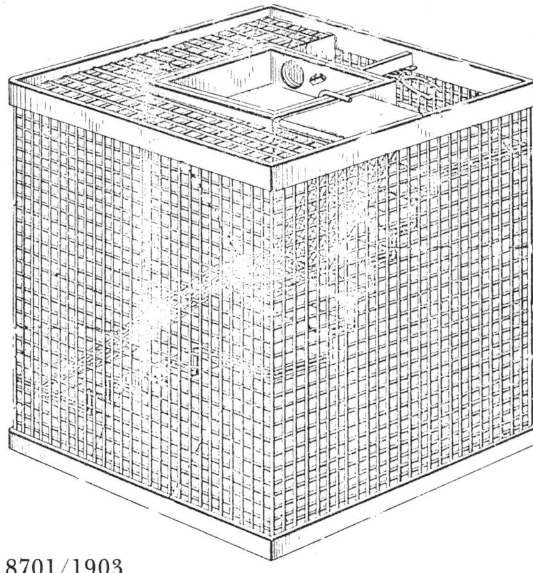

8701/1903
Joseph Bernard Harig *Merchant*
1130 Light Street, Baltimore, Maryland, USA

(Wheatley & Mackenzie)

Bieber's Improved Walking Cane comprises a hollow walking-stick containing two bladders (preferably oiled hogs' bladders) capable of being blown up and so supporting a person in the water, the stick forming a connection between them.

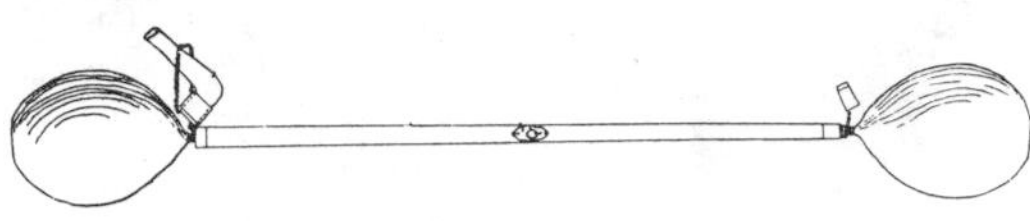

14268/1903
Carl Bieber *Engineer*
Fürstenwall, Düsseldorf, Prussia

(Oskar Arendt)

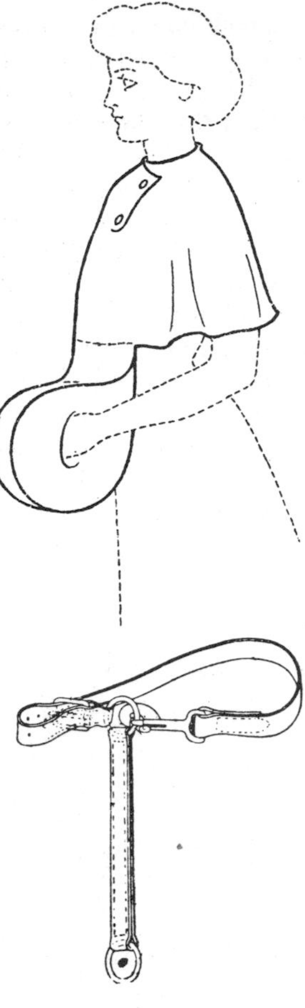

Hendriks's Combined Cape, Stole and Muff for Motorists and the like comprises a cape with a dependent stole, the end of which forms the muff. It is designed to afford a convenient combination of these devices which shall, whilst being effective in a sartorial sense, be easy and speedy to doff or assume.

13629/1903
Marion Hendriks
55 Fordwych Road, West Hampstead, London, England

(Castle Smith)

The Claphams' Device for Preventing a Horse from Running Away with a Vehicle when Unattended is a strap which not only fastens the wheel to the frame, but is also arranged to pull on the reins if the horse starts to move. If the horse should attempt to go backwards, the strap locks the wheel before a pull is put on the reins.

7973/1903
John Newsome Clapham *Hairdresser*
George Spencer Clapham *School Teacher*
Ashurst, New Zealand

(Marks & Clerk)

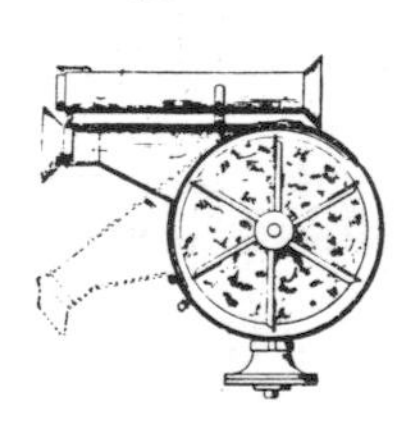

Halliwell's Improved Hair Dryer can direct hot or cold air pumped by an electric fan, and has a flexible extension to aid the user. This is warmed either by electricity or gas jets.

14407/1903
Henry Virtue Halliwell *Inventor*
200 West 14th Street, New York, USA

(Cruikshank & Fairweather)

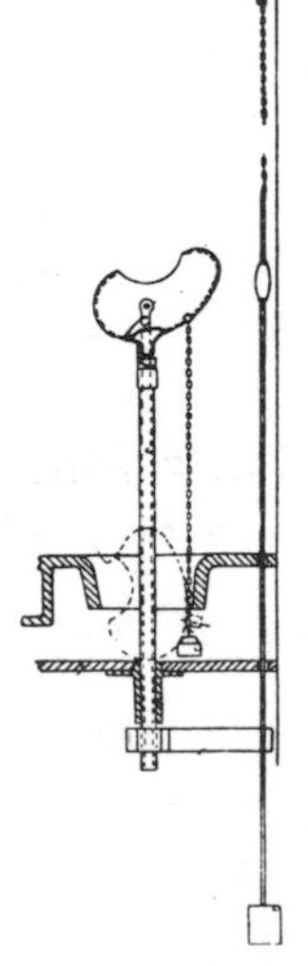

Auerbach's Improved Urinal for the Use of Females obviates the serious inconvenience connected with ordinary closets owing to the necessity of the dress being raised to the level of the closet seat where, moreover, it is easily exposed to the risk of being soiled. In this device, the utensil is designed to be moved upwards and downwards and is so connected that it can be raised and lowered by the person standing over it. Moreover, the utensil is self-tilting so that it turns into position when it has been raised to the appropriate height.

15585/1903
Carl Rudolf Auerbach *Gentleman*
Saalfield on the Saal, Germany

(Wheatley & Mackenzie)

Haller and Ellis's Convertible Cloak, Stretcher, Hammock, Bed or Float is particularly adapted for use by military men, travellers and sportsmen, but from its graceful appearance and great utility will also be worn by ladies and gentlemen for cycling, driving, motoring, fishing, golfing, shooting and walking.

Circular in shape, a strong lattice of webbing inside the cloak is provided to bear the weight of the body when in use as a hammock or stretcher, and is supplied with fixtures for hammock ropes or stretcher poles.

The lateral wings of the cloak fall becomingly to the sides when in normal wear and also cover the body and protect it from dew or rain when in use as a stretcher.

10244/1903
Phillip Haller *Tailor*
John Thom Ellis *Tailor*
South Molton Street, London W, England

Tolcher's Improved Monocycle provides an equilibrant whereby the rider of a monocycle can more readily maintain his balance, whether travelling along the flat or ascending or descending an incline. Briefly, the invention consists in gearing swinging handlebars to a swinging saddleframe so that the weight of the rider can at will be moved forward or backward of the centre of the wheel or directly above it to maintain or regain his equilibrium. For full comprehension reference must be made to the accompanying drawings.

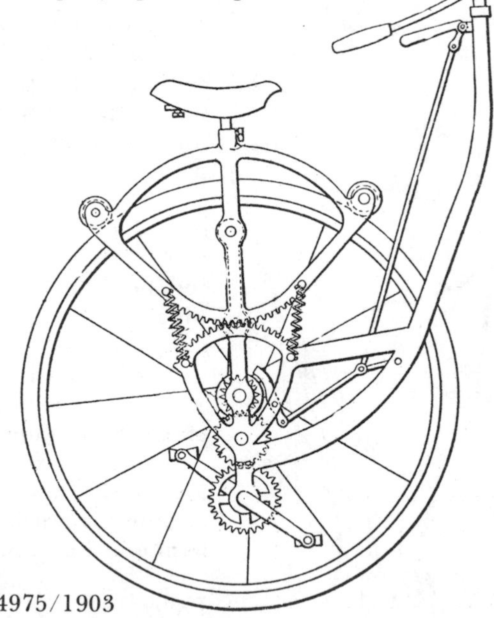

14975/1903
Henry Tolcher *Engineer*
Mare Street, Pretoria, Transvaal, South Africa

(F Wise Howorth)

1903

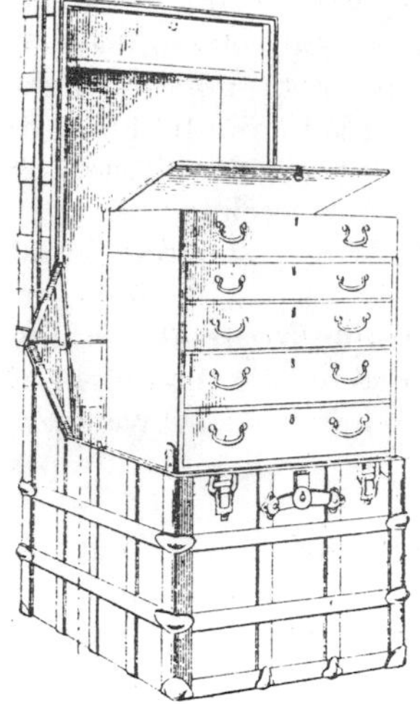

Baruch's Combined Travelling Trunk and Chest of Drawers or Wardrobe provides a receptacle which is suitable for use most particularly by travellers such as actors, salesmen, clergymen etc.

884/1903
Nathan Baruch *Manufacturer*
557 West 124th Street, New York, USA

(A M & WM Clark)

Mayer's Improved Toilet Paper enables relief stamping or printing of advertising matter to be effected on the sheets without prejudicing the hygienic demands. The printed portions may be detached, or are in such positions as not to come into use for toilet purposes.

17066/1903
David Mayer *Manufacturer*
17 Lindenstrasse, Cologne-on-the-Rhine, Germany

(Herbert Haddan & Co)

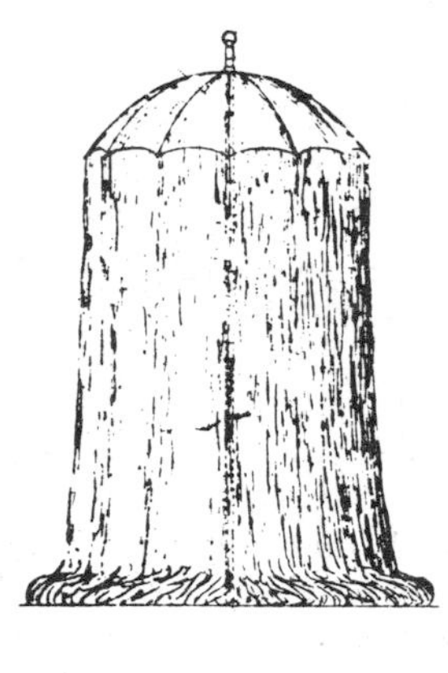

Pardo's Improved Umbrella may be hung from a tree by means of the ring provided, and has a curtain reaching the ground to act as a mosquito net.

18749/1903
Isaac Pardo *Merchant*
67 Neuerwall, Hamburg, Germany

(Boult, Wade & Kilburn)

Lissner's Improved Chalice both obviates the risk of contagion or infection in using the communion cup and makes it possible to allot the sacrament wine in definite portions so that the quantity of wine required may be easily predetermined. The mouth of the chalice is closed and portions of wine are made to flow into the closure (which is funnel shaped) by tilting the cup. The quantity of wine taken by each communicant is separated from the remaining wine by a valve-like arrangment, so that infection of the latter is prevented. In addition, transfer of sickness from the lips of one person to another is rendered entirely impossible by a strip of paper, parchment, or the like pulled off a roll, over the lip, and on to a take-up roll (the rolls are in the base) provided with a winder. A chamber is provided in the base to receive any unimbibed portion of wine.

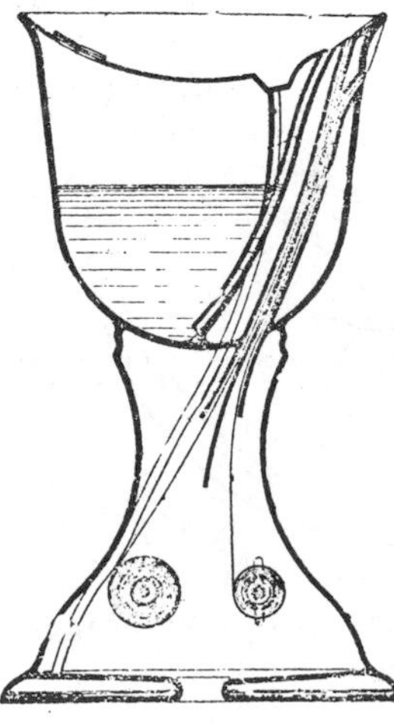

15399/1903
Henrik Lissner *Manufacturer*
Slagelse, Denmark

(Henry O Linck)

Gessmann's Improved Electric Dry Battery adapted to be used in connection with Hat or Head Bands comprises a flexible strip of pure zinc with projections forming the negative zinc-poles. Silver plates, dished and filled with silver chloride mix, and insulated from the zinc by a fabric strip impregnated with calcium chloride, form the positive electrodes. The whole is covered with filtering-stuff and fastened in the hat. To make the battery efficacious it is only necessary to moisten it from time to time with pure water; people who perspire very much need not do this.

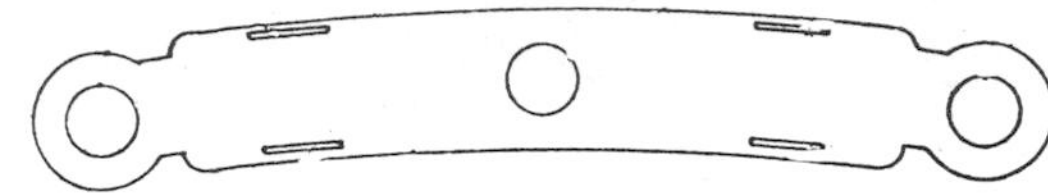

12578/1903
Gustave Wilhelm Gessmann *Inventor*
Landesmuseum, Graz, Austria

(Charles Bauer, Imrie & Co)

The Gannett Development Syndicate's Improved See-Saw Amusement Apparatus is adapted particularly for pleasure resorts such as in parks and other places where people congregate in search of amusement.

A massive see-saw is motor driven, and the cages — in which the passengers sit — have hoods which act as diving bells when the cages are submerged in appropriately placed tanks of water.

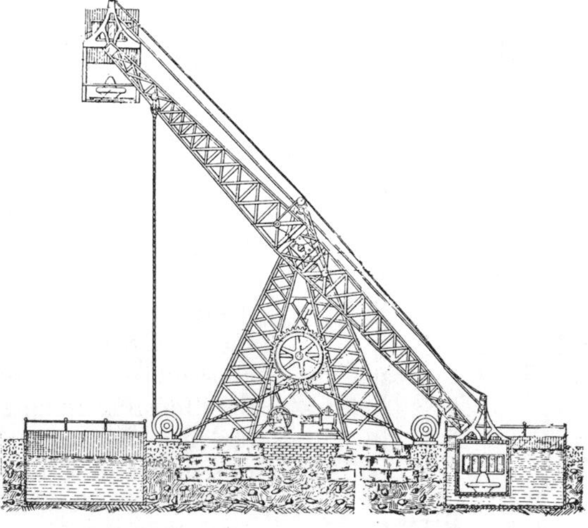

10610/1903
Willard H Gilman
State Street, Boston, Massachusetts, USA

The Gannett Development Syndicate
27 Shoe Lane, London, England

(Haseltine, Lake & Co)

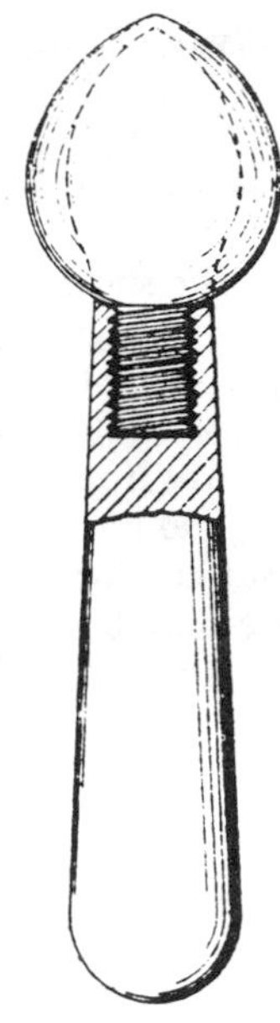

Mahn's Appliance for Smoothing the Edges and Widening the Button Holes of Collars has a head of bone, celluloid, earthenware, glass, india-rubber, vulcanite, ivory or wood, with tapered grooves to suit different collars. The handle may be hollow, and screwed on, so as to serve as a receptacle for studs, pins etc., and may be used for advertising purposes.

For smoothing and rounding collar edges the appliance offers special advantages, seeing that the sharp edges of starched and ironed collars hitherto frequently led to painful inflammations of the skin. Furthermore, a head of disinfecting material may be provided as a preventative against infection.

18471/1903
August Hermann Heinrich Mahn *Gentleman*
30 Hirten Strasse, Hamburg, Germany

(Warwick Henry Williams)

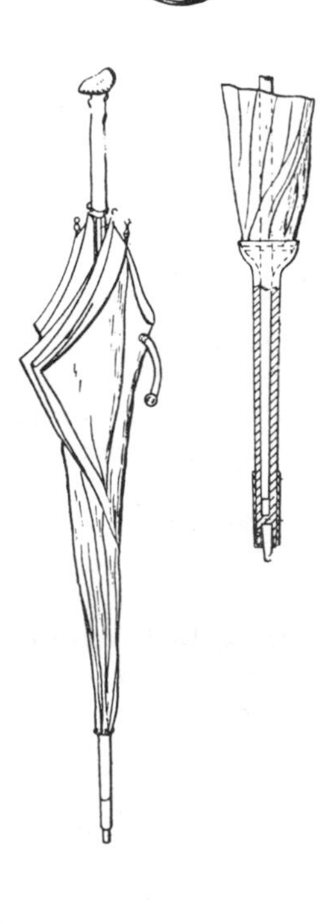

Kern's Improved Insulated Ferrule or Tip for Umbrellas and the like will protect the user from receiving a shock should the end of the umbrella be brought in contact with an electric conductor; it consists of a device which is adapted to entirely cover the extended end of the metal rod, also covering the ends of the ribs and ring to which they are attached.

The device is simple and cheap and can be readily attached to or detached from an ordinary umbrella.

18416/1903
Howard Edward Kern *Inventor*
Allentown, Pennsylvania, USA

(P R J Willis)

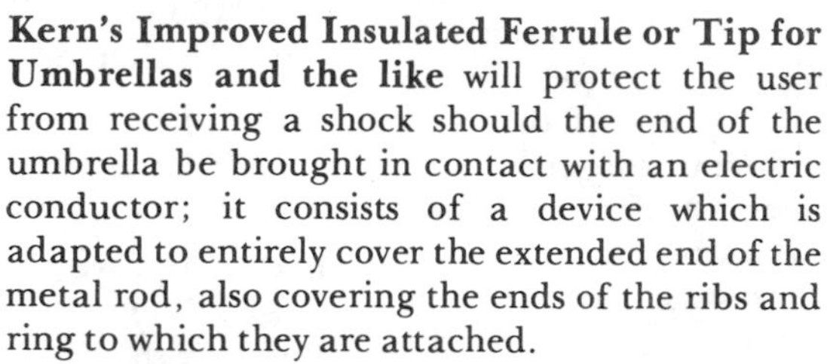

Mniszewski's Improved Cash Register provides a means whereby the employee attending the register must exactly register the amount paid even in case a customer should stand at some distance from the machine and cannot see the registering device. This purpose is attained by connecting a cash-register of any convenient construction with a talking-machine such as a phonograph or gramophone. The cashier who operates the register must then speak each amount into the machine (which remains in action only while the cash drawer is open) so that the amount is not only registered by the till but also by the talking-machine.

19693/1903
Anton Mniszewski *Manufacturer*
13 Ritter Strasse, Posen, Prussia

(Jensen & Son)

Robertson's Improved Medical Irrigator or Syringe provides a simple apparatus for injecting disinfectants or other fluids into the great gut, so as to prevent or check appendicitis or kindred diseases. The appliance is particularly adapted for individual self-treatment. The apparatus comprises a collapsible reservoir for containing the injection fluid, an injection nozzle connected to the reservoir by a bayonet joint, and a cylindrical slide for controlling the flow of fluid when pressure is applied by the person sitting on the vessel.

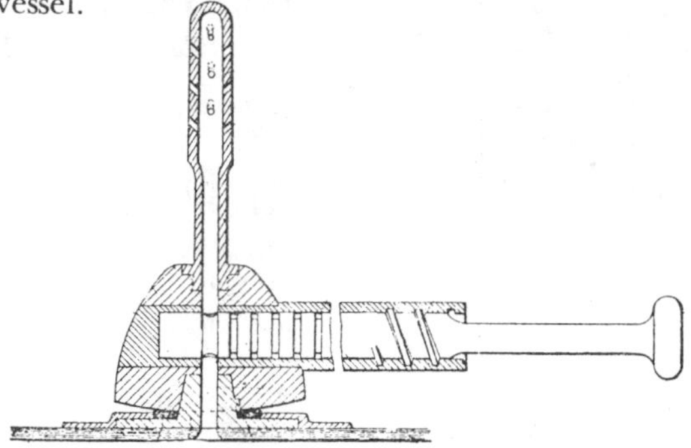

14314/1903
George Washington Robertson *Engineer*
6 Plympton Road, Brondesbury, Middlesex, England

(John P O'Donnell)

O'Higgins's Improved Hair Comb supplies an existing demand for a simply-constructed fountain comb to aid in the sale of an oily hair preparation to persons troubled with dandruff in an ordinary degree. The comb may be conveniently handled, filled without soiling the fingers, cleaned without need for taking it apart and carried about as a toilet article without danger of the contents escaping. It consists essentially in a comb, with hollow teeth and back, connected to a bottle so that liquid preparations may be fed through the teeth. The prongs bear valves at their tips to save waste and mess, and the liquid only escapes when the valves are opened by their touching the scalp.

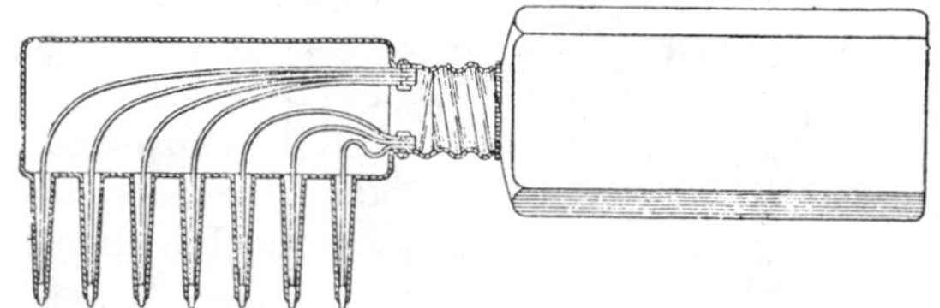

15472/1903
John Bernard O'Higgins *Journalist*
23 West 65th Street, New York, USA

(Boult, Wade & Kilburn)

1903

Callaway's Appliance for Improving the Shape of Finger Nails compresses flat and spreading nails into a more rounded form by an arrangement or press attached to the ends of the fingers. The nails are first softened in hot water, and then the appliance is placed over the finger and the two plates are screwed together sufficiently to compress the nail into the curve.

The appliance is put on at bedtime and removed in the morning.

19689/1903
Henry Clarke Callaway *Gentleman*
4 Norfolk Square, Brighton, Sussex, England

(Frederic Prince)

Schütz's Glass Telegraph Pole obviates the disadvantages of the wood and metal poles now in use, and offers several novel and valuable features.

Iron poles, for example, have the disadvantage that they vibrate considerably and have need of frequent inspection to prevent rusting. Wooden poles, still more used than iron poles, are, in spite of impregnation, apt to rot and succumb to the influence of the weather. The glass pole is hollow, with or without wire insertion, and of circular cross section, this being the easiest form to manufacture and also offering the best resistance to objects thrown against the poles.

Foot brackets may be cast into or bolted on to the pole to facilitate climbing. Instruments may be mounted inside the pole and may be read from without, without a door having to be opened.

16487/1903
Wilhelm Schütz *Architect*
23 Rotenditmolderstrasse, Kassel, Germany

(Paul E Schilling)

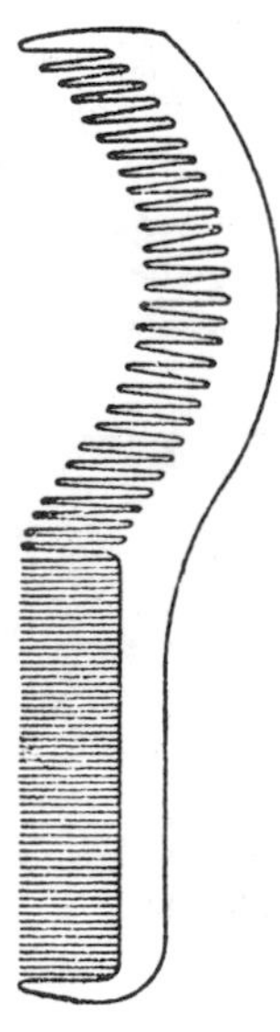

Groedel's Improved Hair Comb is shaped to fit the head; inasmuch as it lies along the scalp throughout more of its length it engages more hair than the ordinary comb so that the hair can be more evenly and quickly dressed, and with less trouble.

The coarse-toothed portion is used first to set the hair in order; the flat portion carrying the fine teeth is used for finally dressing the head.

19145/1903
Ernst Groedel *Merchant*
1A Schäfer Gasse, Frankfurt-on-the-Main, Prussia

(Cruikshank & Fairweather)

Dutrieu's Switchback Roadway is an improved construction wherein the track is made discontinuous, so that a vehicle descending along the inclined track will, on arriving at a gap, bound across the same owing to its acquired momentum. In order to lessen the shock of the impact of landing, the inclination of the track at the point where the vehicle strikes it should be tangential to the descending parabolic curve. Since the point of impact may vary from several causes, that part of the track should be mounted on springs so as to act as a buffer. For further reducing the shock the surface may be covered with loose material such as earth, dry sand, sawdust and the like.

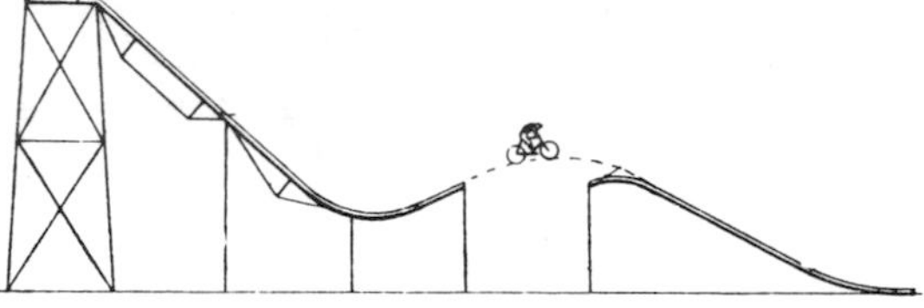

16383/1903
Hélène Dutrieu *Spinster*
5 Rue d'Alexandrie, Paris, France

(Abel & Imray)

Abée's Improved Heart Truss has two pads connected by a spring, one fitting over the lower left wall of the chest, and the other over the right side of the back. This truss offers several advantages over contrivances of the same kind already in existence. In the first place, it is possible to apply the pads to the body in such a way that the heart lies in the line of a diagonal pressure, by which it can be so acted upon that it can not only be lifted, but also shifted laterally. By dispensing with the old mode of fastening, pressure upon the epigastric and hepatic regions can be avoided. Moreover, there is no interference with the breathing, since the connecting spring is able to follow the movements of respiration.

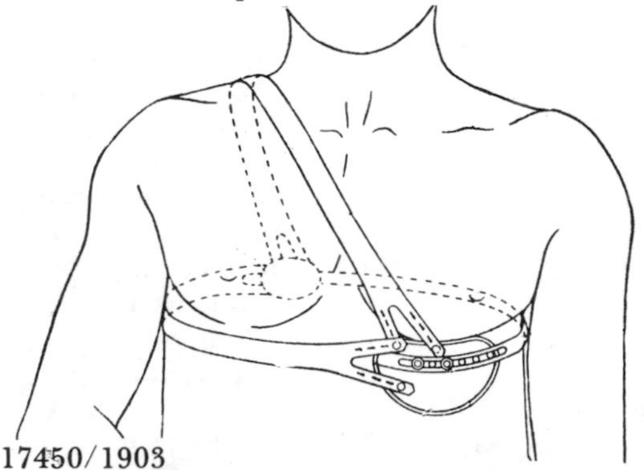

17450/1903
Ernst Abée *Doctor*
Bad Nauheim, Germany

(G F Redfern & Co)

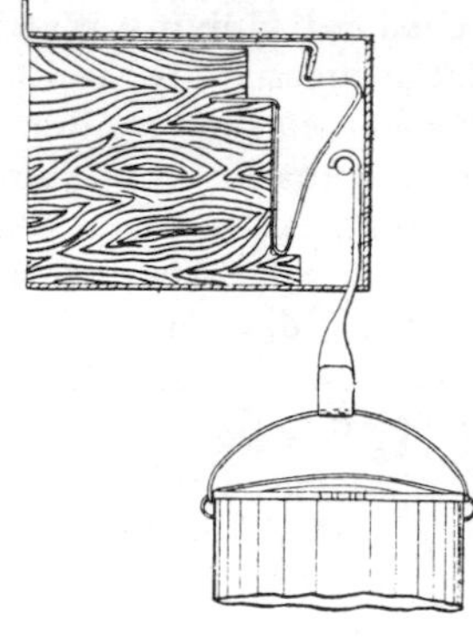

Trimmer's Fastening for Securing Milk Cans to House Doors ensures the greater security of the contents of such small containers as are commonly left at the doors of dwellings in the course of delivery. An automatic metal locking device secures the can to the door such that the can cannot be removed unless the door is open.

1899/1903
James Trimmer *Wheelwright*
14 Princess Street, Chapel Southampton, England

(C: Frederick J Cheesebrough)

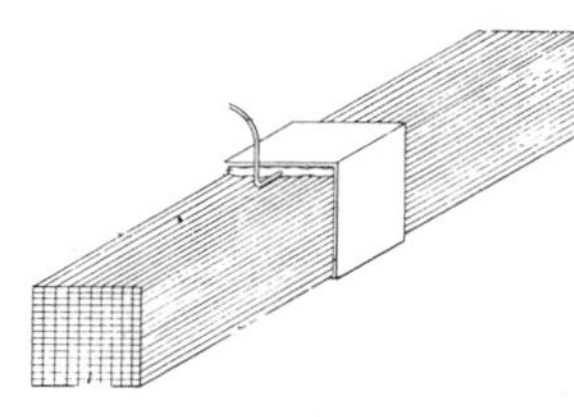

Rolland and Strom's Improved Means for Modelling and Moulding Plastic Forms is of particular advantage to corset-makers, tailors and orthopedists as it provides an exact model or fitting of the client to keep always in their possession.

Rods in a frame are pushed against the subject to be modelled, and tapped with a resilient hammer so that they slide horizontally to assume his/her form; they are then fixed in place by inflating the pneumatic pocket which surrounds them by connecting this up to a compressed air supply.

The employment of pneumatic pockets allows of immediately forming an indefinite number of elements and causes no fatigue to the model. Once the rods are fixed in the frame, a positive may be taken to act as a model of the body.

12512/1903
Charles Rolland
Gustave Adolphe Strom *Manufacturer*
70 Rue de Rivoli, Paris, France

(Herbert Haddan & Co)

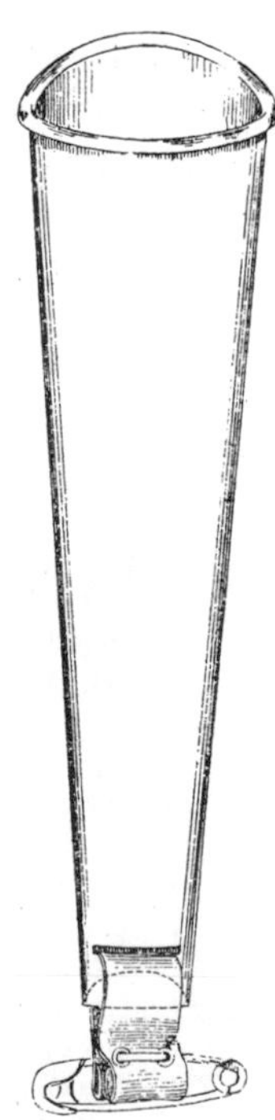

Leatham's Apparatus for Utilising the Breath to Heat the Outer Body comprises a tube shaped like a horn or otherwise placed over the mouth and beneath the chin of the wearer. The other end of the tube is inserted beneath the cloak etc. to which it is secured by a safety pin. The wearer takes in breath through the nostrils and breathes out down the tube.

The device, from 12 to 15 inches long, can be used by either sex when exposed to the cold air, and when worn by cyclists, motorists or the like will have a small whistle attached to the top of the horn.

7806/1903
Annie Elizabeth Leatham *Women's Dress Association*
75 Parade, Birmingham, England

Dickson's Improved Safety Strap for Hansom Cabs and other Vehicles is provided to prevent passengers from being thrown either against the windows, doors, splashboard or even into the roadway in the event of the horse stumbling or falling or of a collision taking place.

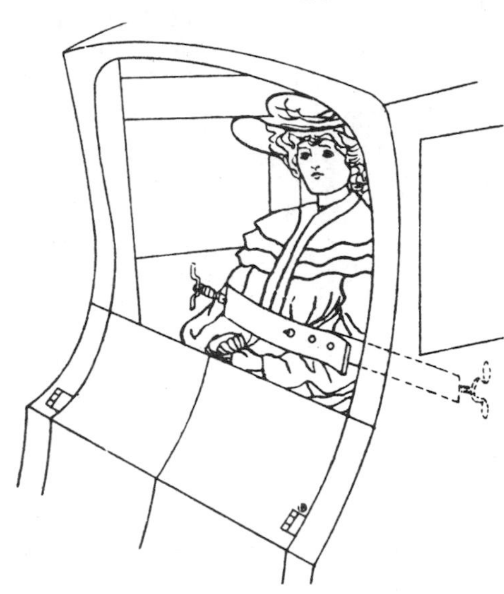

17845/1903
William Dickson *Retired Civil Servant*
15 Dents Road, Wandsworth Common, London, England

Schuller's Improved Umbrella is one in which the usual central stick is replaced by two jointed rods forming a quadrilateral frame when opened, within which there is room for the head and head gear of the user. The purpose of the invention is to simplify the manipulation of such frames and to impart increased strength thereto.

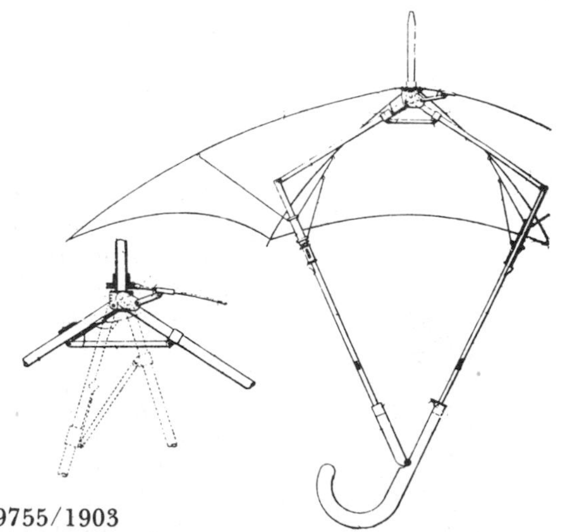

19755/1903
Leonhard Schuller *Manufacturer*
11 Sandwirtgasse, Vienna VI, Austria

(Haseltine, Lake & Co)

1903

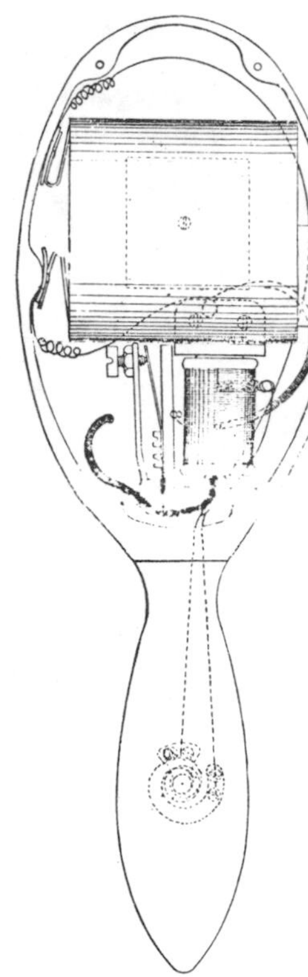

Collins's Improved Electric Hair Brush consists of a wire brush with a cell fixed in the back thereof actuated by a push button in the handle to conduct electricity straight to the scalp.

7742/1903
John Stuart Collins *Perfumer*
North British Station Hotel, Edinburgh, Scotland

(W R M Thomson & Co)

Holmes's Paper Hanging Machine expedites the application of lengths of paper to a wall or ceiling without pursuing the separate operations of first hanging the paper and then brushing or rolling the same, and also provides a simple and effective device which may be easily handled and operated. Once the walls have been covered, the paper can be similarly applied to the ceiling.

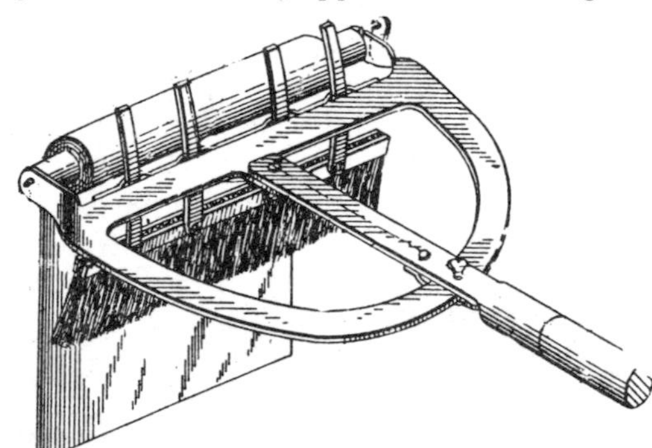

1045/1903
Cuthbert Holmes *Inventor*
Montgomery, West Virginia, USA

(P R J Willis)

Ballin-Oppenheimer's Device for Preventing the Overflow of Milk when cooking is applicable to any vessel, no matter what quantity of milk it is intended to cook in the pot. Its purpose is to arrest the heating when the liquid boils. The boiling vessel stands on a spirit lamp. The device comprises a bowl adapted to be suspended within the cooking vessel above the milk, so that when the latter rises on boiling the bowl fills and sinks and through suitable leverage releases a spring-restrained pivoted arm carrying the lamp cover which arm when released brings the cover over the lamp, descends and extinguishes the flame. If the vessel is heated by gas, the rod is arranged to shut off the gas when the milk boils.

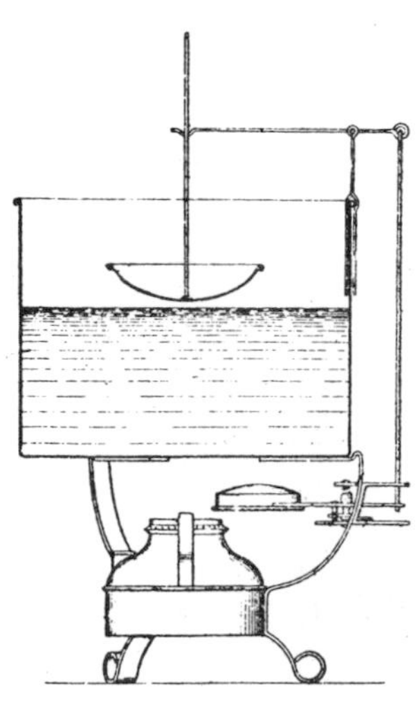

19544/1903
David Ballin-Oppenheimer *Manufacturer*
Heldenbergen, Hesse, Germany

(Marks & Clerk)

von Ehrenberg's Improved Self-driving Water Motor is arranged so that by means of the united action of two levers and of an elevator, water is made to circulate and to automatically produce motive power. The water is raised by means of a chain of buckets driven by a water-wheel on to which the water falls when it is discharged from the buckets.

A power take-off is provided so that the motor may drive a dynamo.

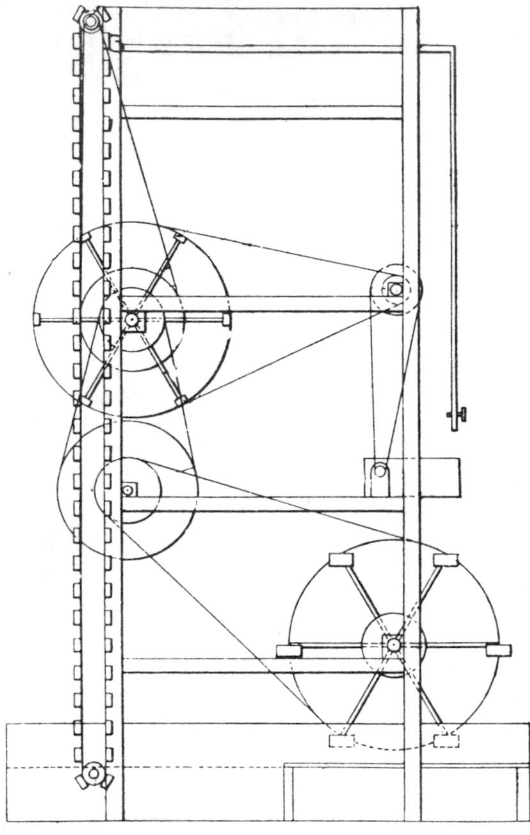

26571/1903
Friedrich von Ehrenberg
Constanz, Germany

(Edward Evans & Co)

Duke's Improved Air Ship can be floated and made to ascend or descend at will without the use of ballast. Accordingly, it has a body made of an alloy of aluminium, magnesium and silver, out of which the air may be pumped to give it buoyancy, and to which air is admitted when it is required to descend.

17651/1903
John Frederick Duke *Metallurgical Chemist*
Mill Tower, Downe, Kent

(Mewburn, Ellis & Pryor)

Scheuer's Improved Muffler is a one-piece, made-up muffler with a pocket in the filling-piece turned over to take a spring clip so configurated as to cause the marginal part of the neck portion to intimately clasp the wearer's neck and maintain the wide faces of the rear and sides of the neck portion and its filling-piece approximately vertical.

3132/1903
William Scheuer *Leather Merchant*
263 West 30th Street, Manhattan, New York, USA

(Allison Bros)

Dumville's Improved Paint Brush has a reservoir in its handle and a channel through which the fluid can flow to the bristles, via a regulating valve. The constant necessity for refilling with paint and concomitant frequent cessation of work is thus rendered unnecessary, this latter facility being at times a great desideratum.

21694/1903
Eileen Isabella Dumville
5, Red Hall Terrace, Vernon Avenue, Clontarf, Dublin, Ireland

(Albert E Ellen)

Roberts's Combined Bookmarker, Paper-knife and Toothpick is to be made in ivory, mother of pearl or other suitable materials.

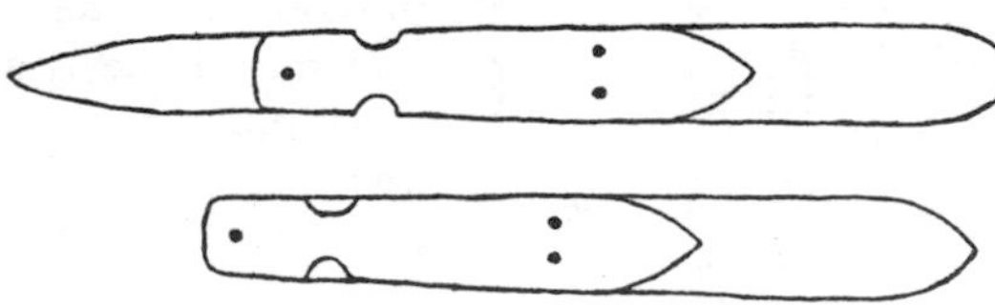

17015/1903
Edwin Roberts *Ivory Worker*
857 Abbeydale Road, Sheffield, England

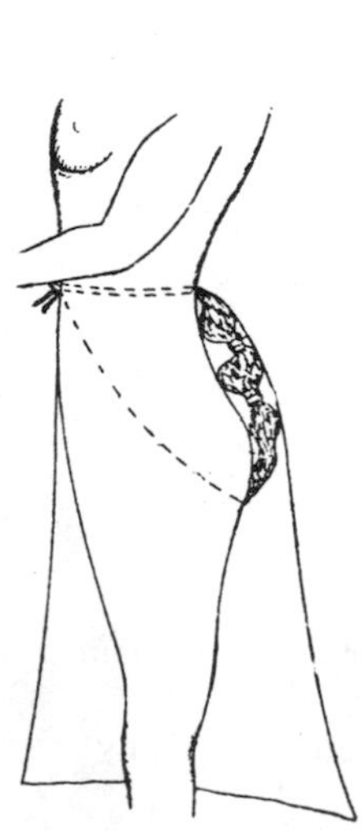

Scott's Improved Bustle and Hip Form produces a more symmetrical figure and causes the skirts to fit more gracefully. Ventilators are fitted to keep the garment free from moisture from the perspiration of the body and render it at all times cleanly and sanitary.

3127/1903
Charles Henry Scott *Ladies' Tailor*
112 South Clark Street, Chicago, Illinois, USA

(H D Fitzpatrick)

The Gebrüder Stollwerck A-G's Improved Phonograph, Especially a Toy Phonograph, with Phonograph Plates of Edible Material makes use of a material known as "the squeaking sweetmeat", which has up till now been unknown but is specially suitable for this purpose, as in a sufficiently warm state it takes on exactly the record of the phonogram by impression of the negative and retains same in true reproduction on becoming cool. Trials have shown, that the idea, to make sweetmeats emit sound, is quite practicable but the material can be rendered even more suitable for the manufacture of edible phonogram reproductions by covering same with metal foil which thus becomes the bearer of the phonogram. And since children are fond of a change, the durability of the phonogram bearer is in any case not of so much importance as the edibility after use.

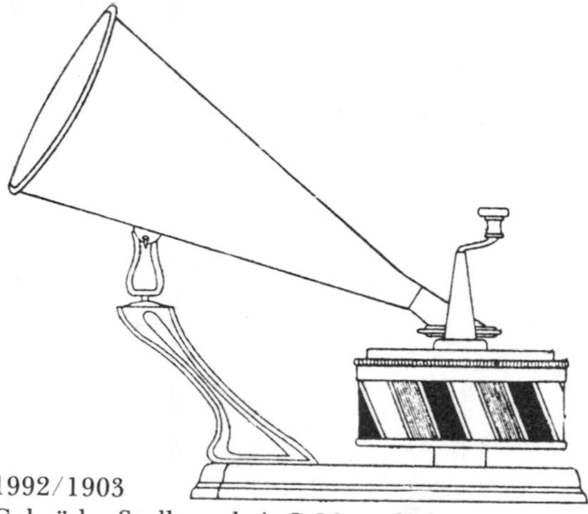

1992/1903
Gebrüder Stollwerck A-G *Manufacturer*
Cologne, Germany

(J W Mackenzie)

The Cooper-Hewitt Electric Company's Improved Luminous Gas or Vapour Electric Lamp employs starting means, based on the observation that, under certain conditions of purity, the tendency of the current is to assume a path between the positive electrode and that portion of the negative electrode which is remote therefrom.

Because it is difficult to start the current in its path, means are provided for bridging the gap between the electrodes and then breaking the contact when the current starts to flow, whereupon it continues to do so through the gas in the tube.

3444/1903
Cooper-Hewitt Electric Company
120 Broadway, New York, USA

(F W le Tall)

1903

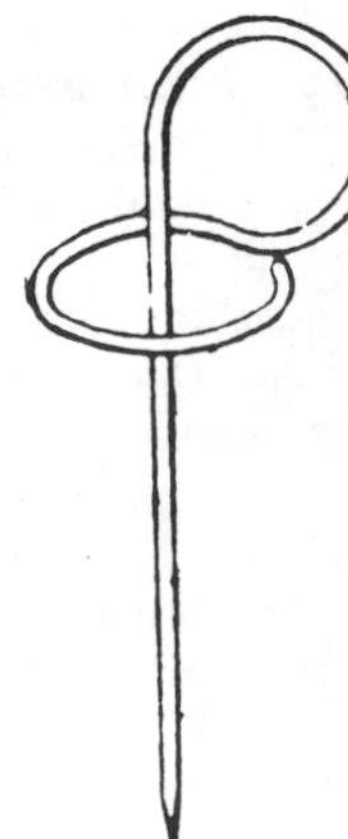

Webb's Improved Skewer for the use of Butchers and others holds the price ticket down and prevents it from being blown away

1388/1903
Stanley Webb *Butcher*
2 Castle Terrace, Tower Street, Winchester, Hampshire, England
(Hughes & Young)

Jephson's Improved Coffin for Indicating the Burial Alive of a Person in a Trance or suffering from a comatose state so that same may be released or rescued, has means for admitting air to the coffin and for giving an audible signal by means of an electric bell, which may be placed either on the grave or in the cemetery house.

There is a glass plate in the lid, and a small shelf attached to one side of the coffin which may hold a hammer, matches and candle so that, when the person wakes, he can light the candle and with the hammer break the glass plate, thus assisting to liberate himself when the earth above the coffin is removed.

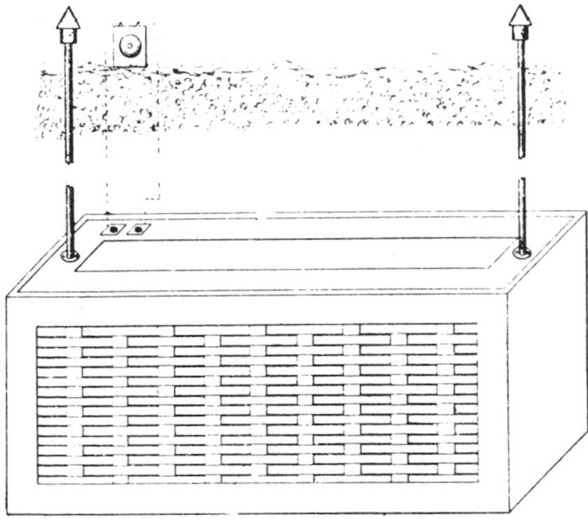

26418/1903
Emily Josephine Jephson *No Calling*
Panton Cottage, Union Road, Cambridge, England
(Cassell & Co)

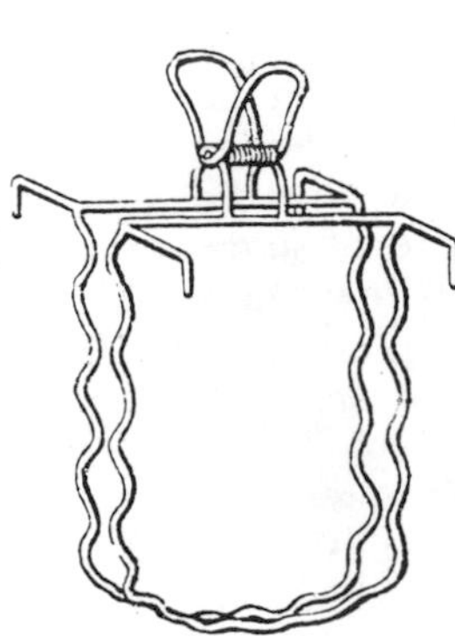

Pickin's Improved Device for Toasting Bread and the like dispenses with the inconvenience of the person having to hold the toaster while the food is being toasted by providing a spring-clamp for the slice of bread which may be hung upon a firebar.

2243/1903
Rowland Owen Pickin *Engineer*
Fledborough, Newark, Nottinghamshire, England

The von Heydens' New and Improved Means for Preventing Coition comprises a guard or shield of plaited leather or the like supported by straps and is designed to be worn by bitches and other female animals during the oestrum to prevent cross breeding. It will be found particularly useful in connection with sporting animals and supersedes both the chemical means used heretofore, which are in most cases so injurious as to remedy the animals permanently sterile, and the undesirable practice of isolating the animals during time of rut, which is also exceedingly destructive to their health.

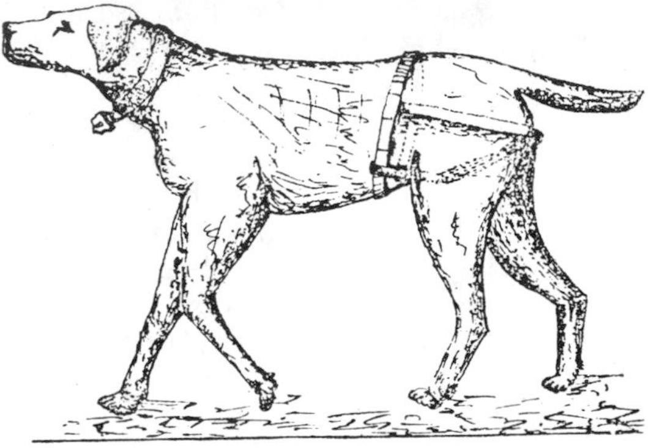

19094/1903
Baroness Margarethe Johanne Christiane Marie von Heyden (née von Levetzow) *Gentlewoman*
Baron Jacob Friedrich Arthur von Heyden *Gentleman*
Westensee, Holstein, Germany
(Herbert Haddan & Co)

Müller's Improved Bicycle is driven by utilising the weight of the rider; it has rests for the feet and the saddle is arranged so that the up and down movement of the bicyclist actuates a pawl and ratchet mechanism which communicates its motion to the driving wheel.

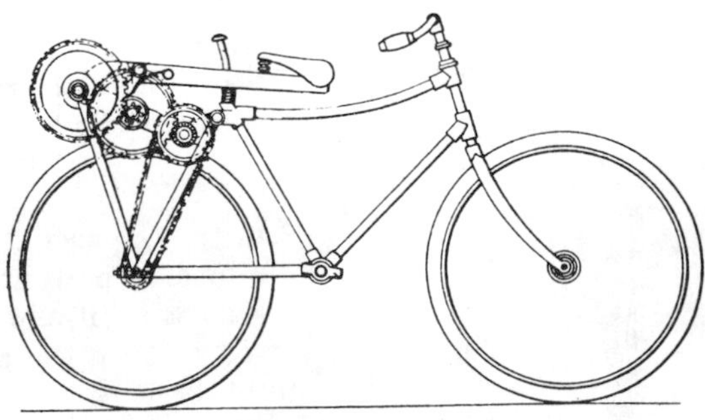

26453/1903
Adolf Ernst Anton Müller *Lieutenant in Kulmer Infantry Regiment*
Military Barracks, Grandenz, Germany
(Dewitz, Morris & Co)

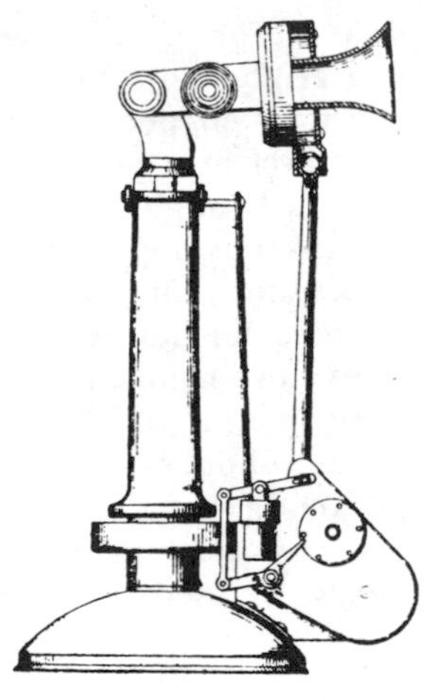

Tubbs, Smith and Hartley's Improved Apparatus for Disinfecting Telephones contemplates the provision of an apparatus for supplying sufficient heat to the transmitter and receiver of a telephone to thereby destroy infectious germs that may collect thereon during use, and to light and extinguish the heating device automatically. The apparatus consists in a gas flame which is ignited when the receiver is placed on its hook so that heated gases are circulated around the parts of the telephone thus sterilising and disinfecting the instrument. A thermostat is provided so that the flame is automatically extinguished when the telephone is sufficiently heated to destroy all germs, and in time for it to cool before the instrument is again used.

21574/1903
Nelson Jerome Tubbs *Inventor*
508 Second Street, Louisville, Kentucky, USA

Robert Harry Smith
Isaac Sidwell Hartley *Printers*
231 Sixth Street, Louisville, Kentucky, USA

(Boult, Wade & Kilburn)

Peschken's Improved Device for Preventing Enuresis Nocturna consists in a U-shaped piece whose shanks act as guides for the pressing bar. Soft padding is attached to the surfaces which act on the member, and the parts are clamped together by means of a threaded nut.

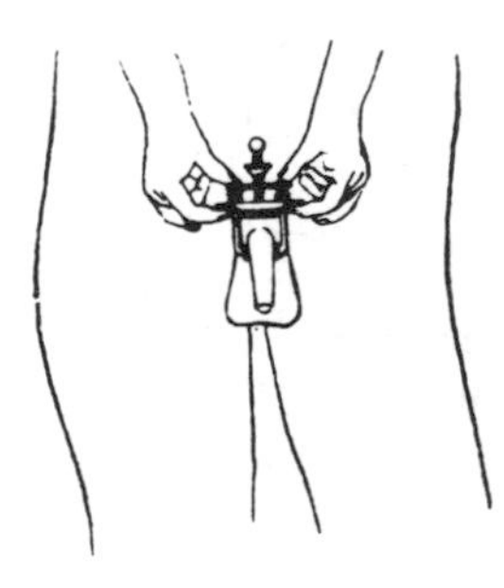

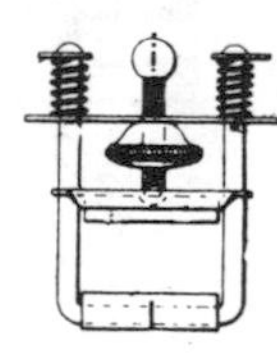

19821/1903
Heinrich Peschken *Pharmaceutical Chemist*
168 Contrescarpe, Bremen, Germany

(P: W E Heys & Son; C: Henry O Linck)

Princess Marie at Ysenburg's Improved Device for Holding up and Retaining Sleeves or the like is especially adapted to the puff-sleeves of ladies' dresses, to prevent their coming into contact with dishes on the table, or household objects when the wearer is at work in the house, so that the wearer will not be obliged to change her blouse or coat.

25773/1903
Princess Marie at Ysenburg, *Princess of Reuss Senior Line*
Castle Gettenbach, nr Gelnhausen, Prussia

Maclean's Rotary Hair Brush having Means whereby the Hair and Scalp can be Electrified consists in the normal type of rotary hair brush as used in a barber's shop, save that copper bristles are interspersed among the ordinary bristles. The copper bristles are connected to one pole of an induction coil and the other pole is earthed. The regulation of the pressure applied is entirely controlled by a switch arrangement on the barber's chair or elsewhere. In the case of the employment of a circuit from an incandescent lamp such switch is arranged to insert suitable resistances as may be required.

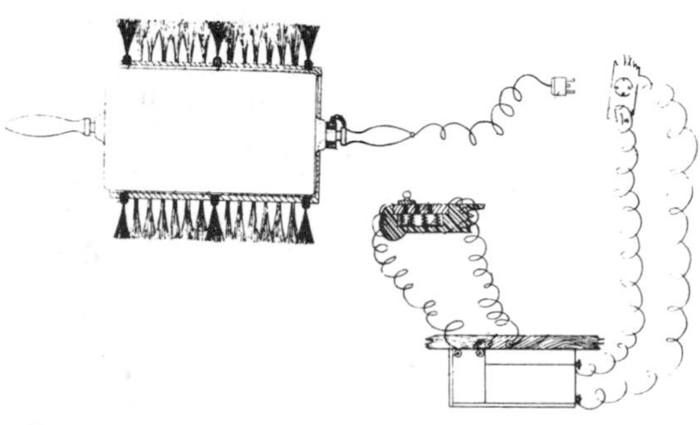

20100/1903
John Maclean *Consultant Engineer*
19 University Avenue, Glasgow, Scotland

(H D Fitzpatrick)

Baden-Powell's Improved Aërial Machine may be used either for studying aërial navigation or for recreational purposes. A car or boat is fitted with aëroplanes and wheels, so that it can run down an inclined track to give it impetus for the glide. A man may sit or lie in the boat and by shifting his weight backward or forward or from side to side or by means of rudders, the movements of the machine while in mid air may be to some extent steered and controlled.

A sheet of water may be used to break the fall in case of accident. Larger patterns of the machine may carry several people and may be fitted with a motor and propellers to prolong the glide.

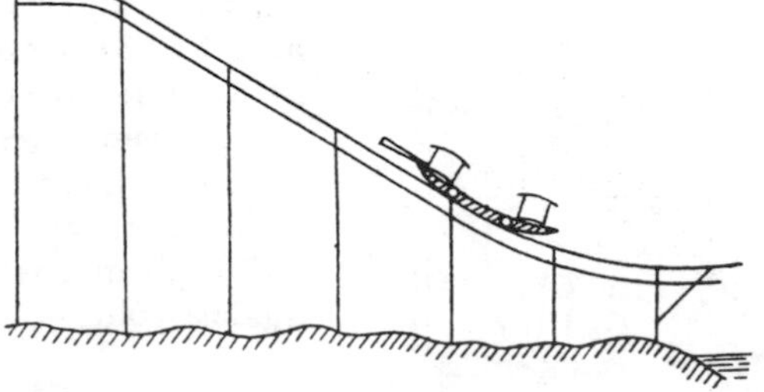

26821/1903
Baden Fletcher Smyth Baden-Powell *Major, Scots Guards*
Aldershot, Hampshire, England

1903

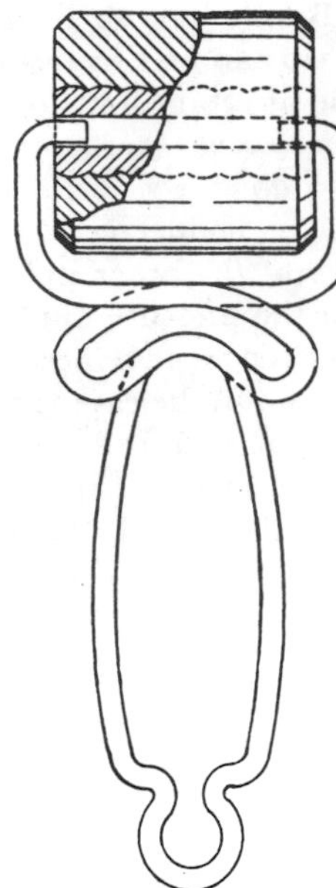

Loring's Improved Soap Device for Shaving applies the soap most economically to the face whilst being particularly clean and easy of manipulation. The soap is in the form of a roller, mounted on a handle, so that it may be rolled over the face to produce lather and a massaging effect at the same time, thus softening the beard and facilitating subsequent shaving.

26738/1903
Frederick Henry Loring *Electrical Engineer*
7 Doughty Street, London WC, England

Gale's Improved Pop-gun may be used in the parlour for projecting darts, peas etc at a target. The missile is placed in the barrel and the cork plug inserted; operating the gun in the usual way causes the missile to be ejected while the cork is retained on its cord. Alternatively, the cord may be detached and the plug used as a missile.

The invention originated in an effort to extend the applicability and popularity of one of the most delightful and yet harmless of toys — the pope [sic] gun — which normally only appeals to very young children because of the faintness of its explosion and because its cork serves no purpose except to cause the explosion. By allowing the discharge of darts, peas, and shot for target practice this improved pop-gun enables the application of personal skill and makes the toy pleasurable to those of more advanced age.

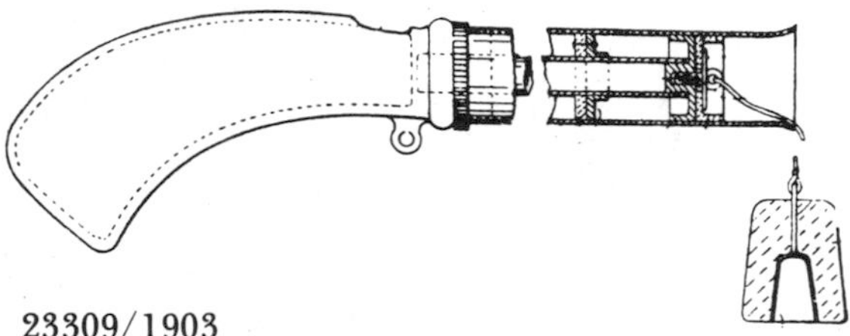

23309/1903
Frederick Gale *Engineer*
Lancefield, Victoria, Australia

(Herbert Haddan & Co)

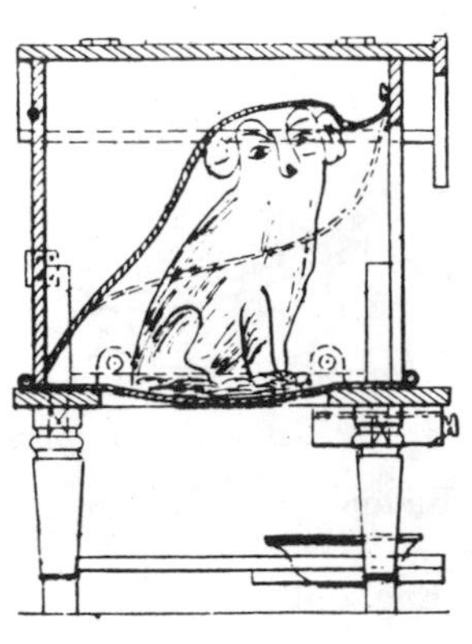

Mann's Improved Kennel or Bench for Dogs and the like has an elevated resilient floor, and a detachable cover, so that the floor and cover adapt themselves to the form of the animal without interfering with its freedom of movement.

By this means the dog is effectively protected against cold.

21577/1903
Karl Friedrich Ernst Mann *Forester*
Neukirchen, Saxony

(Herbert Haddan & Co)

Crist's Improved Apparatus for Conveying Persons or Articles from a Higher to a Lower Level is a conductor down which persons or articles may be caused to travel in a perfectly safe and easy manner and will alight at the bottom free of undue or dangerous impact and with the least amount of shock or sudden stoppage and which when used for amusement may be arranged to give most comical and laughable effects to the manner in which the persons descend. It comprises a shoot or conduit made of canes or laths of wood, which may be used for amusement purposes by attaching it to a suitable structure. It may also be used as a fire escape.

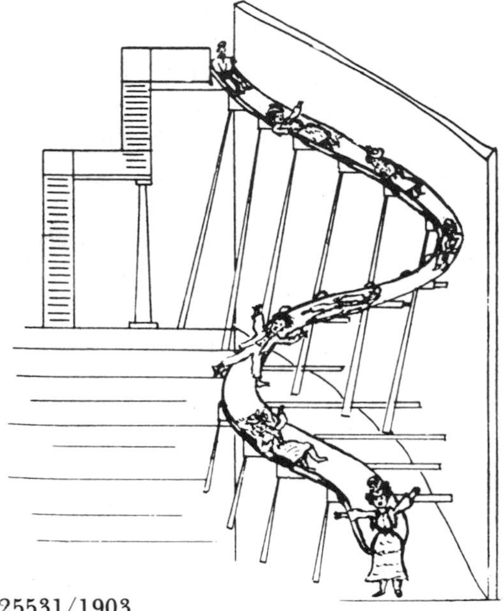

25531/1903
William Eugene Crist *Engineer*
Crystal Palace, Sydenham, Kent, England

(H F Boughton)

Elliston's Improved Bird Cage Perch provides a means whereby the insects which attack a bird are caught. These leave the bird at daybreak, as is well known, and travel along the perch to the bars. Using this invention they can only do so through grooves in the perch, which lead to the cavity, in which they are trapped and unable to get out.

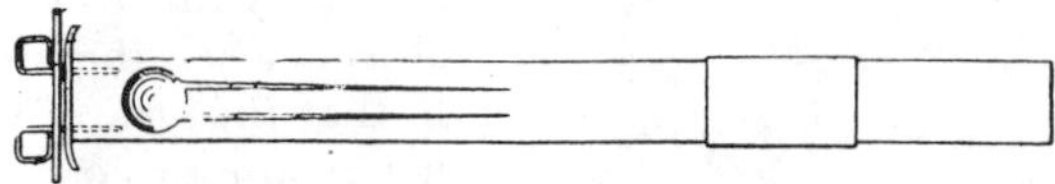

4740/1903
Edward Henry William Elliston *Engineer*
14 Friday Road, Erith, Kent, England

(E Eaton)

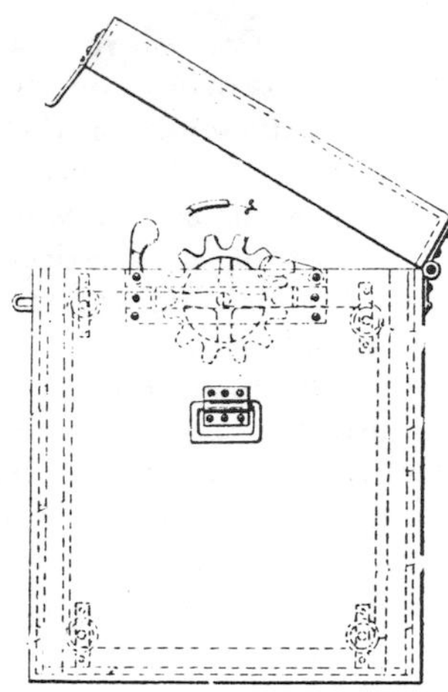

Lankhout and Hertstein's Improved Travelling Trunk is designed to obviate the bending, causing backache, hitherto unavoidable when inserting articles into trunks, and withdrawing them therefrom. This alleviation is obtained by fitting the trunk with an inner box designed to be raised and lowered by means of a windlass.

1/1904
Dr Eduard Lankhout *Advocate*
202 Singel, Amsterdam, Holland

Justus Christoffel Hertstein *Doctor of Medicine*
63 St Annelaan, Nijwegen, Holland

(J Owden O'Brien, *successor to and late of* W P Thompson & Co)

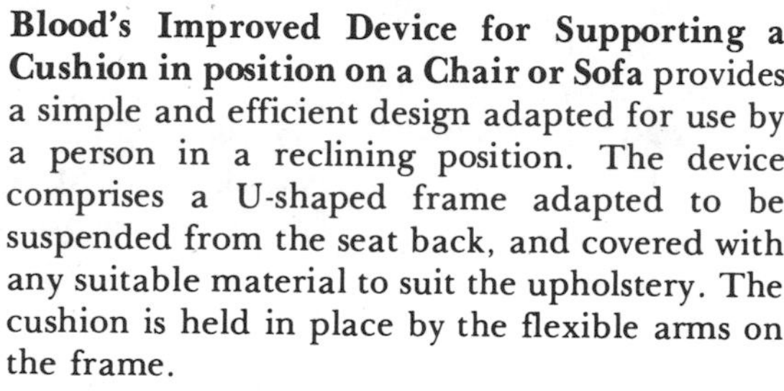

Blood's Improved Device for Supporting a Cushion in position on a Chair or Sofa provides a simple and efficient design adapted for use by a person in a reclining position. The device comprises a U-shaped frame adapted to be suspended from the seat back, and covered with any suitable material to suit the upholstery. The cushion is held in place by the flexible arms on the frame.

741/1904
Fanny Blood
28 Disraeli Road, Ealing, London, England

Gibbons's Galvanic Device for Foot Wear and Bracelets is intended to give relief to persons suffering from rheumatism and ailments of a similar character and from nervous affections. It uses plates of zinc and copper superposed, with the zinc next to the skin on one foot, wrist or arm and the copper next to the skin on the other corresponding member. Acidulated material may be placed between the plates if increased galvanic action is required but for the ordinary strength wanted it can be dispensed with as the salt in the wearer's perspiration will provide all the excitant needed.

288/1904
Robert Pearce Gibbons *Timber Merchant*
Kopu Thames, Auckland, New Zealand

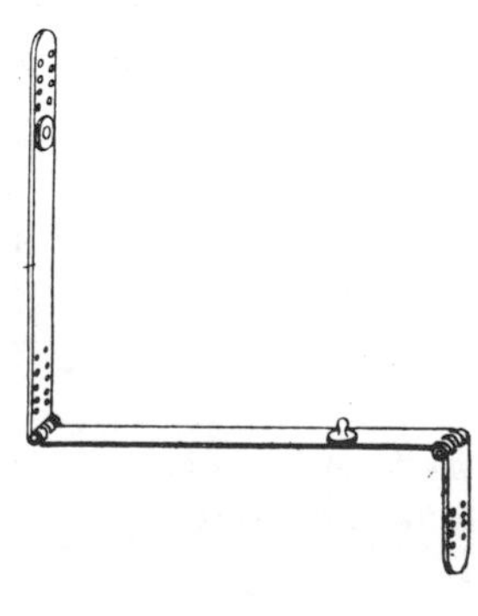

Brune's Device for lifting Ladies' Skirts comprises three hinged bars sewn to the skirt and is specifically designed so as not to cause creases in the material.

1174/1904
Paul Brune *General Agent*
19 Friesenplatz, Cöln, Germany

(Ferdinand Nusch)

Guillot's Improved Corset renders the body more slim without impairing its suppleness so that the wearer is able to bend in any direction without hurting the body or damaging the corset.

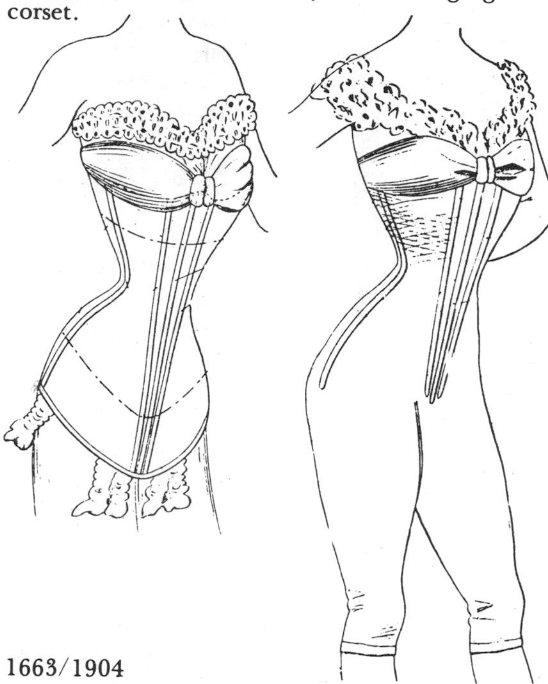

1663/1904
Charles Guillot *Manufacturer*
10 Rue de la Paix, Paris, France

(Boult, Wade & Kilburn)

Winans's Improved Alarm Call for Sportsmen will be particularly useful when shooting in a wild region where the parties may become separated by considerable distances in the course of a day's sport. The invention comprises a horn or reed mouthpiece adapted to fit the barrel of a gun, so that a sportsman may give signals by inserting the mouthpiece into his gun, and blowing to produce a sound like a coach-horn.

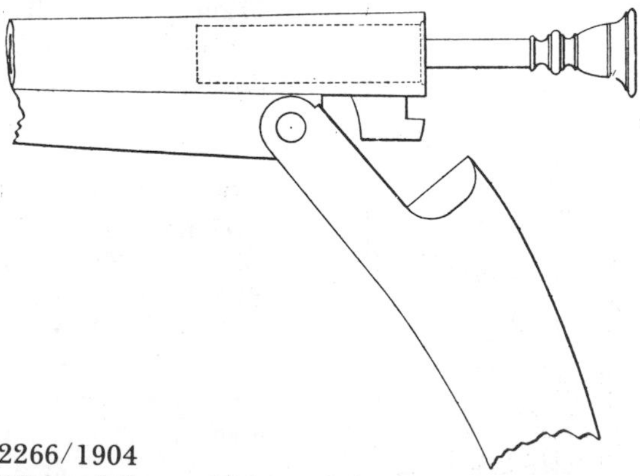

2266/1904
Walter Winans *Citizen of the US and Gentleman*
Surrenden Park, Pluckley, Ashford, Kent, England

(Newton & Son)

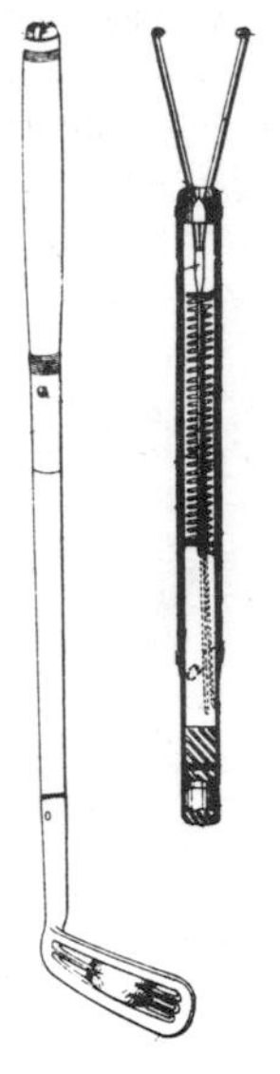

Brown's Improved Golf Club has a dual-purpose device fitted to the handle to enable the player to remove the ball from the hole once it has been putted without his having to stoop to pick it up.

1045/1904
James Ross Brown *Shipsmith*
10 River Street, Montrose, Forfarshire, Scotland

(W R M Thomson)

Llewellin's Fireproof-paper Curtains are so inexpensive that, when dirty, they can be thrown away. The curtains are made of any suitable kind of paper, embossed, stamped, perforated or decorated as desired and treated with any suitable material to render them non-inflammable. Such a process may consist in soaking them in a solution in water of sulphide of ammonia, boric acid and borax.

1568/1904
Louisa Llewellin *Lady*
10 Albemarle Mansions, Holloway Road, London N, England

(Hughes & Young)

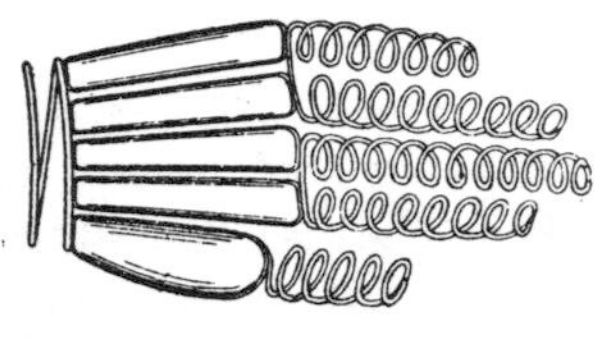

Chell's Hand Guard for use in Cutting Bread and like purposes protects that hand which holds the loaf etc as opposed to that which holds the knife. It consists in a metal wrist-shield, to which are attached wire helices to protect the fingers and thumb. The guard may be reduced so as to protect only the thumb and finger liable to be cut, and may have roughened surfaces to lessen the risk of the hand slipping while holding the loaf.

2703/1904
Francis John Chell *Ironmonger*
194 Brighton Street, Seacombe, Cheshire, England

(W P Thompson & Co)

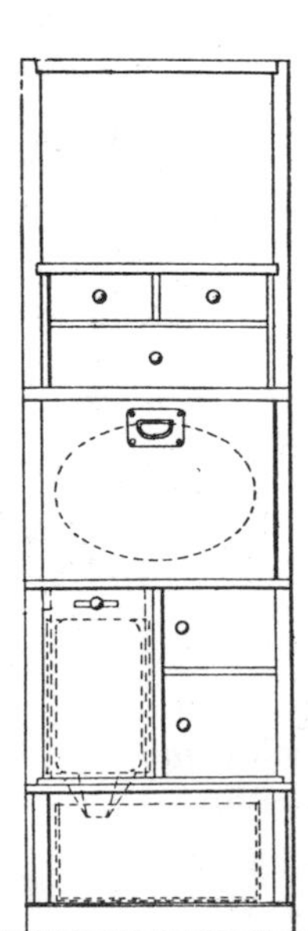

Heymann's Improved Folding Washstand for use on Shipboard, in Railway Sleeping Carriages, Consulting Rooms and the like has a second basin for use as a bidet, or for slops, or for the use of sea-sick passengers and is designed to economise on space by making only one hygienic structure necessary.

1201/1904
Alfred Theodor Heymann *Manufacturer*
42 Neuerwall, Hamburg, Germany

(A M & WM Clark)

MacCallum's Apparatus for Travelling on Water effects flotation by means of an endless band passing over drums at the ends of the vessel. Intermediate wheels and stretchers keep the band in shape, and projections at intervals prevent backslip. This mode of progression practically eliminates skin frictional retardation, and largely reduces wave-making resistance.

Two band vessels of this kind may be placed side by side, with any desired space between them, and bear a framework for the accommodation of passengers and machinery, and the arrangement may also be used to construct a pleasure craft of the kind known as a water velocipede.

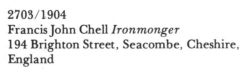

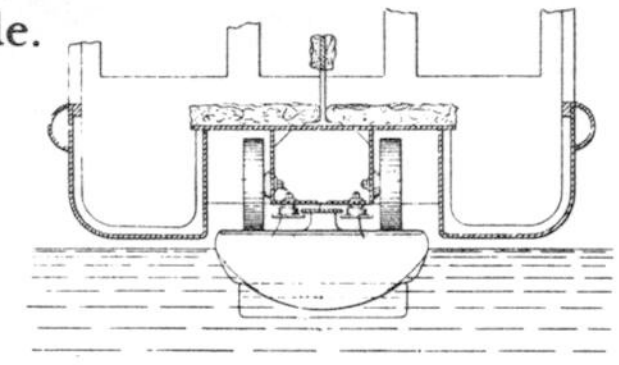

165/1904
Patrick Fraser MacCallum *Engineer*
Fairbank, Helensburgh, Dumbartonshire, Scotland

Vickery's Improved Jewel Case, or Dressing and other Bag Box and the like is fitted with an alarm bell which rings if an unauthorised person lifts the handle.

This invention is effected by pivoting one end of the handle to the case or bag and constructing the other end so that an electric circuit is closed when the bag is lifted, causing a bell to ring. The battery and bell mechanism are arranged within a false bottom or false top so as to cause no inconvenience. A suitably disguised switch, which may be controlled by a lock and key for additional security, is provided so that an authorised person may lift the bag without connecting the circuit and ringing the bell.

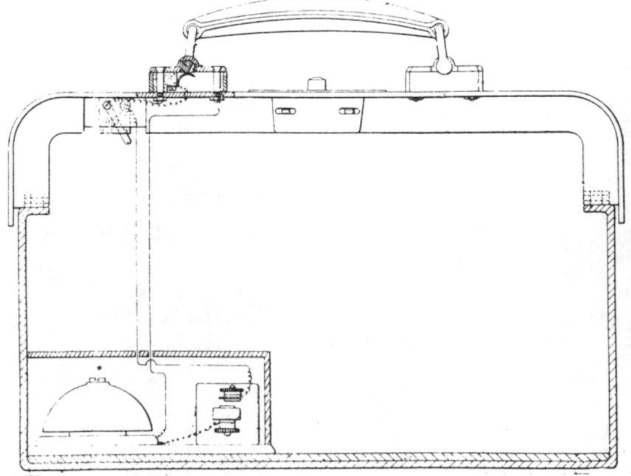

2465/1904
John Collard Vickery *Jeweller*
179 Regent Street, London, England

(Harris & Mills)

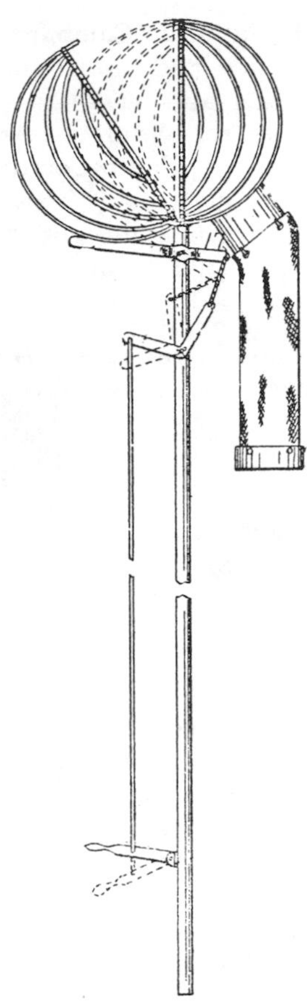

Rohr's Improved Fruit Picker makes it possible for persons standing on the ground to pick apples, peaches etc without bruising or injuring the same and without the use of ladders.

1681/1904
Frederick Rohr *Miner*
Irwin, Pennsylvania, USA

(Allison Bros)

Llewellin's Improved Gloves for Self-defence and other Purposes are provided with sharp steel talons, and are designed especially for the use of ladies who travel alone and are therefore liable to be assailed by thieves and others.

In use, the gloves could be worn during the whole journey or put on when required and by drawing them over a person's face it would be so severely scratched as to effectually prevent the majority of people from continuing their molestations.

The invention can also be used by mountain climbers to enable them to catch hold of whatever they pass over during a fall.

1567/1904
Louisa Llewellin *Lady*
10 Albemarle Mansions, Holloway Road, London N, England

(Hughes & Young)

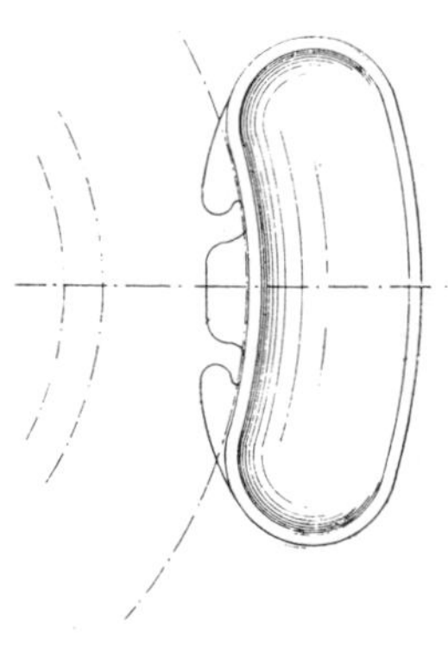

Huysmans's Improved Receptacle for Waste is designed to be detachably fixed to dinner plates and the like for receiving bones and scraps.

1406/1904
Lievin Huysmans *Gentleman*
14 Rue de Danemark, Brussels, Belgium

(J S Lorrain)

Drake and Gorham's Improved Ray Bath is an apparatus that can be made at minimum cost and easily erected without necessitating skilled labour. Portable vapour and cabinet baths are converted to take Nernst or ordinary lamps mounted on triangular reflectors suspended in the corners of the cabinets, with Nernst lamps being preferred since these have the property of emitting a quantity of actinic and other rays, and so have greater beneficial effects than ordinary incandescent lamps.

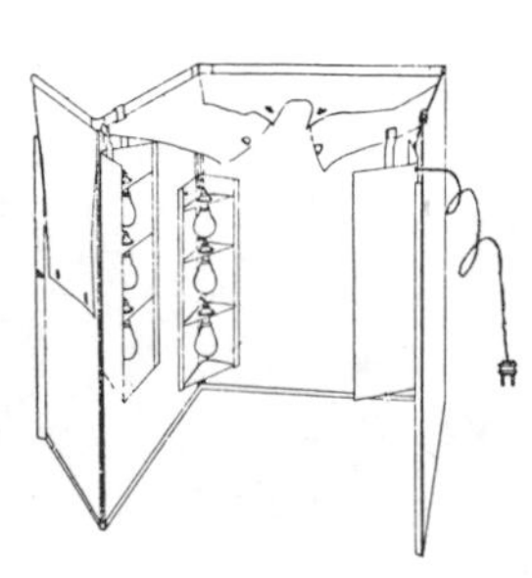

1069/1904
Bernard Mervyn Drake
John Marshall Gorham *Electrical Engineers*
66 Victoria Street, Westminster, London SW, England

Compton's Improved Wave Power and Tidal Motor is an apparatus which imparts a continuously rotating motion, derived from the action of sea waves or the tide, to a shaft. A boat-shaped float drives a system of ratcheted wheels in order to convert its reciprocating movements into unidirectional rotation. In this way electricity may be generated and conveyed to the shore by a submarine cable but other arrangements may be utilised as found convenient.

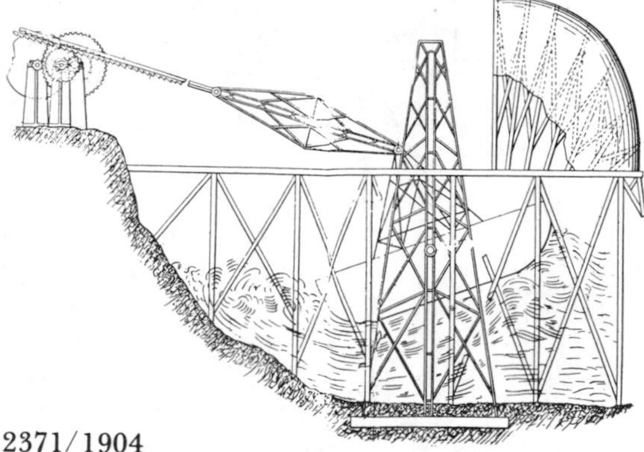

2371/1904
Melvin David Compton *Engineer*
16 Park Place, New York, USA

(Boult, Wade & Kilburn)

Rushworth's Improved Tyre for Cycles and Vehicles consists in a spring-steel outer rim which runs on the road, connected to the inner rim of the wheel by means of a series of circular, curved or bent metal springs; this provides a tyre that will not be liable to puncture and will not require such frequent repairs as the pneumatic tyres now in use and yet will possess a similar resiliency.

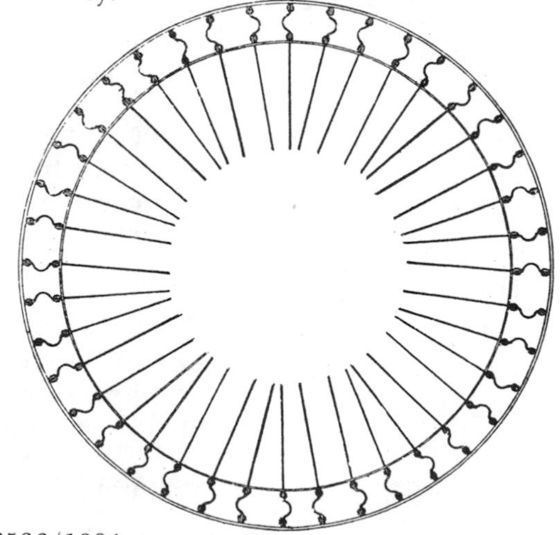

3536/1904
Joseph Rushworth *Civil Engineer*
35 Beechfield Road, Harringay Park, Tottenham, Middlesex, England

1904

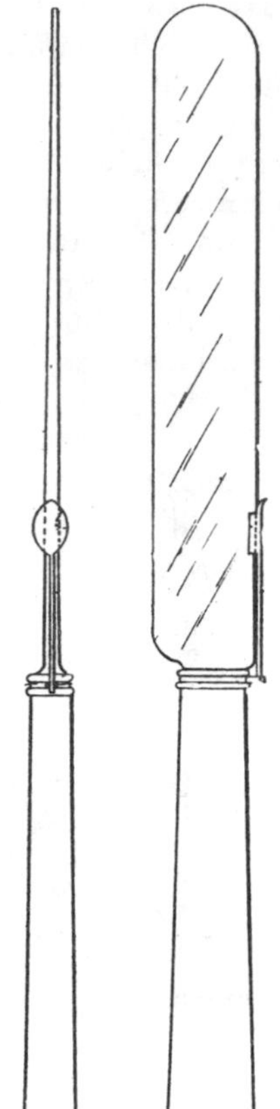

Klipp's Improved Table Knife obviates the defects of ordinary knives which, by constant cleaning and sharpening, become sharp at the back and are liable to mark the user's fingers if pressed thereon. In order to prevent this, a finger piece is arranged to clip to the bolster. When the knife has to be cleaned etc, the finger piece may be turned back on its pivot.

2897/1904
William George Klipp *Picture Restorer and Mount Cutter*
64 Glendowar Road, Mutley, Plymouth, Devonshire, England

(Hughes & Young)

Edmundson's Device for Peeling Oranges can also be used as a paper-knife. When employed for its primary object, *viz:* peeling fruit, the hook portion is inserted into the rind and worked round, simultaneously removing the rind and keeping the hands free from juice.

3850/1904
Joseph Edmundson *Manufacturer*
102 St Hubert's Road, Great Harwood, nr Blackburn, Lancashire, England

(George Davis & Son)

Gerlach's Spring Action Net for Bathing Children facilitates the bathing of infants so that one person can see to it alone without requiring help. The child is placed in a net supported over the bath on elastic straps, and immersed by pressing the net down.

2059/1904
Irene Gerlach *Independent Lady*
142 Friedbergerlandstrasse, Frankfurt-on-the-Main, Germany

(Ferdinand Nusch)

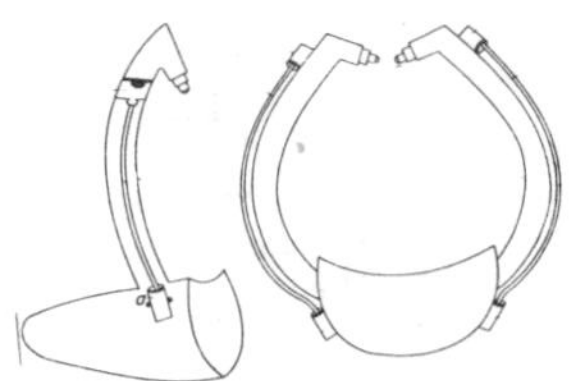

Kahn's Improved Ear Trumpet is binaural, and is to be worn on the head, or on a hat or bonnet. The earpieces are spring-loaded by means of steel wires or whalebones so as to remain firmly fixed in their places in the ears.

2659/1904
Charles Kahn *Acoustic Instrument Maker*
t/a F C Rein & Son, 108 Strand, London WC, England

The American Hard Rubber Company's Improved Comb is provided with an advantageous disposition of teeth with undulatory or sinuous edges, the undulations on one edge of a tooth alternating with those on the other edge and on those of the next tooth. Designed for combing or straightening out the hair, this improved comb is particularly desirable to females.

2437/1904
American Hard Rubber Company
9 Mercer Street, New York, USA

(Page & Rowlinson)

Bungard's Improved Undershirt relates to that class of knitted cotton or wool garment which is made to fit rather tightly, and, as a consequence, after having been washed a few times and shrunk a little and become still narrower and also shorter, they have the unpleasant drawback that they creep up on the body, especially during a walk, in hot weather, and on corpulent persons, which causes a very unpleasant feeling and is very troublesome to corpulent persons when attempting to pull it down again especially on the back and may even lead to a cold in the abdomen.

This undershirt is designed to overcome these problems by providing a tail which may be drawn between the legs and fastened to the front. It is of value to all those whose occupation compels them to run about in the open air, to those suffering from inflammations of the bladder or blind-gut and to travellers who are compelled to always sleep in hotels in beds with fresh sheets and sometimes even half damp sheets and easily catch cold as a result.

It can also act as a body-supporter.

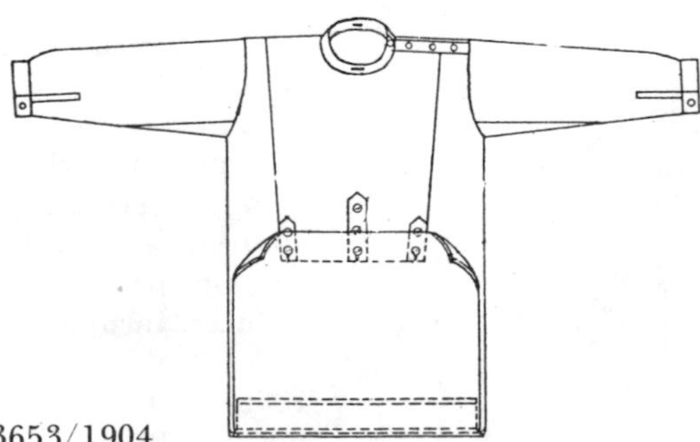

3653/1904
Johann Mathias Bungard *Merchant*
59A Perlengraben, Cologne, Germany

(B Brockhues)

1904

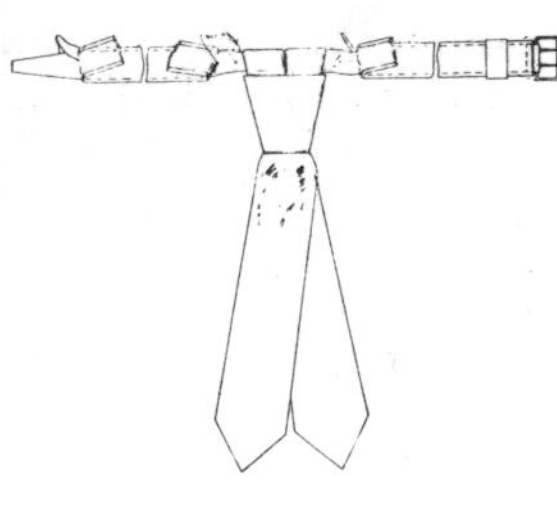

Schröder's Improved Scarf or Necktie obviates the fact that, as is well known, a scarf is rendered unsightly and worthless the moment its band is worn out by the collar of an overcoat or otherwise, even if the knot or bow has remained intact.

In contradistinction to ordinary neckbands, this improved scarf is made up of several superposed layers of fabric so that when the outermost becomes soiled and worn it can be torn off and a fresh surface exposed to wear.

Preferably each layer is sewn separately, and removed by unpicking the seam.

2931/1904
Ignaz Schröder *Merchant*
49 Danzigerstrasse, Neustadt, Germany

(J G Lorrain)

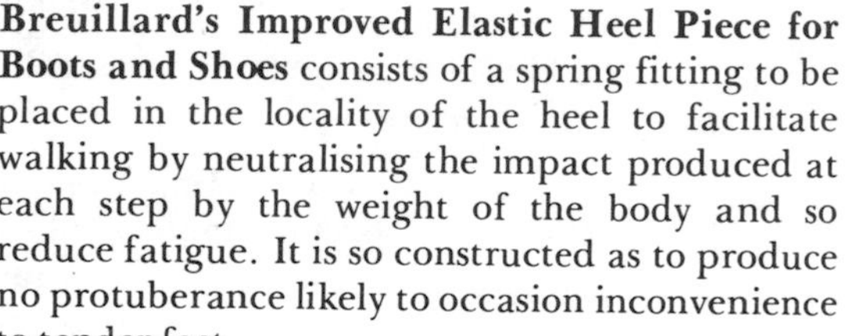

Breuillard's Improved Elastic Heel Piece for Boots and Shoes consists of a spring fitting to be placed in the locality of the heel to facilitate walking by neutralising the impact produced at each step by the weight of the body and so reduce fatigue. It is so constructed as to produce no protuberance likely to occasion inconvenience to tender feet.

3600/1904
Jean François Breuillard *Doctor of Medicine*
90 Rue de Rennes, Paris, France

(F Wise Howorth)

Gabriel's Appliance to Assist Animals with their Breathing — useful also to humans — consists in a clip which distends the nostrils and enlarges the air passages, so permitting freer breathing and the admission of a larger volume of air into the lungs. Among other uses the appliance may be worn during exercise, during sleep or during an operation when an anaesthetic has been or is being administered.

The device may take several forms: a cross-shaped spring clip constructed so that as the pressure on the outer limbs is released the spring causes the two inner limbs to grip the inner cartilage while the outer limbs press against the inner surface of the nostril; or it may take the form of a pince-nez; or it may take the form of a spring bow, the tendency of which is to maintain the ends at a distance apart greater than that to which the nostrils can be conveniently distended.

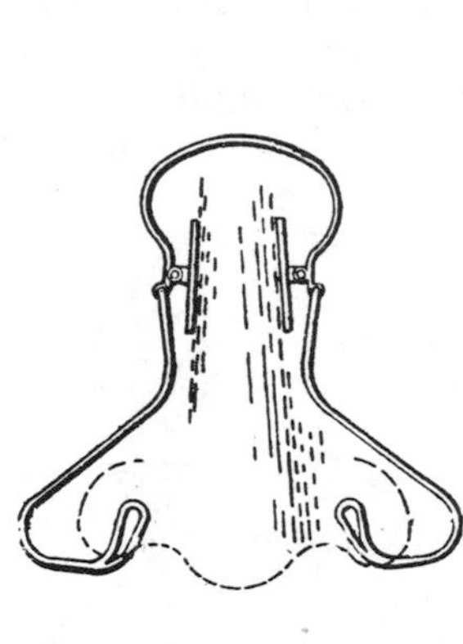

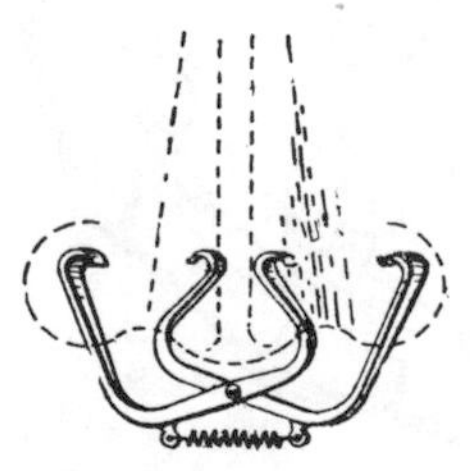

3516/1904
Joseph Gabriel *Merchant*
61 George Street, Manchester, England

(W E Heys & Son)

Sax's Improved Picture provides means for giving the impression of fiery rays issuing from pictures and the like employed for ecclesiastical or mundane purposes.

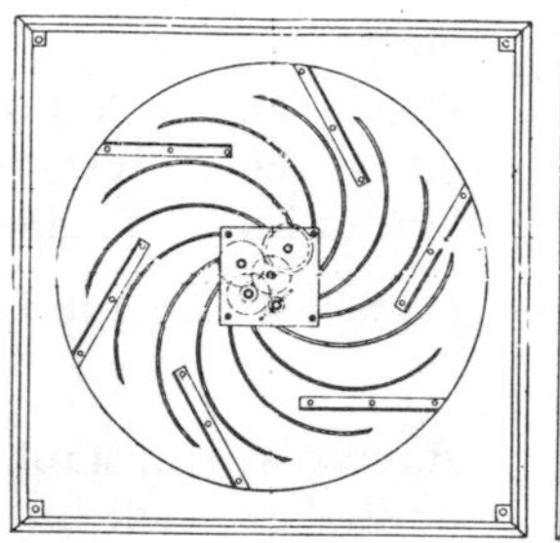

2089/1904
Henry Sax *Merchant*
Tegelen, nr Venloo, Holland

(Wheatley & Mackenzie)

Tomlinson's Improved Galvanic Curative Appliance for Human Use consists in a disc of copper stitched into the left sock nd a disc of zinc stitched into the right sock, or *vice versa*.

The invention is more especially directed toward the cure or alleviation of rheumatism, although it is calculated to benefit the wearer generally.

3276/1904
Frank Tomlinson *Traveller*
13 Newhampton Road West, Wolverhampton, Staffordshire, England

(J A Coubrough)

Ader's Improved Vessel Adapted to Slide on the Surface of Water Instead of Being More or Less Immersed Therein has two lateral wings and a transverse tailpiece, through whose undersides compressed air is supplied to form a cushion between the surfaces and the water.

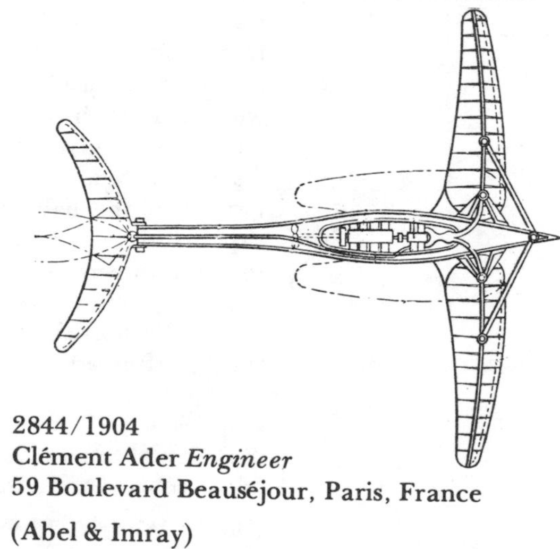

2844/1904
Clément Ader *Engineer*
59 Boulevard Beauséjour, Paris, France

(Abel & Imray)

1904

Perks's Automatic Cart Loader uses the forward motion of the cart, transmitted via the rear wheel and a lever, to raise a load and discharge it into the cart.

3057/1904
George Henry Perks *Coal Merchant*
40 Hawkins Street, Hill Top, West Bromwich, Staffordshire, England

(P: Self; C: George T Millard)

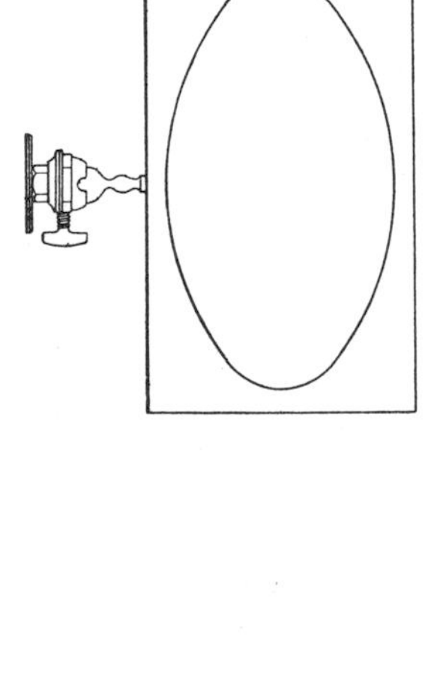

Barratt's Adjustable Reflector for Vehicles is devised especially in view of the recent legislation requiring every driver of a van or similar vehicle in London to be able to see the traffic on each side of him. By this invention, a mirror in combination with a ball and socket joint permitting of adjustment in all directions, it becomes possible for the driver not only to observe cross traffic but also to see what is approaching from behind without changing his seat or turning his head.

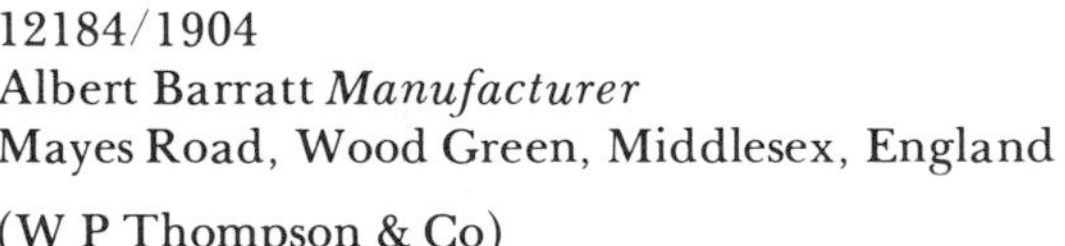

12184/1904
Albert Barratt *Manufacturer*
Mayes Road, Wood Green, Middlesex, England

(W P Thompson & Co)

Lichtmann's Improved Cap for Promoting the Growth of Hair prevents baldness where some hair is still left, even though it may be very sparse. The loss of hair is stopped after the cap has been used for a short time, owing to the gradual re-establishment of the normal circulation of the blood, promoting a strong growth of hair in the scalp.

The cap must at first be worn every night but afterwards only from time to time. The cap, in shape like an ordinary sleeping or skull cap, has an inner absorbing layer and an outer waterproof layer. The cap is slit at the back and front and provided with strings for securing to the head. In use, the inner layer is moistened with a restorer containing, for example: 25 alcohol; 10 glycerine; 10 cinchona tinct; 2 resorcin; 53 water and scent as desired. The growth of the hair is assisted by the alcoholic vapours developed (owing to the heat of the head) between the scalp and the cap. Said vapours promote the circulation of the blood so that badly nourished hair roots are fed and, so to speak, refreshed.

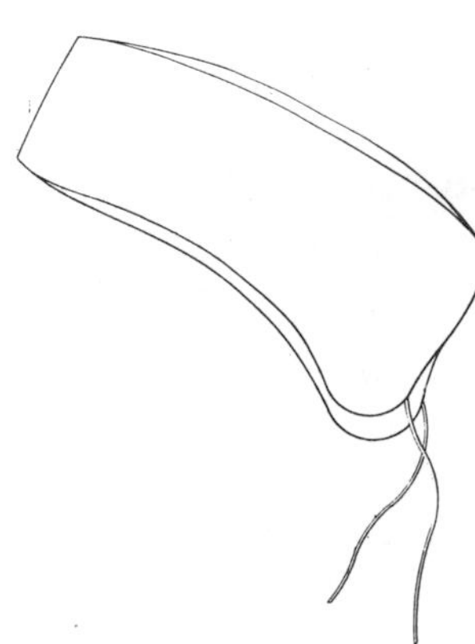

376/1904
Victor Lichtmann *Doctor*
54 Erzsébet Kórut, Budapest VII, Hungary

(Boult, Wade & Kilburn)

Phillips's Improved Foot-warmer for Motor Vehicles uses the hot exhaust gases from the engine to warm the feet in such a way that the warmed chamber does not interfere with access to the parts of the vehicle.

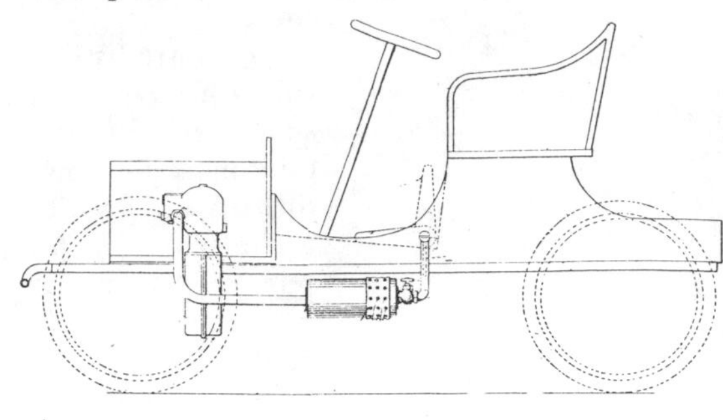

2901/1904
Walter Phillips *Manager*
Humber Ltd, Motor and Cycle Manufacturers
Coventry, Warwickshire, England

(Boult, Wade & Kilburn)

Bond's Improved Skipping Rope is designed to rotate more easily than the ordinary skipping rope and is also provided with a stiffened centre portion so that the distal end assumes a rectangular section when swung and the feet are thus less liable to become entangled in it. This makes it more convenient for use by adults as well as children and encourages skipping as a valuable form of gymnastics.

3718/1904
Francis Thomas Bond *Doctor of Medicine*
3 Beaufort Buildings, Gloucester, England

Miller and Bunnell's Armoured Pneumatic Tyre employs a series of curved sheet-metal shields, preferably of spring steel, to prevent the cutting or puncturing of the vulnerable rubber.

5014/1904
Major Miller *Inventor*
James Andrew Bunnell *Traveller*
Clyman, Wisconsin, USA

(Wheatley & Mackenzie)

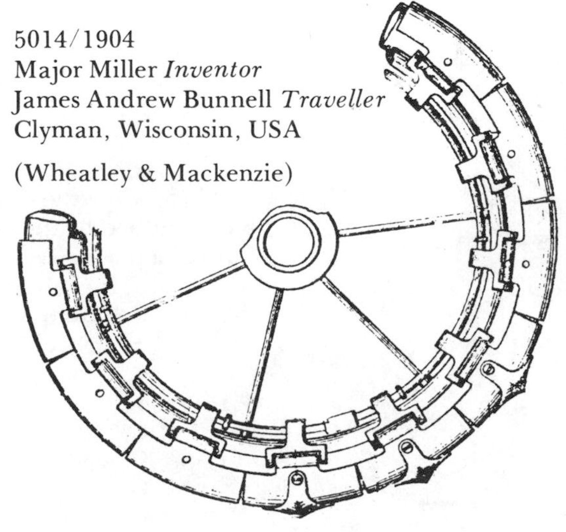

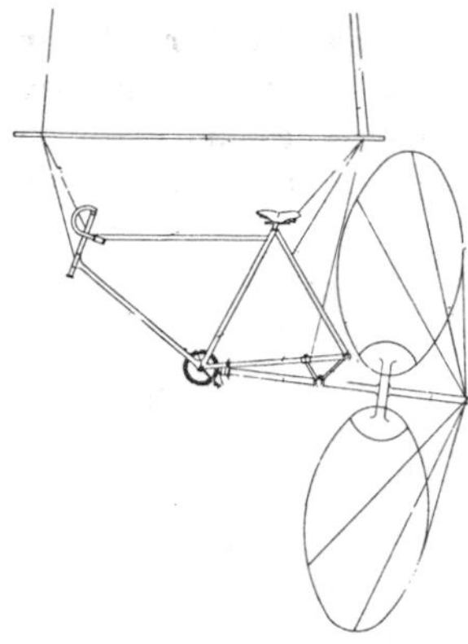

Crawford's Machine Propelled through the Air is for exercise, recreation or amusement and may be suspended from a building, tree, kite, or captive balloon. A cycle frame provides both support for the user, and a means for turning the propellor.

3929/1904
Robert Lindsay Crawford *Advertising Agent*
3 Hilberry Avenue, Tue Brook, Liverpool, England

(John Hindley Walker)

Lee's Improvement in Telephonic Communication connects every house supplied with electricity from a central station for lighting or transmission of power to that station telephonically via the very wires which convey the power. To achieve this purpose, the vibrations caused by speech or pressure are introduced into the electric circuit of the system at any desired place, in he wires leading to an incandescent lamp for example, by a telephonic transmitter, and picked up at any other suitable place via a telephonic receiver connected therewith. Any suitable resistances may be introduced between the telephonic instruments and the wires carrying the electricity, to reduce the voltage of the current acting upon the instruments as may be necessary or desirable.

By connecting each house with the central power station telephonic communications and emergency messages such as those calling the police, fire brigade, messengers, hackney carriages and so on can be passed on on more economical terms than if they were connected directly with the telephone system.

7676/1904
John Hodgson Lee *Lieutenant RN (Retired)*
4 Elm Grove, Peckham, Surrey, England

(W H Beck)

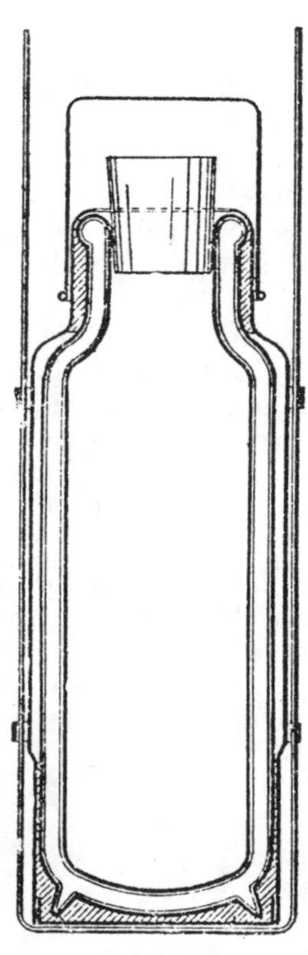

Burger's Improved Bottle consists in a glass vessel for workmen, sportsmen, etc constructed with double walls, the space between then being exhausted, for preventing the cooling of water, milk, coffee etc. The inside space between the walls of the vessel is coated with a mercury layer.

The bottle is enclosed in a case and its neck sealed with a cork. The tubes for exhausting the bottle are arranged safely at the sides of the bottom.

4421/1904
Reinhold Burger *Manufacturer*
2E Chausseestrasse, Berlin North, Germany

(Herbert Haddan & Co)

Pfeiffer's Improved Roundabout has cars made to resemble submarine vessels. These are rotated with an undulating motion over a circular basin of water so as to become successively submerged and raised.

The upper parts of the cars are of glass to facilitate observation. The cars are watertight and also fitted with a ventilator so that air enters the interior of the boat at all times. If there is a leakage in a boat, an electromagnetic valve empties the basin. A suitable landing stage is provided for the embarkation of passengers.

1098/1904
Hans Pfeiffer *Dentist*
9/2 Fraunhoferstrasse, Munich, Germany

(Herbert Haddan & Co)

Lange's Improved Double Bicycle for Looping the Loop in Circuses is of use if a performer should wish to perform the feat of looping the mutilated loop. He is provided with a second cycle attached above his head which enables him to complete his journey head downwards. The upper cycle may have more than two wheels, and be furnished with such rests and padding for the rider as may be required.

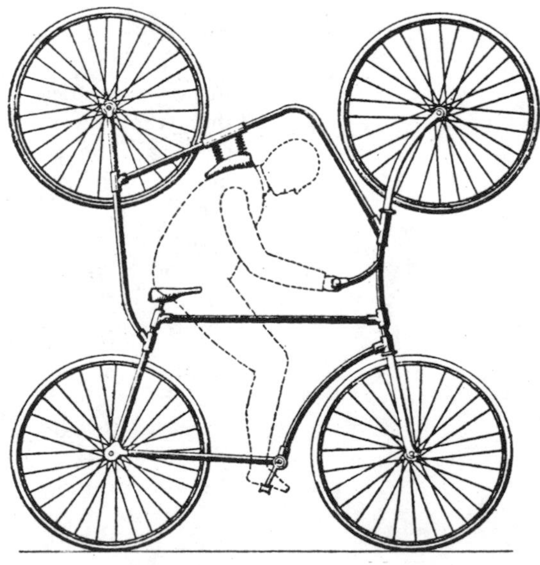

8706/1904
Karl Lange *Merchant*
66 Warschauer Strasse, Berlin, Germany

(G F Redfern & Co)

1904

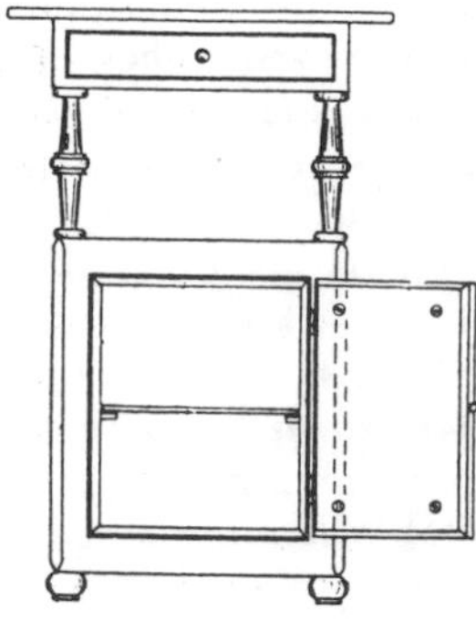

Countess Borcke-Stargordt's Pedestal Cupboard is provided with a lining, of glass, fayence, china or other suitable material, to prevent the absorption of evaporating urine, or gases or odours therefrom, by the wood of the pedestal. Notwithstanding of the greatest cleanliness it is at present impossible to avoid this drawback with ordinary cupboards, as the evaporation takes place in a single night and a nauseous odour becomes sensible when the door is opened.

12280/1904
Magdalena, Countess Borcke-Stargordt, *born* Magdalena, Countess Lehndorff
Stargordt, nr Regenwalde, Pomerania, Germany

(J G Lorrain)

Anderson's Device for Holding Down Ladies' Dress Skirts when the wearer is cycling, walking, playing tennis or taking other exercise consists in a ring which is sprung around the leg above the ankle and means for attaching it to the inside of the skirt, also sprung so that it will be automatically released if subjected to excessive strain.

14788/1904
Thomas Law Anderson *Mate*
Friendship Terrace, Keadby, Lincolnshire, *late of* The Keel *Faith,* Stanforth, nr Doncaster, Yorkshire, England

(Louis E Kippax)

Lorenzen's Horse Velocipede is so constructed that the draught animal walks not on the road but on an endless platform, whose movement is then transmitted to the road wheels by means of gearing. The pull exercised by the animal is utilised in this horse velocipede as much as it is in ordinary road vehicles, an increase in efficiency being besides obtained by the utilisation of the friction of the hooves upon the platform, which friction is not utilised with vehicles of ordinary construction

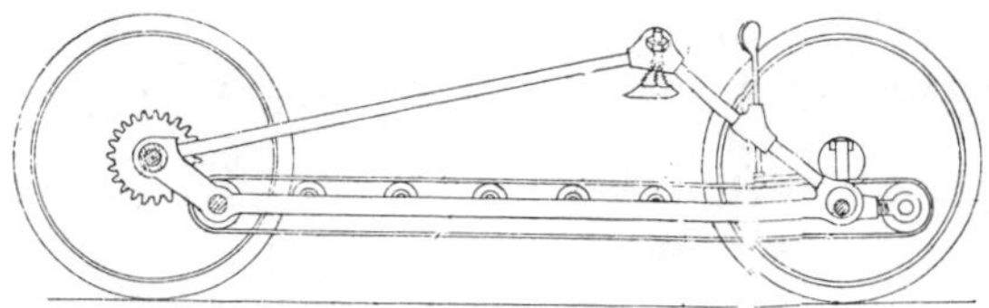

8488/1904
Christian Lorenzen *Engineer*
Roselea, Spencer Road, Wealdstone, Harrow, Middlesex, England

(Ferdinand Nusch)

Lowcock and Grave's Trick Cycle Riding Apparatus provides a wheel wherein a performer may display his skill and judgement.

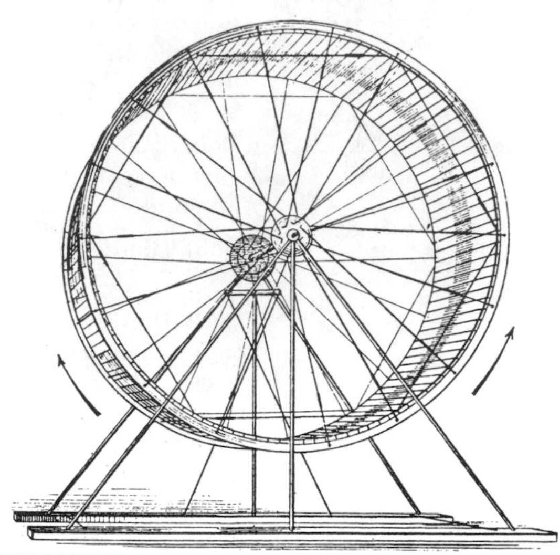

12831/1904
Tom Lowcock *Cycle and Motor Manufacturer*
96 Gisburn Road, Barrowford, Lancashire, England

Thomas Henry Grave *Twister*
6 Forest View, Barrowford, Lancashire, England

(Tasker & Crossley)

Skorzewski's Improved Ballon overcomes the drawbacks of ordinary apparatus constructed on the "lighter than air" principle: that they are only able to remain in the air for a short time owing to the permeability of the envelope, that the decrease in lifting power during the ascent must be compensated for by throwing out ballast, and that it is also necessary to let out air to descend to earth so that the balloon having once landed cannot raise itself again without being once more inflated.

A further disadvantage of the ordinary flexible envelope balloon is that it loses its shape under the pressure of air, and that dangerous currents can be set up in the gas inside it. Moreover, a tear in the envelope means a loss of gas which it is impossible to stop and which necessitates a more or less rapid descent.

In Skorzewski's balloon, this state of things is remedied by using impermeable chambers, for instance of glass, filled with a light gas, preferably pure hydrogen, and by condensing water ballast from the atmosphere in order to descend rapidly.

16752/1904
Comte Vladimir Skorzewski
Czerinejewo, Bromberg, West Prussia

(Carpmael & Co)

Fischer's Head-washing Apparatus promotes the health and beauty of the hair whilst at the same time being exceedingly practical and easy to handle. The person wishing to have the head washed need not, as formerly, take off the collar and neck-tie, and bend down his head, thereby often having the water run into the eyes and ears. The said person may remain in the chair, reading and smoking at will while washing and douches are proceeding.

The apparatus comprises a ring or gutter of aluminium or similar material formed to receive the shape of the head, with its inner edge shaped so as to receive an india-rubber ring which holds it pressed against the head, and is sometimes held tighter by a band, spring or similar device. At the lowest part of the gutter is a funnel-shaped exit with an extension for the attachment of a rubber tube.

12475/1904
Paul Fischer *Coiffeur*
Köttelbrücke, Königsberg, Prussia

(Henry O Linck)

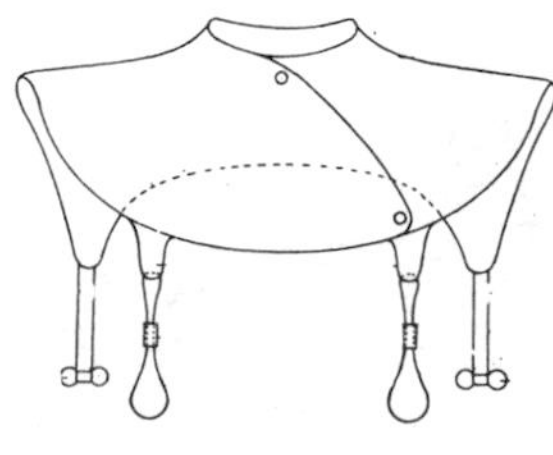

Buley's Appliance for Retaining Blankets, Rugs and similar Coverings in Position on the User consists in a yoke resting on the shoulders and carrying straps with knobs at the back and extension pieces at the front provided with loops and catches.

14765/1904
Olga Emilie Buley *Married Woman*
144 Askew Road, Shepherds Bush, London, England

(P: W D Rowlingson; C: Hughes & Young)

Leupold's Neck-compress or Appliance for External Treatment of Throat Complaints is a heating or cooling device, formed of metal bottles hinged together and filled with a suitable heating or cooling medium. It is designed to obviate the disadvantages of present methods of treating illnesses or complaints from "catching cold" (namely, steam-compresses of woven fabrics dipped into boiling water and then applied to the part affected by the disease) by providing an appliance which supplies heat only where desired and which may be adjusted to suit various sizes of neck and so not interfere with breathing.

12430/1904
Karl Rudolf Leupold *Engineer*
45 Nordstrasse, Zwickau, Saxony

(Fairfax & Wetter)

Mortet's Improved Means for Producing Spectacular Effects on Railways for Recreation enables an electric car to travel a few metres in space, this passage in the air being obtained through the weight of the automobile and the acquired speed.

13339/1904
Marie Mortet *Married Woman*
13 Rue de Poissy, Paris, France

(Hy Fairbrother)

Delory's Improved Comb for Applying Dyes, Lotions and Pharmaceutical Products to the Roots of the Hair or of the Beard obviates the drawbacks of methods of applying or removing hair colour at present in vogue, where it is very difficult to apply the lotion near the roots of the hair or only on a particular spot and in any case the operation can only be performed by a skilled operator and not by the person himself. This invention overcomes these problems and enables any person to perform the operation of dyeing or removing the whole of his hair or beard in a very easy and uniform manner. It consists in a special toothed, channelled and threaded comb which is allowed to glide very flatly on the head or face and which is arranged such that each of its teeth distributes the liquid uniformly.

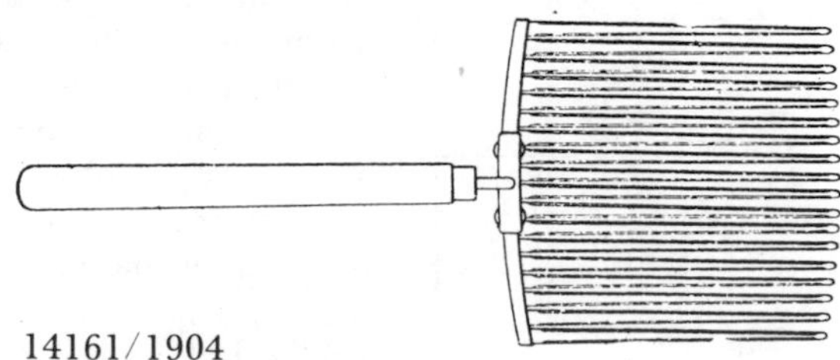

14161/1904
Gustave Delory *Manufacturer*
31 Rue de Maubeuge, Paris, France

(Jensen & Son)

Smith's Hair Net Case is in pocket book or wallet form, and bound like a book in leather, silk, paper or cloth. It is supplied with an accordion-shaped extension so that it can be opened for inspection and selection of a hair-net to be removed.

12601/1904
Crissie Mary Smith
14 Hartington Gardens, Edinburgh, Scotland
formerly of 27 Lexham Gardens, Kensington, London W, England

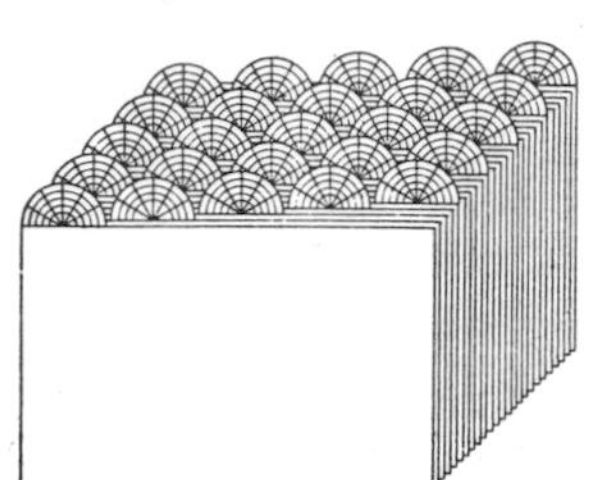

Cooper's Recording Tabs or Appendages so Designed as to Facilitate the Recording of Dates and Transactions or Events are placed on the upper edges of semicircular cards with divisions to indicate dates, periods, transactions or events. The system may be used, for example, to indicate the incidence of epidemic or zymotic diseases:

Number of radial division	*Diseases represented by colour or shading*		*Diseases represented by crosses*
1	Smallpox	: Green	Diarrhoea
2	Measles	: Blue	Phthisis
3	Scarlet Fever	: Red	Typhus Fever
4	Diphtheria	: Yellow	Erysipelas
5	Whooping Cough	: Purple	Cholera
6	Typhoid Fever	: Brown	Bubonic Plague

Deaths from the disease are indicated by black marks on the tabs, and the number of deaths from each disease in each house in each year can be shown on tables on the cards below, for example:

Year	*House referred to by card* e	f	g
1900	Measles Bubonic Plague (1 death)	Smallpox Cholera (1 death)	Diphtheria
1901	Diarrhoea (1 death) Scarlet Fever (3 deaths) Whooping Cough Cholera (1 death)	Diphtheria Erysipelas Typhoid	Phthisis Scarlet Fever

14209/1904
Charles Hamlet Cooper *Civil Engineer*
15 Dora Road, Wimbledon, Surrey, England

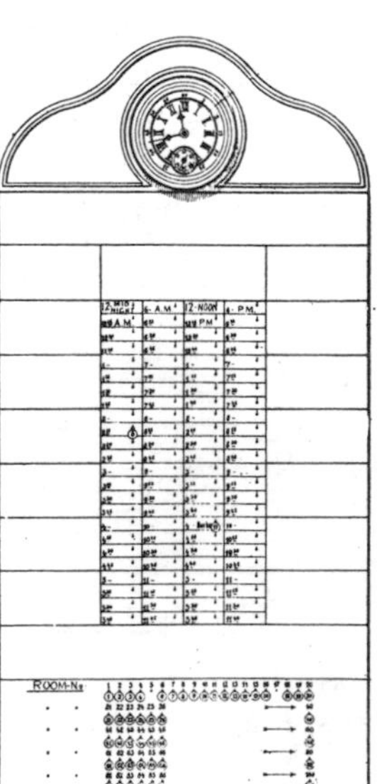

Steinhardt's Appliance for Indicating the Times when Persons Require to be Woke at Hotels and the like, and for Affording other Indications of a similar nature enables a rapid and reliable observation or register of the requirements of the travellers or occupiers of the rooms to be made.

13287/1904
Charles Christian Steinhardt *Merchant*
8 Tauenzienstrasse, Berlin, Germany

(Abel & Imray)

Hutchings's Apparatus for the Generation of Motive Power by Floating Bodies such as Ships and Docks. In propelling ships it is customary to use steam or other engines driven by burning coal, petroleum or other fuel. This is a costly means of generating power, owing to the great consumption of coal or petroleum and the expense involved in the employment of a large number of men to attend to the stoking of coal and watch the working of the machinery.

This invention provides means for generating power either for propulsion or for loading, unloading, lighting or like operations by the action of the ship and apparatus in contact with the water through which the ship or other floating body may be moved; the motive power is generated by utilising the rising and falling wave motions of the water. This is effected by a specially designed turbine screw, a combination of floating compartmented vessels with pistons within them operated by the rising and falling of water within them, so that relative movements of the vessel are directly conveyed, by a system of pipes and valves controlling the flow of reciprocating fluid, to any suitable means for utilising the same.

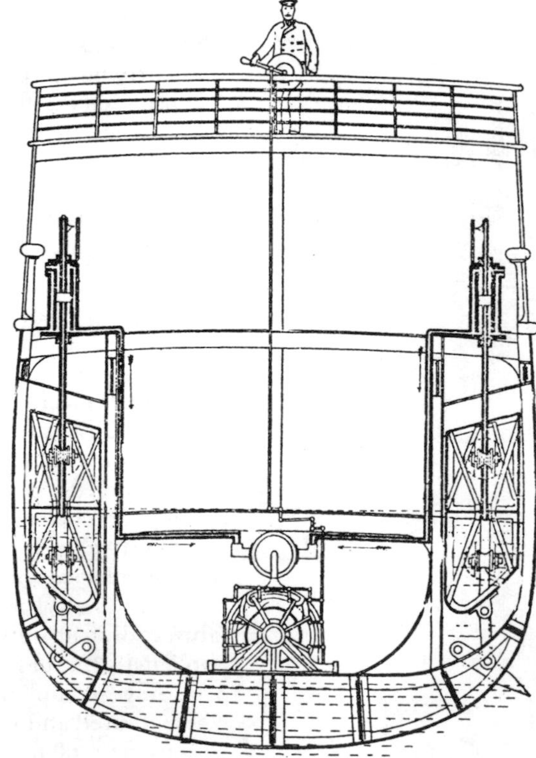

12099/1904
John Hutchings *Engineer*
210 Moorgate Station Chambers, Moorfields, London, England

(William Brookes & Son)

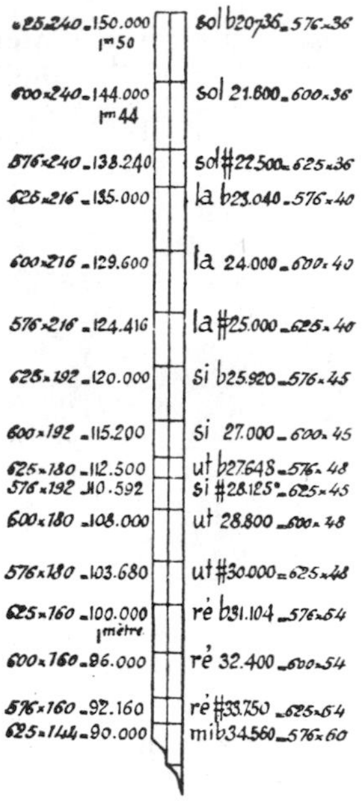

625×240–150.000 1m50	sol b20736–576×36
600×240–144.000 1m44	sol 21.600–600×36
576×240–138.240	sol #22500–625×36
625×216–135.000	la b23.040–576×40
600×216–129.600	la 24.000–600×40
576×216–124.416	la #25.000–625×40
625×192–120.000	si b25.920–576×45
600×192–115.200	si 27.000–600×45
625×180–112.500	ut b27.648–576×48
576×192–110.592	si #28.125°–625×45
600×180–108.000	ut 28.800–600×48
576×180–103.680	ut #30.000–625×48
625×160–100.000 1 mètre	ré b31.104–576×54
600×160–96.000	ré 32.400–600×54
576×160–92.160	ré #33.750–625×54
625×144–90.000	mi b34560–576×60

de Saint-Yves's Means of Designing Scales for Applying the Musical Scale to Architecture, the Fine-arts, Graphic and Plastic Trades and Industries is archeological and archeometrical, an instrument of precision, a cyclic and encyclopedic protractor, a synthetical code of higher religious, scientific and artistic studies, using the zodiac of the cosmological language, the involute zone of the twelve morphological letters termed Vattanes, a trigram of schematic languages expressing the eternal Divine life, the cosmological, planetary and inner letters and the enharmonics and angles of the primary colours from which the solstices and arithmetical expression of the colours can be generated. It can reveal the origin of the astral zodiac, signify the Divine constitution, verify the cosmological positions of the Egyptian Ptolemy, and provides a means whereby the mathematic reasoning displayed in aesthetical proportions can be applied to architectonics.

14377/1904
Joseph-Alexandre de Saint-Yves *Founder and Director of L'Institut International des Hautes Etudes*
9 Rue Colbert, Versailles, France

(Wheatley & Mackenzie)

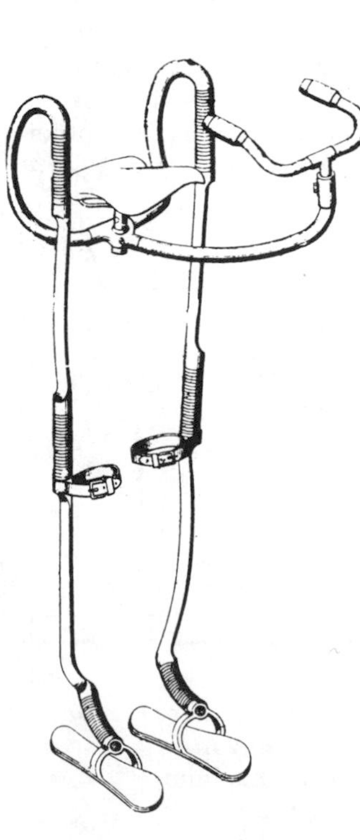

Skorzewski's Apparatus for Facilitating Walking or Running avoids one of the main causes of muscular fatigue, namely the necessity of a man's lifting his centre of gravity a certain height at every step. During the period of descent there is no compensation for the energy expended. The muscles have to remain strained in consequence of the necessity of supporting the weight of the body during this movement.

This invention has for its object an arrangement which allows very high speeds to be attained by the ordinary motions of walking and running. It is adapted to store up, and systematically to give up at the required moment, the energy resulting in the alteration in height of the centre of gravity of the body.

The user sits on a saddle, connected to handlebars and two extensions to strap to the legs, and made to conform to their shape. Soles are provided. The joints of the frame are preferably made of armoured rubber, and the whole is filled with compressed air through a suitable cock or valve.

14477/1904
Comte Vladimir Skorzewski *Count*
Czerinejewo, Bromberg, West Prussia

(Carpmael & Co)

Stoltz's Improved Tug of War Apparatus provides means whereby two or more persons contest with each other to determine superiority of strength, skill of manipulation, or power of endurance.

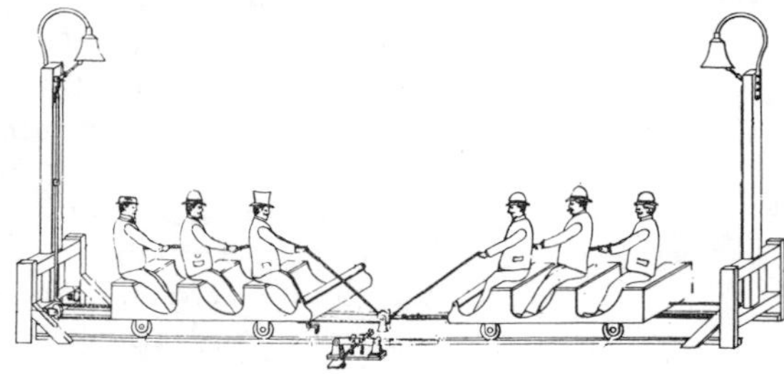

11067/1904
Melville Stoltz *Machinist*
80 Nicholas Avenue, New York, USA

(Haseltine, Lake & Co)

Boddy and Bottomley's Medicinal Compound for the Treatment of Piles is made of 8 ounces of Red Armenian bole, 4 ounces of Crab's Eye and 1 ounce of Dragon's blood, rolled out or ground to a fine powder and then formed into suitably sized pills by admixing a quantity of Venice turpentine.

16317/1904
Joseph Boddy *Confectioner*
1 St Thomas' Terrace, Stanningley, Leeds, Yorkshire, England

William Bottomley *Printer*
Fairfield House, Rodley, Leeds, Yorkshire, England

(Brewer & Son)

Benston's Improved Flying Machine or Navigable Ballon is constructed in such a manner as to admit of the necessary precautions being taken for safeguarding the occupants against accidents or injury. It is shaped like a torpedo, and divided into two compartments, the lower for containing passengers and machinery, the upper for increasing the buoyancy.

When descending, the rush of air through vertical tubes drives fans which cause parachute wings to be wound out of the sides of the body.

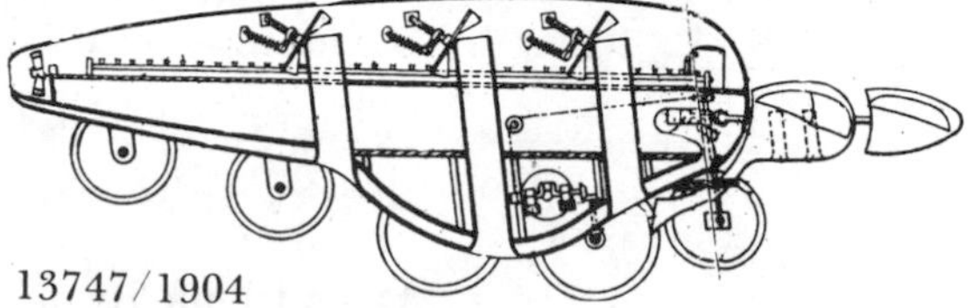

13747/1904
Edward Benston *Gentleman*
2 Waterloo Road, Cranbrook, Kent, England

(Cassell & Co)

1904

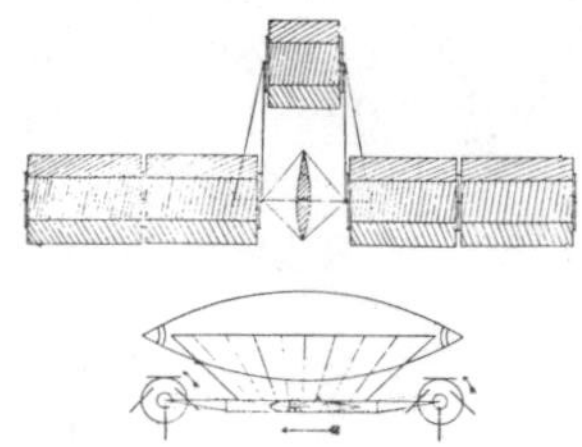

Hawkins's Improvement for Aërial or Flying Machines is applicable also to submarines of the type lighter than water.

It consists in giving stability to such machines by providing an arrangement of fanwheels such that if the axis of the machine is tilted from the horizontal the lifting power of the fans is altered to restore equilibrium. Machines employing this type of propellor will rise from the ground without a long run, will remain stable laterally, will rise or fall on an incline according to the will of the aëronaut, and will retard horizontal speed powerfully when within a few feet of the ground.

1543/1904
Edward Caesar Hawkins *Gentleman*
Shelton Hall, Long Stratton, Norfolk, England

Sarason's Improved Medicated Bath contains a liquid aërated by liberating nascent oxygen containing ozone from peroxides and their salts using catalytic agents such as extract of malt.

Its purpoe is to replace the effervescent carbonic acid baths heretofore used, which have the disadvantage that the acid is liable to become partly reabsorbed by the skin; at the same time the gas rising from the water might cause spasms to the bather. In this aërated oxygen bath however, the effect of the gas on the bather is both invigorous and beneficial. Oxygen *in statu nascendi* has a great avidity towards the surface of the body such as the skin and capillary vessels and will, by being reabsorbed, be of specific effect on the tips of the nerves as also on the composition of the blood.

4500/1904
Leopold Sarason *Physician*
48 Hubertus Allee, Grimewald, nr Berlin, Germany

(Haseltine, Lake & Co)

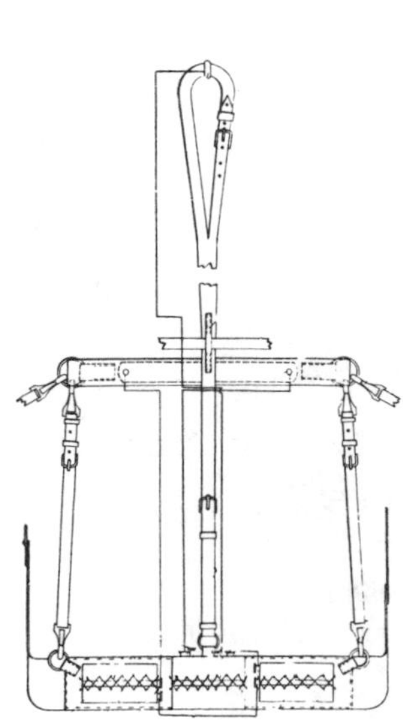

Huish and Stevens's Device for Preventing Self Abuse in Horses is so constructed that when the act is attempted an electric shock is administered and at the same time an electric bell rings.

The battery is disconnected if the horse should lie down by means of a mercury switch.

48/1904
Charles Henry Huish *Surgical Instrument Maker*
12 Red Lion Square, London WC, England

Frank Stevens *Electrical Engineer*
4 Princess Road, Kilburn, London, England

(J E Evans-Jackson & Co)

Kwiatkowski and Stefanski's Improved Water Power Engine is operated by a waterwheel which, via cranks and lazy tongs, pumps water to itself.

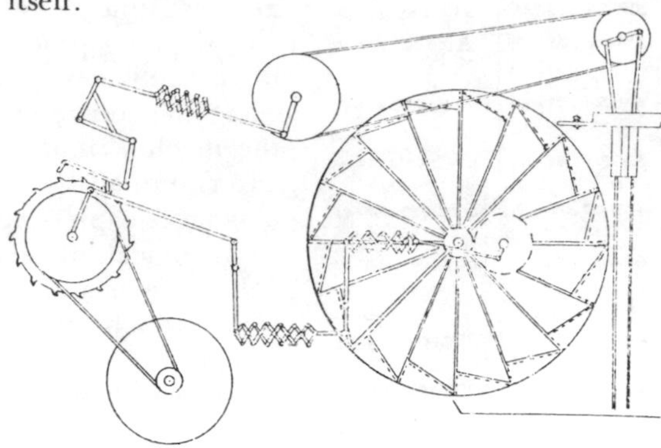

5723/1904
Franz Kwiatkowski *Tileworks Foreman*
76 Kronprinzenstrasse, Posen, Germany

Feliks Stefanski *Innkeeper*
Jarschonnkowo, nr Gnesen, Germany

(Jensen & Son)

Schwedler's Improved Electric Bath is specially intended to influence the action of the heart. Whilst the patient is under treatment, a rotating electric current is made to pass through his body along a direction determined by its immersed parts and to disperse in the region of the heart. The patient is seated as if in an ordinary armchair and footbath, with his arms and feet immersed in the glass troughs connected with the electrodes. It is thus possible to treat patients without their being obliged to undress entirely, which is a notable advantage, as the risk of catching cold and being subject to other inconveniences is thereby avoided.

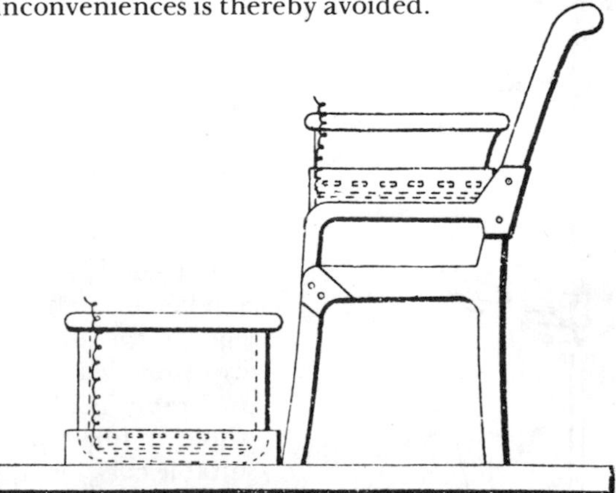

10393/1904
Hermann Immanuel Willy Schwedler *Director of The Sanitas Electric Company Ltd*
7A Soho Square, London W, England

(Wheatley & Mackenzie)

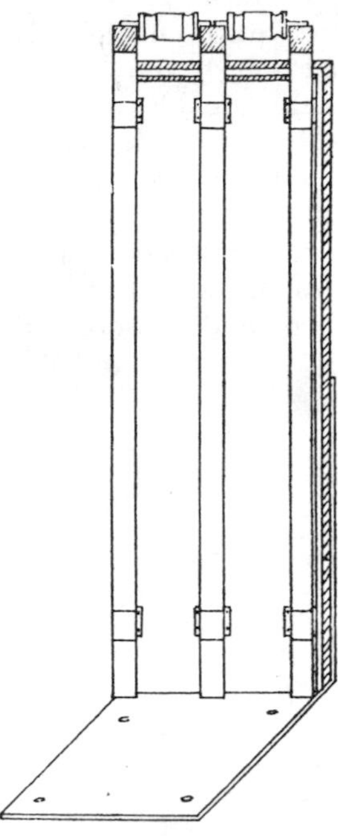

Ryan's Improved Wicket for the Game of Cricket — whether played indoors or out — provides electrical means whereby a bell is rung if the wicket is struck.

3618/1904
James Ryan *Resident Engineer*
St Brides Institute, Bride Lane, London, England

Hülsmeyer's Hertzian-wave Projecting and Receiving Apparatus to Give Warning of the Presence of Metallic Bodies such as ships, wrecks, or submarine boats projects waves which are reflected by the body into a receiver at the same place as the projector. The system is adapted to give a visible or audible signal of reflection. It may be turned so that the waves are projected to each point of the compass in turn, and so arranged that the officer in charge knows at once the direction from which the warning comes.

13170/1904
Christian Hülsmeyer *Engineer*
3 Grabenstrasse, Düsseldorf, Germany

(Charles Bauer, Imrie & Co)

Fuchs's Improved Portable Folding Seat is suspended from the user by its four corners by adjustable braces, or fixed to the clothing. The seat part is of pliable material, lined or quilted, fastened to legs which are folded by a spring when the user stands up. When the seat is in use, the legs are forced apart by the pull of the spring, and a large surface is presented so that the seat does not sink into the ground, while the user can bend a little forward or backward, if needful.

The seat is particularly useful in gardening or other occupations, and also as a camp stool, when shooting etc, in fact whenever it is necessary to assume what would otherwise be an inevitably fatiguing squatting position.

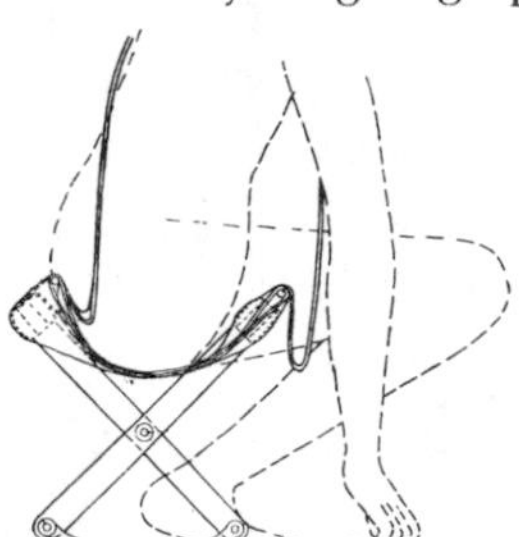

5110/1904
Rudolph Fuchs *Actor*
1 Werra Strasse, Meiningen, Germany

(W P Thompson & Co)

Stern's Improved Means for Allowing the Passing of one Vehicle by Another relates to systems of transport on single line or narrow routes whereby collisions may be avoided and travel carried on without interruption. It is applicable to railway or road vehicles, but is more particularly for amusement or recreation cars of about the size generally employed in scenic railways. It consists in passing one vehicle over the top of the other, and of modifying the vehicle roofs and tyres to allow of such passage.

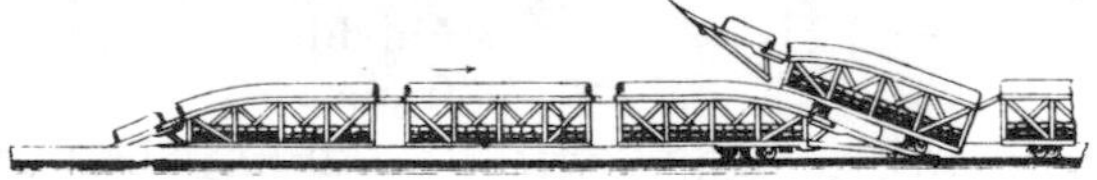

13095/1904
Phillip Kossuth Stern *Electrical Engineer*
130 Fulton Street, New York, USA

Wulff's Improved Apparatus for Throwing Animals to take a Somersault is a device for projecting horses, elephants, monkeys etc into the air so that they perform the so-called salto-mortale. The animal is supported by means of a body belt, fastened to vertical posts on a plate, so that its feet are just in contact with the plate. The plate is connected to a base by means of a hinge at one end, and springs are provided which tilt the plate when a catch is released.

When the animal is strapped into position and the trigger pulled, the plate tilts and the animal turns a somersault, the rings on the body belt readily disengaging themselves from the hooks on the supporting posts.

The apparatus is designed to ensure that the animal does not cling with the legs, which would be objectionable.

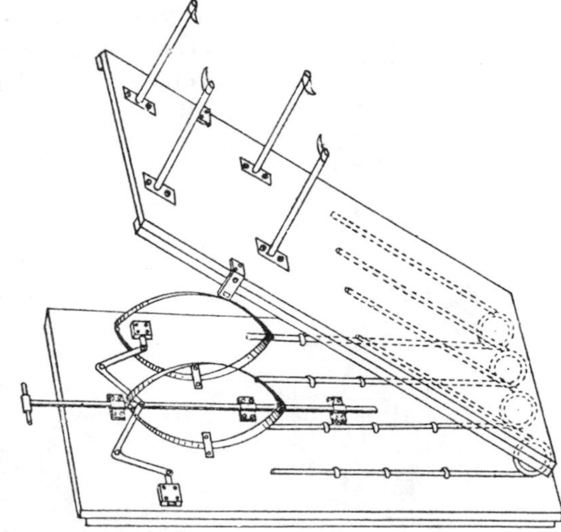

8713/1904
Eduard Wulff *Circus Manager*
10 Rue du Moniteur, Brussels, Belgium

(W P Thompson & Co)

1904

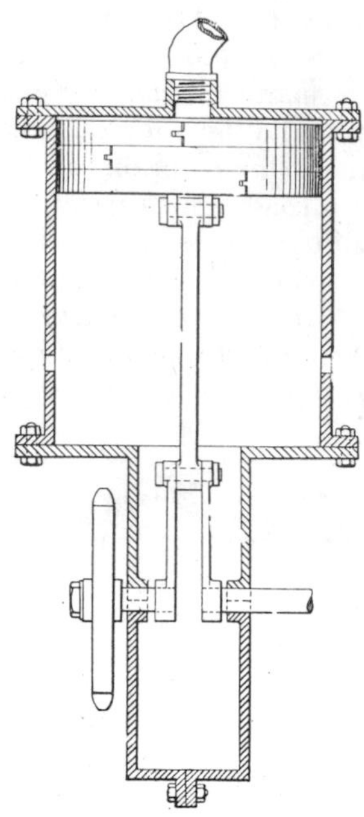

Tollemache's Improved Means of Inflating the Tyres of Motor Cars and the like is a small air pump which may be driven by the engine of the vehicle when the car is at rest.

3670/1904
The Hon Bentley Lyonel John Tollemache
Siddington Hall, Cirencester, Gloucestershire, England

The Wright Brothers' Improved Aeronautical Machine comprises two planes of light framework covered with canvas and so arranged that they can be warped by pulling on ropes.

A forward horizontal rudder is carried on projecting struts; springs attached to the frame curve the surface of the rudder when inclined and increase its effect.

There is a vertical rudder at the rear attached by universal joints to two parallel pairs of struts, hinged to the rear edges of the planes and actuated by ropes.

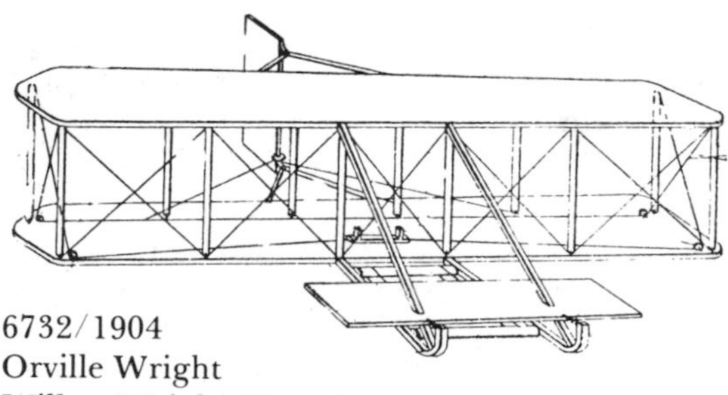

6732/1904
Orville Wright
Wilbur Wright *Manufacturers*
1127 West Third Street, Dayton, Ohio, USA

(Herbert Haddan & Co)

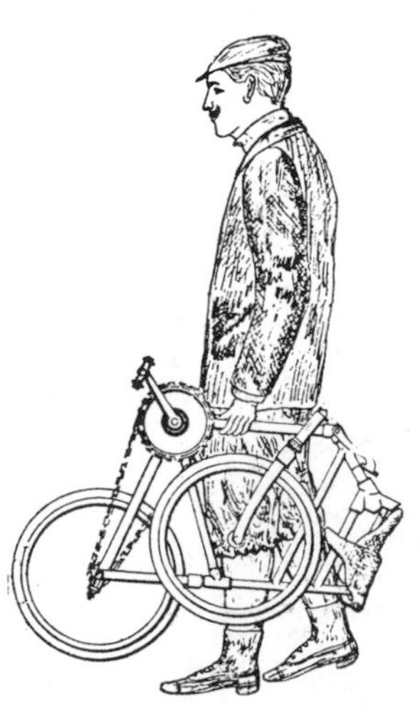

von Bülow's Improved Guiding Device for Bicycles, Motorcycles and the like dispenses with the ordinary handle-bar and uses instead a saddle-back and arm-rest so that any action of hand is done away with. Both hands of the rider are therefore free for any other purpose whatever.

5508/1904
Joachim von Bülow *Gentleman*
10 Schoenerberger-Ufer, Berlin West, Germany

(Henry O Linck)

Bishop's Improved Hat Guard enables a cord to be pulled from the hat quickly and secured to the coat should a wind suddenly arise, and equally quickly dispensed with and hidden from view should its services no longer be required. It will be obvious that such an ingenious contrivance (meeting as it does a want long felt, especially by wearers of straw hats) will be eagerly patronised as a boon and comfort to all.

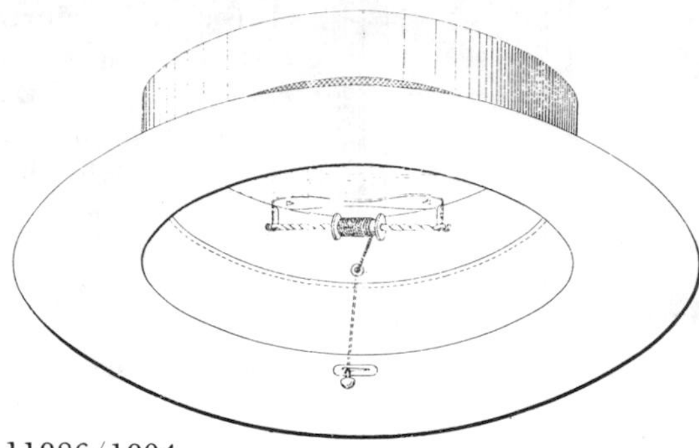

11986/1904
Robert Hodges Bishop *Commission Agent*
Hartwell, Great North Road, Highgate, London N, England

(J Owden O'Brien)

Lee's Improved Toast Rack is designed to keep the toast in the most desirable state, namely that which freshly toasted bread possesses. To this end it has a hollow base and partitions to receive hot water and corrugations on the partitions to keep the toast from touching them. This ensures that the radiant heat is such as to keep the toast warm, whilst not allowing it to become soft on the one hand nor hard and brittle on the other.

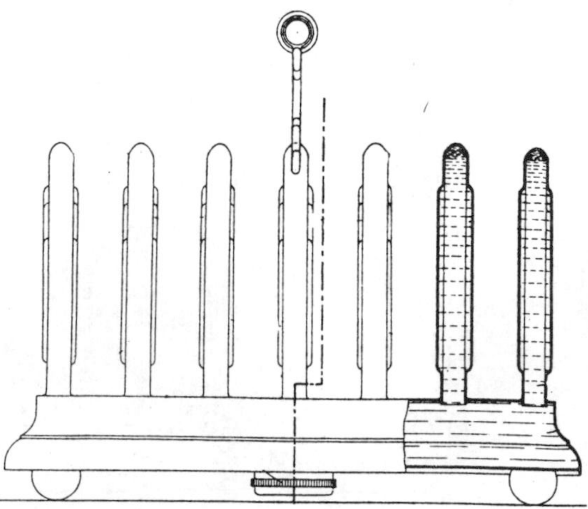

19426/1904
Thomas Russell Lee *Gentleman*
Springwood, New Ferry, Chester, England

(Cheesebrough & Royston)

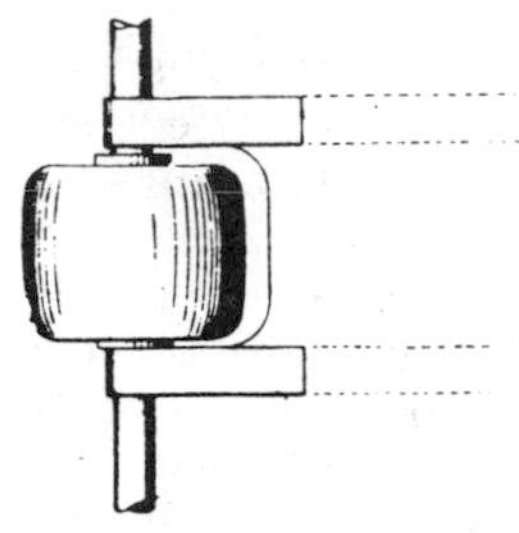

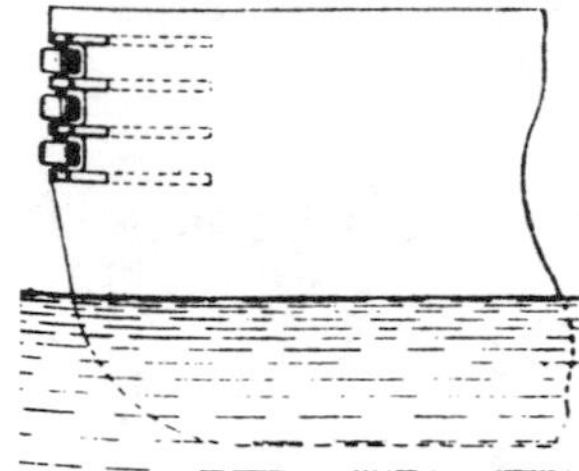

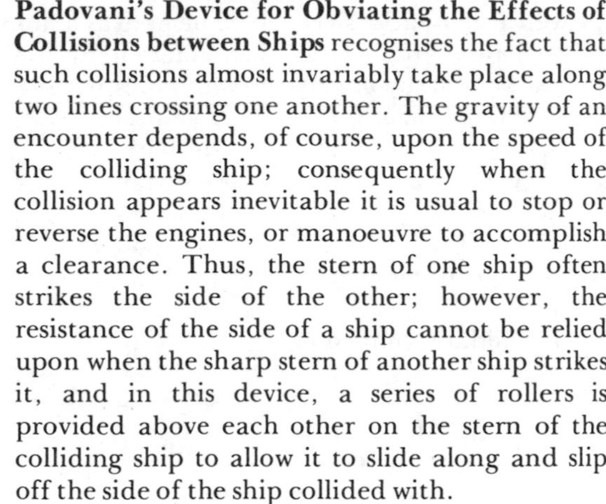

Padovani's Device for Obviating the Effects of Collisions between Ships recognises the fact that such collisions almost invariably take place along two lines crossing one another. The gravity of an encounter depends, of course, upon the speed of the colliding ship; consequently when the collision appears inevitable it is usual to stop or reverse the engines, or manoeuvre to accomplish a clearance. Thus, the stern of one ship often strikes the side of the other; however, the resistance of the side of a ship cannot be relied upon when the sharp stern of another ship strikes it, and in this device, a series of rollers is provided above each other on the stern of the colliding ship to allow it to slide along and slip off the side of the ship collided with.

16265/1904
Mathieu Padovani *Engineer*
4 Rue de l'Intenance, Bastia, Corsica

(A F Spooner)

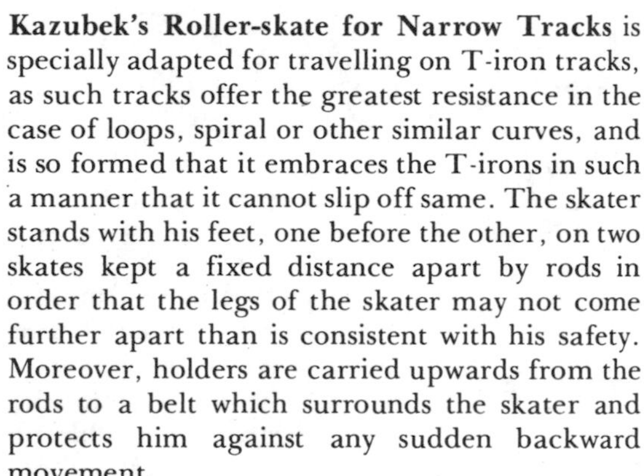

Kazubek's Roller-skate for Narrow Tracks is specially adapted for travelling on T-iron tracks, as such tracks offer the greatest resistance in the case of loops, spiral or other similar curves, and is so formed that it embraces the T-irons in such a manner that it cannot slip off same. The skater stands with his feet, one before the other, on two skates kept a fixed distance apart by rods in order that the legs of the skater may not come further apart than is consistent with his safety. Moreover, holders are carried upwards from the rods to a belt which surrounds the skater and protects him against any sudden backward movement.

16441/1904
Adalbert Kazubek *Electrical Engineer*
2 Josefstrasse, Berlin, Germany

(Wheatley & Mackenzie)

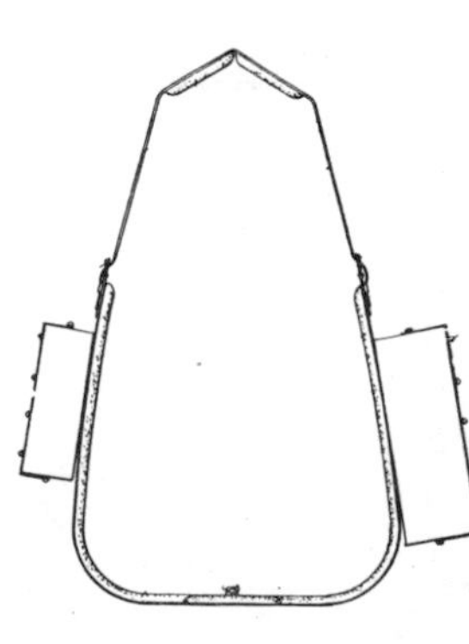

Stevens and Huish's Improved Device for Preventing "Crib Biting" and "Wind Sucking" in Horses comprises a band to affix to the animal's neck, provided with a battery and an induction coil. A contact is so arranged that, when the horse extends the muscles of its neck upon attempting either of the above actions, a shock is administered.

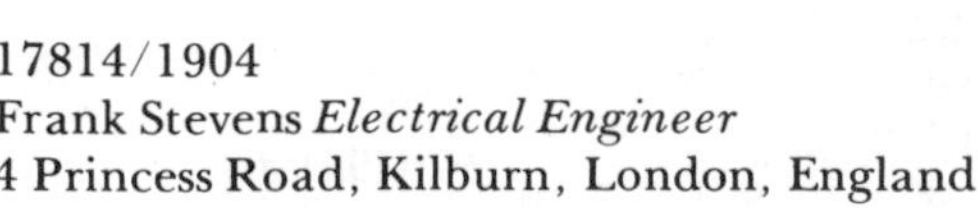

17814/1904
Frank Stevens *Electrical Engineer*
4 Princess Road, Kilburn, London, England

Charles Henry Huish *Surgical Instrument Maker*
12 Red Lion Square, London EC, England

(J E Evans-Jackson & Co)

Buggé's Improved Brake for Vehicles raises the vehicle off the ground at the pull of a lever, thus bringing it to a halt. By acting independently of the wheels, Buggé's brake relieves them of the strain and wear to which they are subject under the present plan of applying brakes to their rims, and also saves the tyres from injury in that they are kept from being forced over rough ground by the impetus of the car. By locking the brakes in action, the car can be left by the occupant without any fear of it being maliciously set in motion.

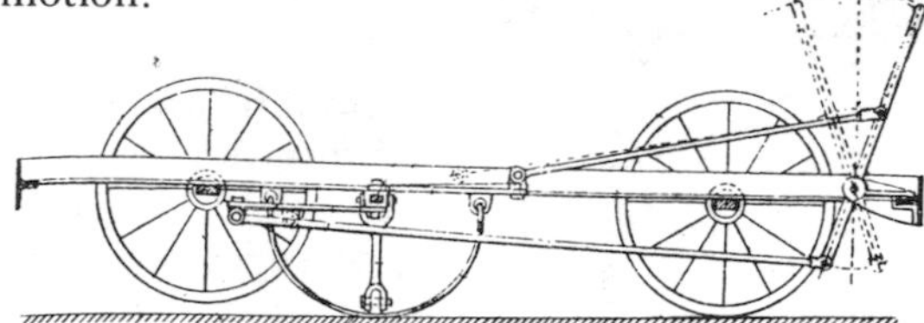

7517/1904
Rasmus Buggé *Master Mariner*
12 Annis Road, Hackney, London NE, England

(Reginald W Barker)

Kleinbach's Improved Orthopaedic Exercising Apparatus is for stretching the vertebral column and exercising and strengthening the muscles of the neck and the upper portions of the body.

It comprises a cap bearing a pulley on the top, which is strapped to the patient's head. Handles are attached to a cord which is so arranged that by pulling down on them the patient raises himself from the floor by the upward pressure exerted upon the head.

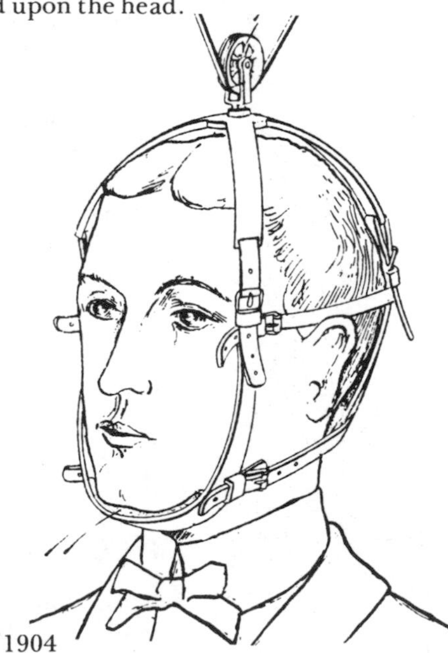

18258/1904
John Kleinbach *Gentleman*
Spokane, Washington, USA

(Marks & Clerk)

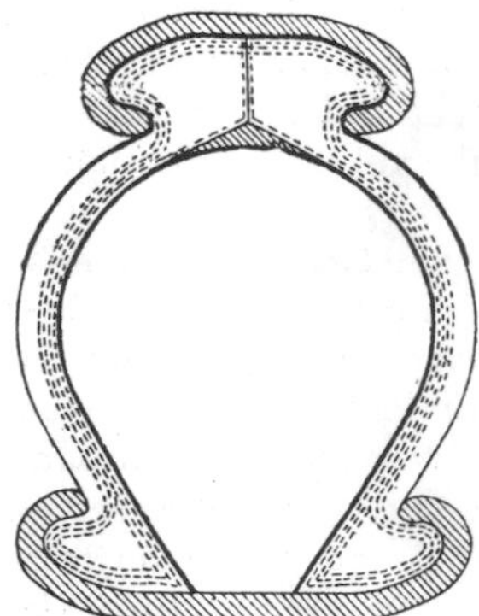

Somerville-Large's Improved Pneumatic Tyre Cover is made in two rings, joined around the circumference which touches the road by means of metallic clips and can be self-attached to the outer periphery of tyres to protect them from injury without impairing their elasticity.

5314/1904
Philip Townsend Somerville-Large *Member of the Institute of Civil Engineers*
Carnalway, Kilcullen, County Kildare, Ireland

Schmidt and Sharp's Improved Means for Protecting the Air Tubes of Pneumatic Tyres uses a shield made of a number of thin metal plates loosely joined together to allow for expansion and contraction, and preferably enclosed in a canvas envelope.

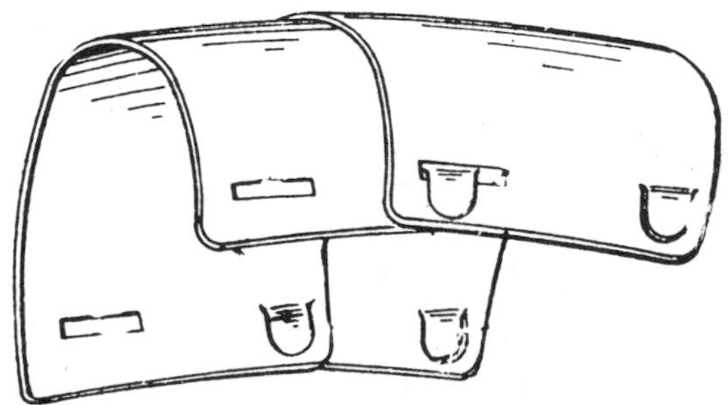

5615/1904
Francis Schmidt *Engineer*
1 Brook Road, Itchen, Hampshire, England

Charles Sharp *Grocer*
Winton House, Obelisk Road, Woolston, Hampshire, England

(Browne & Co)

Dubuis's Improved Bath is fitted with appliances to enable it to be used as an electric light, steam, or Turkish bath. A hinged cover is provided to enclose the patient, from which suitable glow lamps may be hung as desired.

End pieces are also provided, one of which may have an opening to allow the patient's head to pass through and rest on a pillow while he breathes the fresh air. Another opening, with a slide, is provided in he top in case the patient desires to sit up, in which case a chair may be placed in the bath.

An openwork mattress can be hung from hooks along the sides of the bath, and a stronger mattress or frame placed below for safety. An additional horse-hair or other mattress may be provided for the patient's use when cooling down. A winch and cords may be provided for lifting or straightening the mattress. The bath may be used as a light bath, or a water bath, or a combination of both, and, by fitting it with steam pipes, it may be used as a vapour or Turkish bath.

The bath is preferably enclosed in a wooden case to retain the heat, suitably treated to render it fireproof. Designed for private use, the construction and joining of its parts enables the bath to be opened from inside in case it becomes too hot.

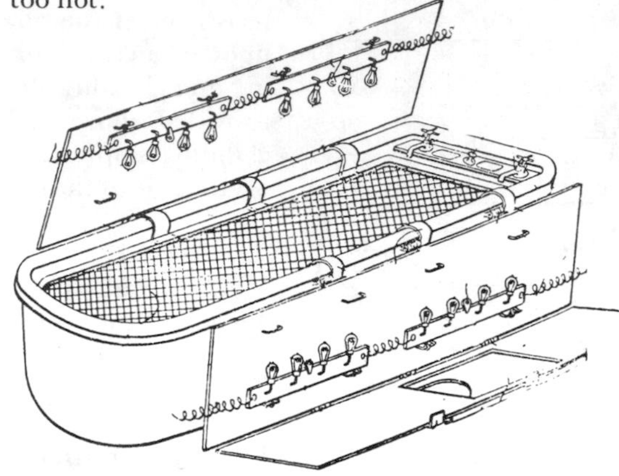

3345/1904
Gabriel Dubuis *Medical Vibration Specialist*
26 Hugh Street, Pimlico, London, England

(Hy Fairbrother)

Walliker's Improved Device for Picking up Pins, Needles and the like is a simple and effective appliance by which the said articles can be lifted from the floor without the use of stooping or having to bring the hand into contact with them. A horseshoe magnet is provided with a telescopic handle and a hook and chain to suspend it from the wearer's waistband; users of the appliance can so conveniently carry it attached to the person and while having it always handy in case of emergency will at the same time have their hands entirely free for other employment. If desired, a suitable glove or mitten may be provided for the user to wear as a protection when removing the pins and needles from the magnet.

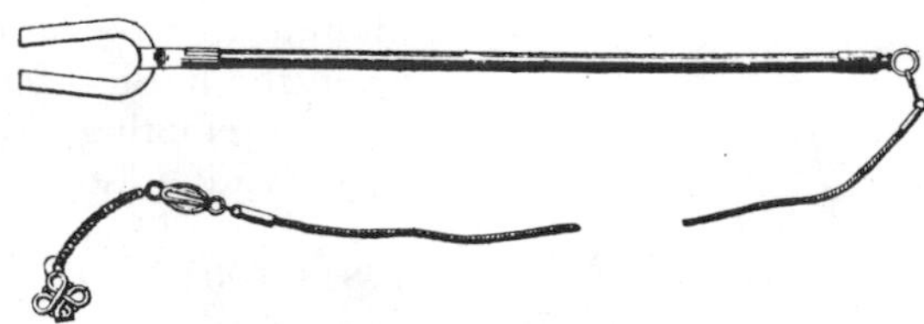

7549/1904
George Steward Walliker *Engineer*
57 Selhurst New Road, South Norwood, London SE, England

(Hughes & Young)

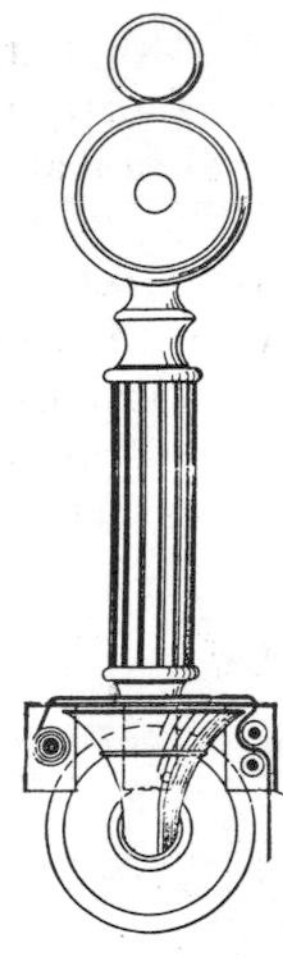

Churchill-Otton's Sanitary Rotating Diaphragm for Telephones ensures that the transmitter mouthpiece is always covered by a strip of paper fed from a roll, so that a fresh surface can be provided for each user, the previous part of the strip having been pulled forward and torn or cut off.

9683/1904
Sidney Churchill-Otton *Commercial Traveller*
Rothschild Chambers, Collins Street,
Melbourne, Australia

(Wheatley & Mackenzie)

Reynolds and Seward's Improved Flying Machine employs the means used by birds in flying. It is operated by the aëronaut strapping himself in, grasping the hand staples and working the stirrups with his feet. To make a start, the man stoops over, and runs forward, operating the silk and whalebone wings.

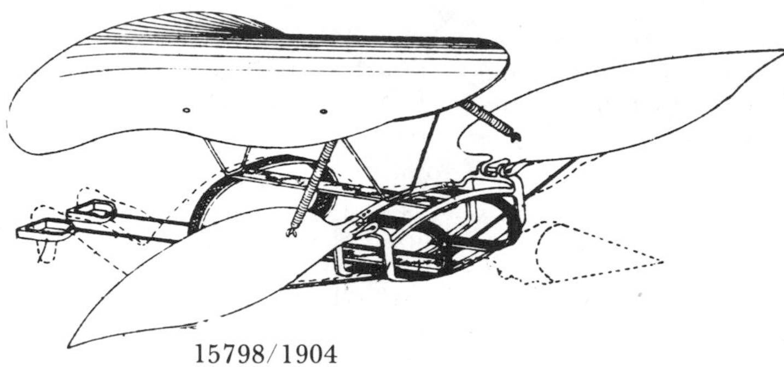

15798/1904
Alva L Reynolds *Mechanic*

Henry I Seward *Agent*
Los Angeles, California, USA

(W P Thompson & Co)

Dr Lorenz's Cap for Assisting Maintenance of an Erect Posture of the Body indicates when the wearer deviates from an upright position and is intended to be worn by school children to prevent stooping. It is made in the form of an inverted bottomless saucer with holes in the side for lightness and ventilation. In as much as the cap only remains on the head so long as the wearer remains in an erect posture its falling off calls attention of both the wearer and of the teacher or other attendant to the fact that normal erect posture is not being maintained. The appliance does not, in itself, force the wearer to assume a definite position but rather mechanically aids the exercise of the will-power.

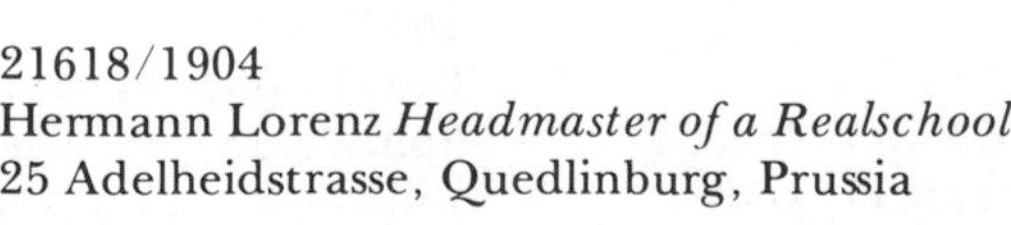

21618/1904
Hermann Lorenz *Headmaster of a Realschool*
25 Adelheidstrasse, Quedlinburg, Prussia

(Henry O Linck)

Boyce's Apparatus for Amusement is designed to interest onlookers as well as entertain users, and is particularly well-adapted for summer resorts or parks, where there are generally many pleasure-seekers.

Persons slide down a wooden incline, the object being to lodge themselves in pockets which may be differently numbered so that the apparatus may be used for competitions. Obstructions are placed here and there, and the slide is wider at the bottom than at the top, making it more difficult to reach the pockets. The persons start sitting from the top. For arresting motion, the incline is dished or ridged at the bottom and guarded with a net and cushion. The slide ends above the ground line so that sliders may more readily recover their feet. The whole may be housed in a structure ornamented with a landscape or the like.

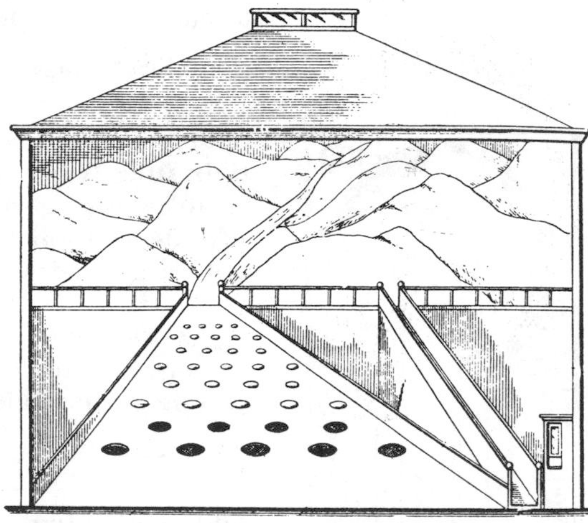

18724/1904
Edward Clarence Boyce *Contractor*
302 Broadway, Manhattan, New York, USA

(Ernest de Pass)

Young's Device for Teaching Penmanship keeps the fingers and thumb in the correct position for writing.

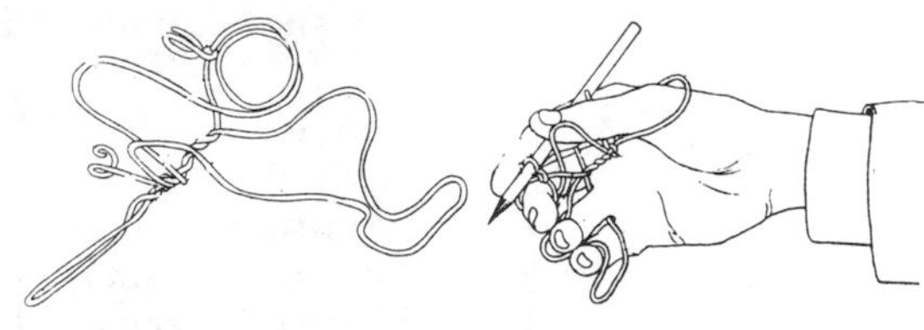

22928/1904
Frank Charles Young *Inventor*
42 Church Street, New Haven, Connecticut, USA

(H D Fitzpatrick)

1904

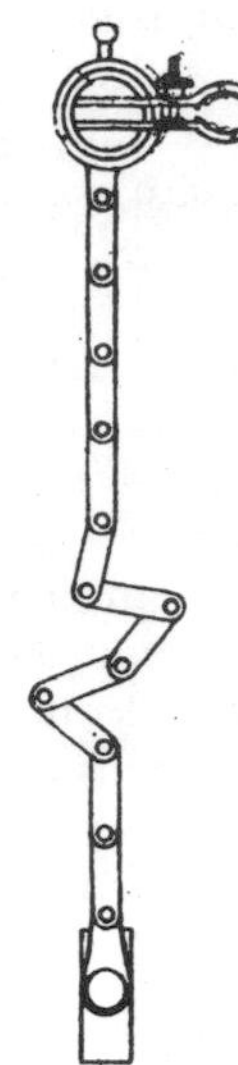

Nootbaar's Improved Device for Holding Expanded Umbrellas can be employed everywhere with the greatest ease and advantage, and is designed to overcome the inconvenience of holding an expanded umbrella when sitting in a open carriage or motor car, or riding a bicycle, since the arm soon becomes tired, especially when driving against the wind. It is also troublesome to have to hold the umbrella when one is driving oneself, or when employing a telescope, or reading or otherwise using the hands.

This new device securely holds the umbrella and enables it to be fixed in any position, at any desired angle: erect, inclined forward to protect the face or rearward to shelter the back.

17559/1904
Ernst Nootbaar *Actor*
6 Feuerstenplatz, Dresden, Germany

(Paul E Schilling)

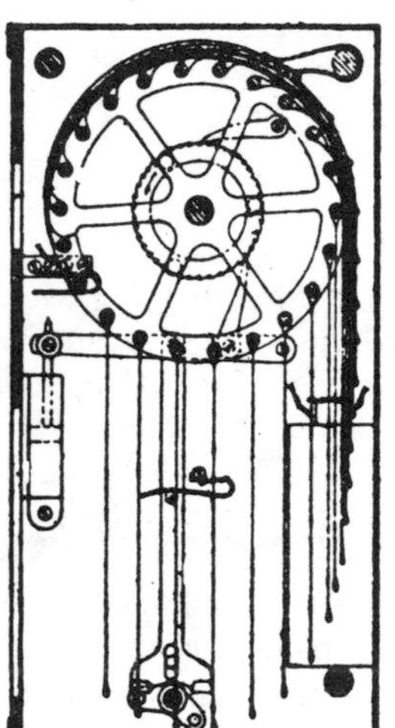

Coupette's Improved Advertising Device for Indicating Names of Stations is a simple and reliable means of displaying station names and advertisements by flipping them over a roller, a bell being sounded at each change.

12768/1904
Paulin Coupette *Chief Engineer*
11 Kaiser Wilhelm Ring, Cologne, Germany

(Abel & Imray)

Anderson's Improved Medicated Powered Sanitary Paper combines a drying curative powder with sanitary paper.

The invention consists in the combination of powder and paper, and not in the powder itself.

Any suitable known means of applying the powder to the paper may be used.

18807/1904
(Mrs) Ada Parr Anderson
Pergola, Leicester Road, Wanstead, Essex,
late of The Laboratory, St Thomas's Road, Hackney, London NE, England

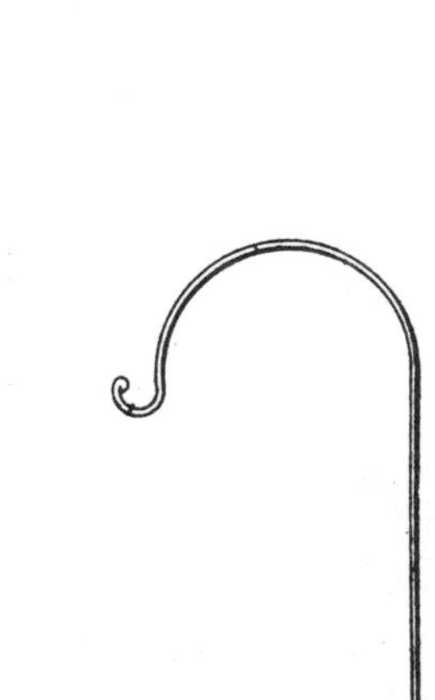

Simon's Improved Hook enables a watch to be easily fixed to a bedrail, looking-glass, picture, gas-bracket or other article in order to avoid placing the watch on its back and also to ensure its being seen without trouble and without knocking about the watch.

17951/1904
John Simon *Accountant and Estate Agent*
6 Court Row, Guernsey, Channel Islands

(Hughes & Young)

Bublitz and Scheel's Improved Vehicle for Carrying Flowers, particularly for use at Funerals is designed to display floral offerings advantageously, to carry a large number of same, and to protect them against frost in winter and overheating in summer. It has hinged side and rear-end glass doors, display racks inside and shelves bearing trays of artificial flowers should there be insufficient floral tributes to cover the racks.

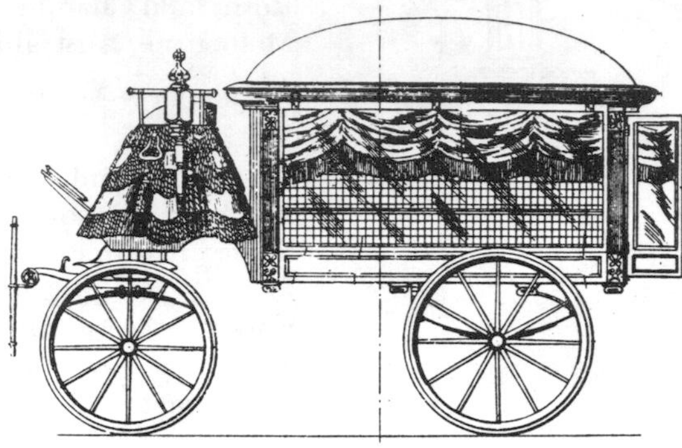

15570/1904
Frank Bublitz *Innkeeper*
419 North Fairfield Avenue, Chicago, Illinois, USA

William Scheel *Motorman*
438 North Fairfield Avenue, Chicago, Illinois, USA

(Haseltine, Lake & Co)

Roberts's Improved Road Locomotive is adapted to run on an endless track consisting of a pair of chains carrying wood blocks. The use of the tracks described obviates the defect of heavy road vehicles as now constructed, which are limited in their use by reason of their wheels sinking to too great an extent when travelling over soft or sandy ground and over surfaces of great irregularity.

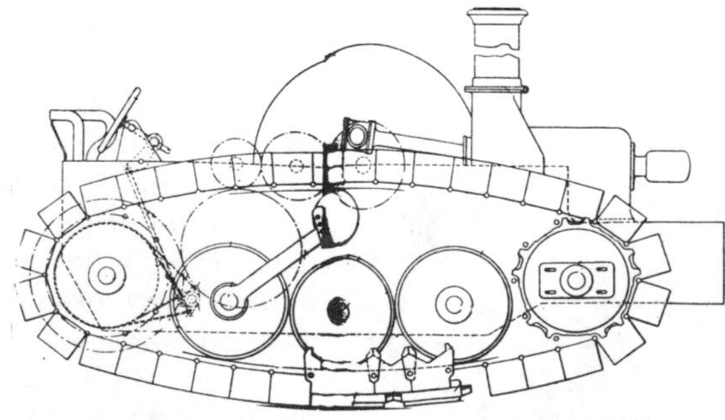

16345/1904
David Roberts *Engineer*
Spittlegate Iron Works, Grantham, Lincolnshire, England

(G F Redfern & Co)

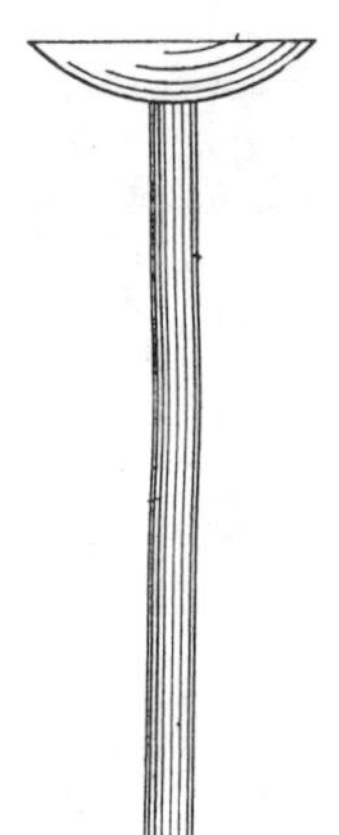

Pikler's Improved Appliance for Playing at Ball is characterised by its peculiar shape; a handle terminating in a dish for throwing and catching the ball.

The game is preferably played by four persons and its essential feature consists in catching and throwing the ball by means of the flat surface of the dish. The dish may be mounted on either end of the handle.

20805/1904
Charles Pikler *Official*
25 Külsö Kerepesiút, Budapest, Hungary

(W P Thompson)

Berger's Improved Trousers Protector keeps the back part of the trouser legs out of the mud and prevents fraying at the edges. It consists of a hinged ring sewn to the trouser bottoms in such a way that the rear semicircle may be lifted and retained, thus preventing soiling of the cloth and overcoming the disadvantages of the conventional trouser protector in which the bottoms are raised by cords or laces, thus causing creasing in the middle or upper part of the trouser and representing an objectionable appearance. An auxiliary bow may also be fixed to the lower front part of the trouser leg to prevent it becoming torn or frayed by friction with the boot.

20897/1904
Hugo Berger *Foreman*
37 Ludwigstrasse, Chemnitz, Germany

(Herbert Haddan & Co)

Turner's Improved Thimble has an open side to relieve the pressure and heat on the user's finger that arises from the finger and nail being entirely encased.

20139/1904
William Turner *Commission Agent*
56 Belvidere Road, Liscard, Cheshire, England

(Chas Coventry)

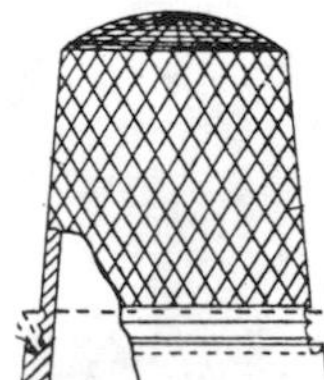

Griffith's Improved Thimble affords protection to the hand of the operator against the liability of the needle to slip upon the surface of the thimble. It has a rim with holes all round it so that, should the needle slip, it shall be caught in one of the said holes.

23236/1904
Frederic Griffith *Jeweller*
Bankfield, Arden Road, Dorridge, Warwickshire, England

(Charles T Powell)

The Bollingers' Improved Knife for slicing Bread, Cake, and other Light Spongy Materials is made with a plurality of blades arranged in step fashion on a common handle. The difference in height of the step is adjustable so as to cut thick or thin slices.

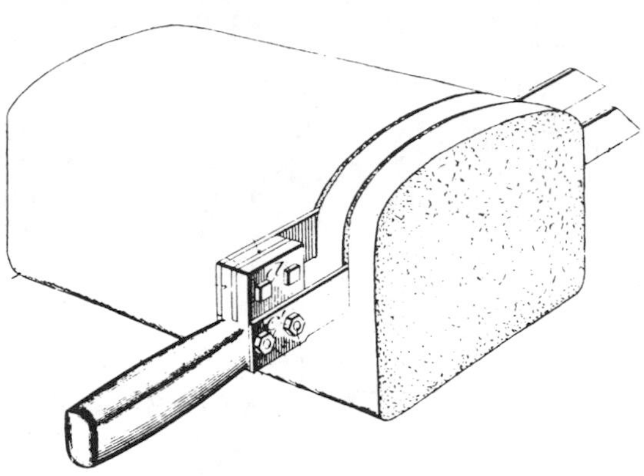

563/1904
John Calvin Bollinger *Labourer*
George Edward Bollinger *Labourer*
Olympia, Washington, USA

(Herbert Haddan & Co)

England's Improved Spoon has been devised as a result of a matter of common observation; that the ordinary spoon, when used for taking such substances as treacle or honey from a jar, becomes coated with sticky matter on the handle, unless great care is exercised in laying it down apart from the jar and in that case treacle or honey is deposited on some other article, the portion so deposited being usually wasted.

The object of this invention is to construct a spoon which shall be capable of resting on the jar without slipping or overbalancing and without the handle becoming soiled with its contents and which shall also effect a saving to the contents of the jar. Accordingly, this improved spoon is provided with a draining hole in the bowl, and a hook at the end to support it. Moreover, the bowl many be weighted to counter the effect of the spoon's having a long handle.

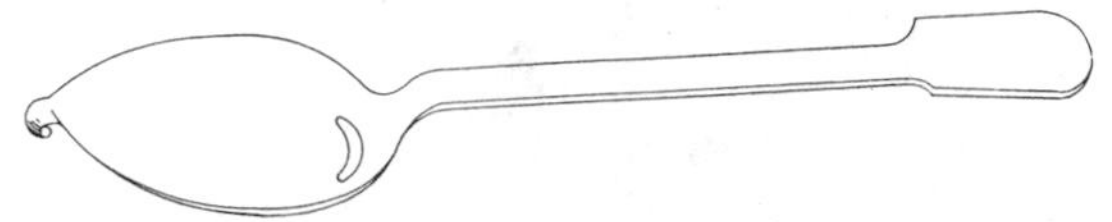

20718/1904
Annie Flora Catharine England *Married Woman*
26 Leith Mansions, Elgin Avenue, Maida Vale, Middlesex, England

(Page & Rowlingson)

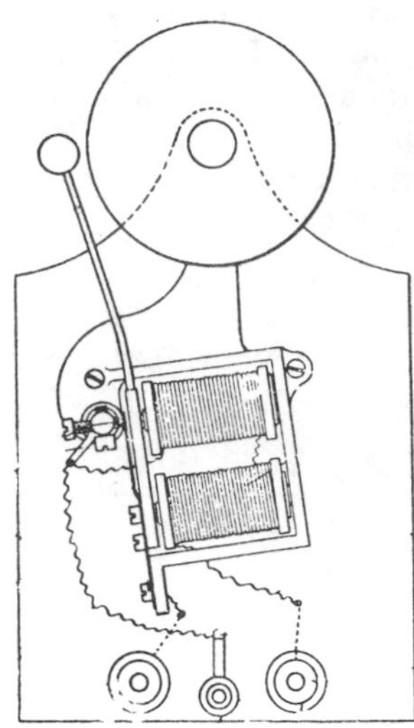

Page's Arrangement whereby Ordinary House Electric Bells can be used for Electro-medical Purposes employs an ordinary electric bell to provide a make-and-break for supplying an intermittent current to the patient from a bichromate or other cell. The circuit consists of a patient grasping two handles, one of which goes to the cell and the other to the bell armature. The other side of the cell is connected to the bell pillar via a third terminal.

A special switch is provided for switching on and off the current from the bell battery, and the bell may be silenced when in use by means of a suitable arrangement operated by a push knob, or by removing the gong.

8378/1904
Alfred Page *Engine Keeper*
112 Albert Buildings, Burnbank, Lanarkshire, Scotland

(H D Fitzpatrick)

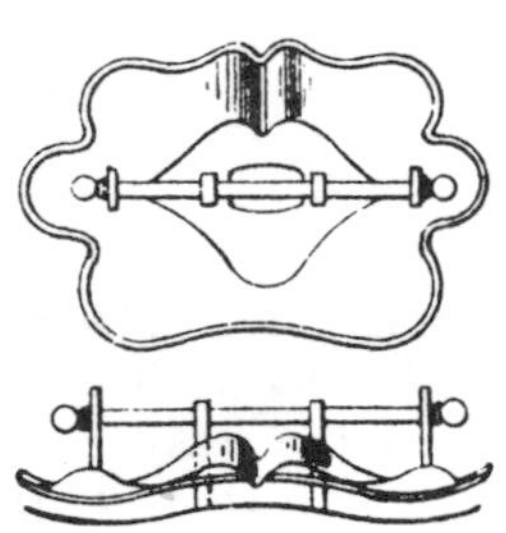

Leroux's Modifying Apparatus for the Mouth comprises two uprights or racks, an inner plate and an outer body-part, and a connecting bar for screwing the plates together. The different parts may be of celluloid. It is only necessary to pout out the lips forcibly whilst bringing them a little forward to introduce them very accurately into the apparatus and to fix the whole solidly by means of the bar, which may be used to tighten the apparatus to the degree required. The tooth at the upper edge of the apparatus is strongly prominent, to accentuate the natural curve of the upper lip, with the concavities and convexities very prominent in the proper places in order to give to the contour of the worst-shaped mouth the harmonious curves of a well-modelled mouth.

Two small notches are provided to allow room for the internal ligaments adhering to the gums of the upper and lower lips.

26012/1904
Veuve Anna Marie Leroux
26 Rue St-Georges, 26 Rennes, France

(Edwards & Co)

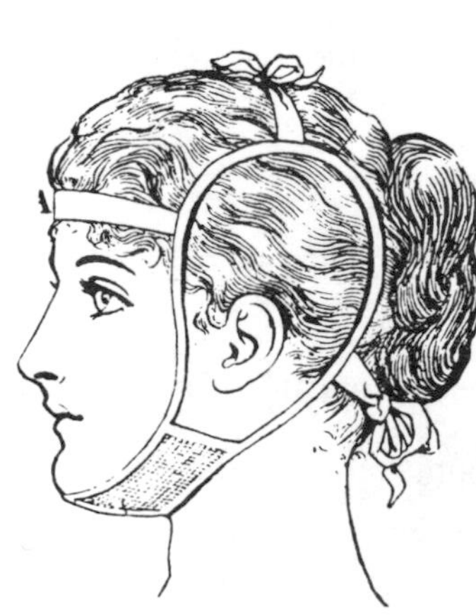

Turner's Appliance for Curing "Double Chin" is intended to be worn at night and may be fitted to the head with perfect comfort. It consists of a pliant frame carrying a piece of non-elastic ventilated material, the whole being tied about the forehead by ribbons.

19381/1904
Adelaide Sophia Turner *née* Claxton Ye Denne
28 Bath Road, Chiswick, Middlesex, England

Botermans's Stocking Supporter, preferably for the use of ladies, has a garter placed over the bare leg below the knee and carrying two spring supporting forks pointing upwards. The prongs of the forks carry means for securing the stocking-top.

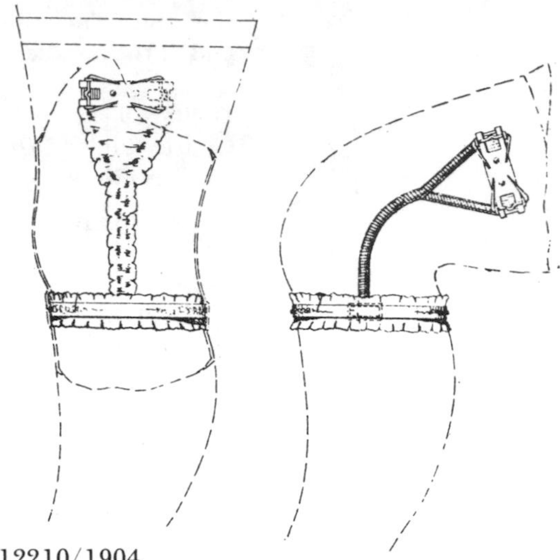

12210/1904
Abraham Botermans *Manufacturer*
97 Oranjeplein, 'S-Gravenhage, Holland

(Gerson & Sachse)

Hogben's Improved Game or Apparatus for Providing Amusement is known as "croquet-billiards" and is essentially a modification of the game of billiards, so adapted that it can be played on any level surface such as a floor or lawn and so that croquet mallets can be used to play the balls instead of cues. Unlike billiards the apparatus for playing the game is inexpensive and can be easily stored.

It consists of lengths of springy material stretched between vertical supports driven into the ground which act like the cushions of a billiard table when struck by croquet balls. The mallets are modified so that the handles may be used to strike the balls when the latter are against the cushions.

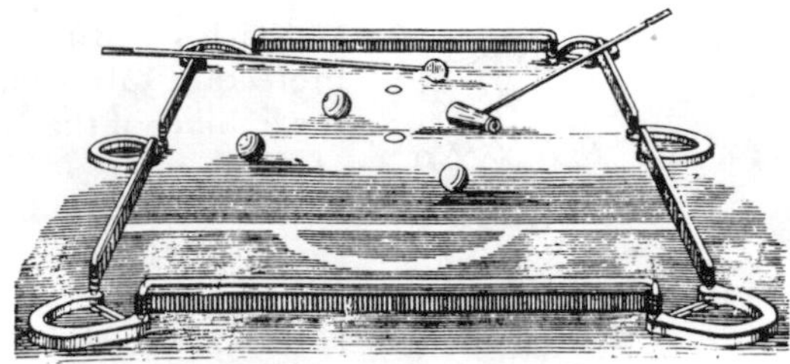

25223/1904
Walter James Hogben *Tobacconist*
35 Broad Street, Canterbury, Kent, England

(Stanley Popplewell & Co)

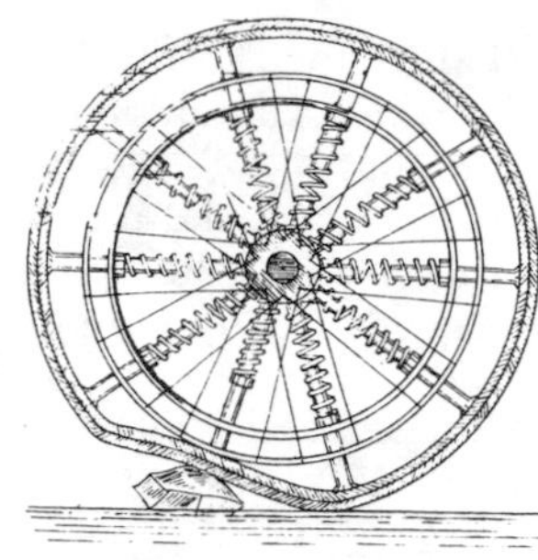

Wood's Arrangement of the Structure of Wheels to Absorb Vibration and Supersede Pneumatic Tyres for Motor Cars, Motor Cyles, Ordinary Cycles and all other Vehicles has a leather tyre supported on spring-loaded spokes.

On a practically smooth road there is not much movement of the spokes, the strength of the springs being adjusted to the load, but any sudden jerk from impact with the road drives in the spoke and spring, so absorbing the shock before it reaches the vehicle. In the remote case of a spring breaking, a new one can be fitted in almost immediately and without taking anything to pieces.

20993/1904
Henry Hopkinson Wood *Insurance Clerk*
27 Queen Anne's Road, York, England

Allen's Improved Shaving Apparatus is designed to be simple, inexpensive, safe, efficient, and automatic in its operation. It moves from chair to chair in a barber-shop, and consists essentially in a rotary razor and a fan to remove the lather etc.

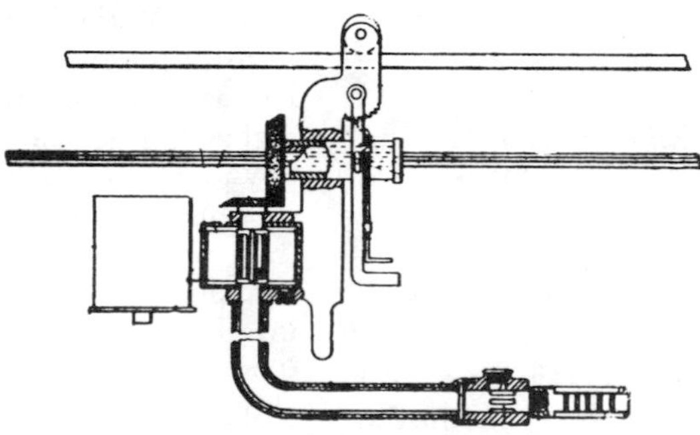

23207/1904
William Robert Allen *Inventor*
Hoquiam, Washington, USA

(P R J Willis)

Claypole's Improved Folding or Retractile Seat is especially designed for parks, tramcars or other outdoor purposes and has a hollow back into which the slatted seat is retracted by means of a spring. As long as the occupant sits thereon the apparatus remains in position, but as soon as he rises from the seat the spring causes it to retract into its cavity, thus protecting it against the weather.

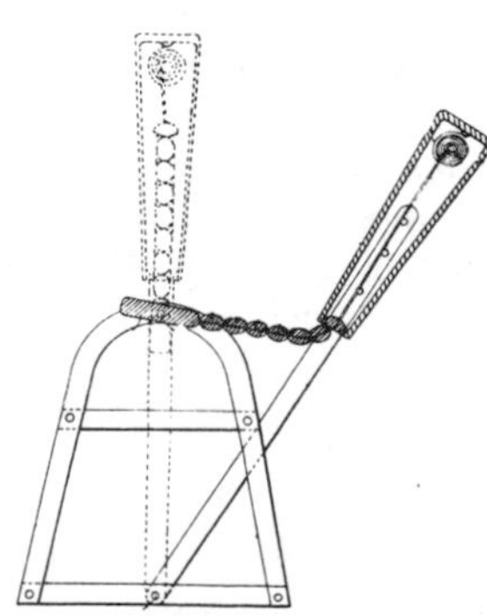

25906/1904
Caroline Lucy Claypole
t/a The Armstrong Automatic Manufacturing Company
38 Wilson Street, Finsbury, London EC, England

(W P Thompson & Co)

Hülsmeyer's Improved Hertzian-wave Projecting and Receiving Apparatus to Give Warning of the Presence of Metallic Bodies provides means whereby the distance of the reflecting object may be calculated. When the direction of the reflecting object has been ascertained, the projector is tilted in a vertical plane until the reflected Hertzian effect is a maximum; from the angle the distance may be calculated.

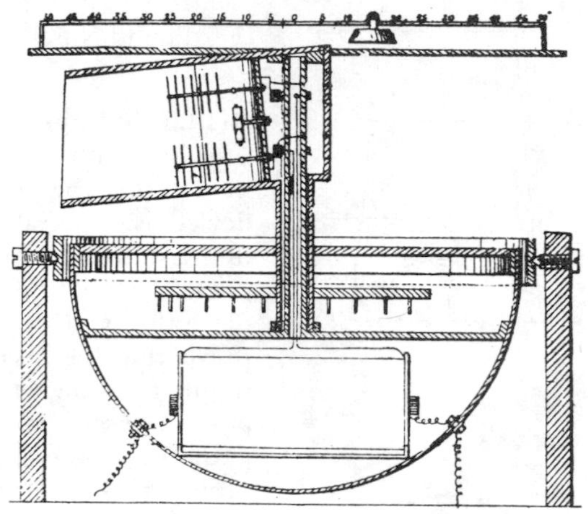

25608/1904
Christian Hülsmeyer *Engineer*
3 Grabenstrasse, Düsseldorf, Germany

(Charles Bauer, Imrie & Co)

Rosenfeld's Improved Illusion Apparatus does away with the risky false bottom and cover arrangement earlier described (19063/1904), and the essence of the invention now consists in providing a hiding place on the side of the vessel away from the spectators for the person who is to enigmatically appear. In the performance of the illusion a cloth is held before the vessel, behind which cloth the person emerges from concealment and mounts quickly over the edge of and into the glass reservoir. The cloth must be held high enough to hide the person's movements but at the bottom it need not reach below the glass vessel.

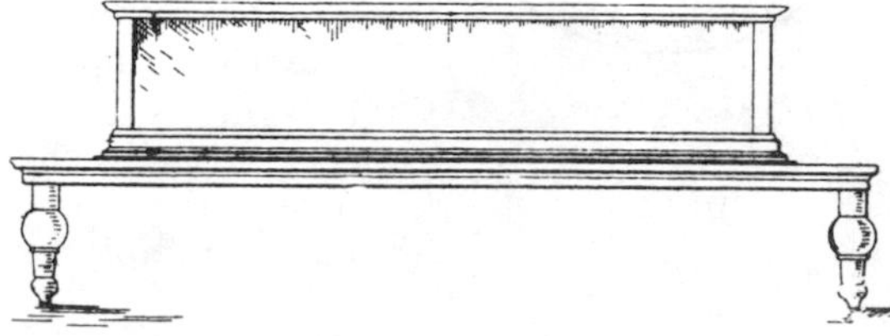

20629/1904
Carl Rosenfeld *Director*
52 Behrenstrasse, Berlin, Germany

(Marks & Clerk)

1904

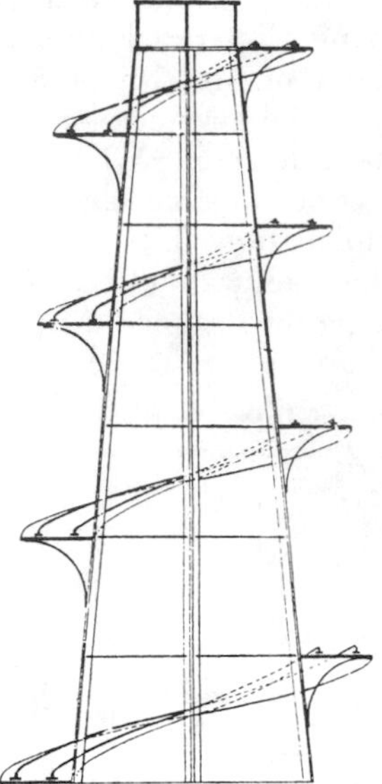

Mortet's Improved Luminous Amusement Apparatus consists in an electrically illuminated ball containing a seat for a person performing different amusements. It travels along a helical road formed by rails fixed around a tower, revolving as it runs.

25104/1904
Marie Mortet *née* Ripert *Authoress*
13 Rue de Poissy, Paris, France

(Ferdinand Nusch)

Brittain's Improved Sanitary Dining-plate is provided with hemispherical receptacles round its rim for the reception of condiments. The device is especially applicable for service at dining-tables in public places as a means for providing each diner with an individual supply of salt and pepper, thereby effectually preventing the insanitary custom of allowing a number of diners to use a single receptacle. It is also very useful for service at receptions or other gatherings where plates of food are passed around to diners located away from the source of food supply.

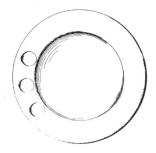

29157/1904
Frank Morris Brittain *Mining Engineer*
Cleghorn, Anerley Road, Croydon, Surrey, England

(Wheatley & Mackenzie)

Fillatre's Improved Nose Bag is fitted with internal springs so that as the contents are consumed by the horse, the remains are raised to the level of its mouth so that it is able to empty the bag completely without the inconvenience of being obliged to shake its head and catch the food flying upward.

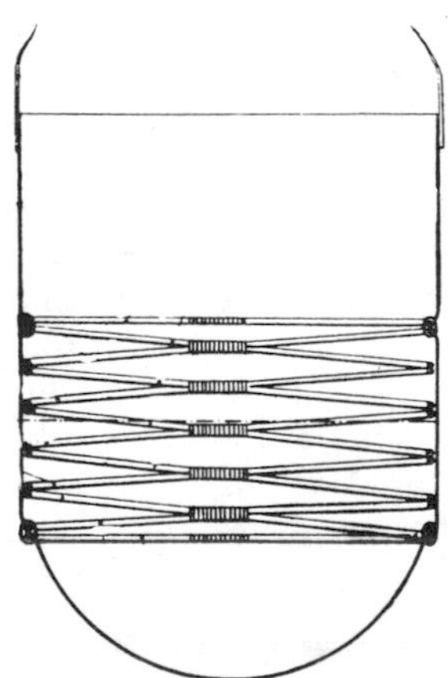

28298/1904
Paul François Ladislas Fillatre *Agent*
75 Rue Saint Sauveur, Paris, France

(F Wise Howorth)

Squier's Improved System of Wireless Telegraphy obviates the use of an aërial by using a tree. The leaves of the tree serve to increase the capacity of the system, and the roots make a suitable "earth".

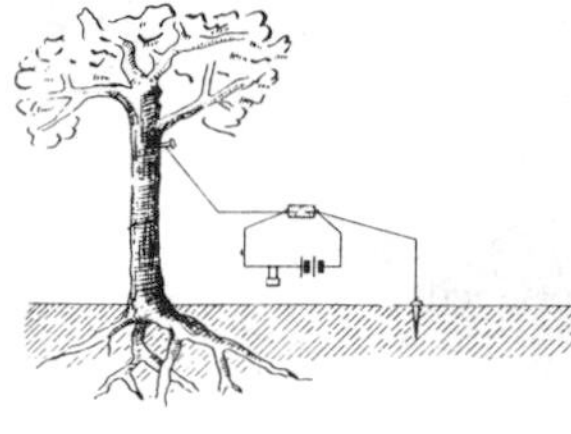

25610/1904
George Owen Squier *Officer, Signal Corps, US Army*
San Francisco, California, USA

(F W Howorth)

Mitchem's Improved Table for Displaying Goods is designed to obviate the accumulation of dust and attendant damage of goods. Accordingly, instead of its having a solid top, it consists in a normal table frame with wires passed across on which the goods are supported.

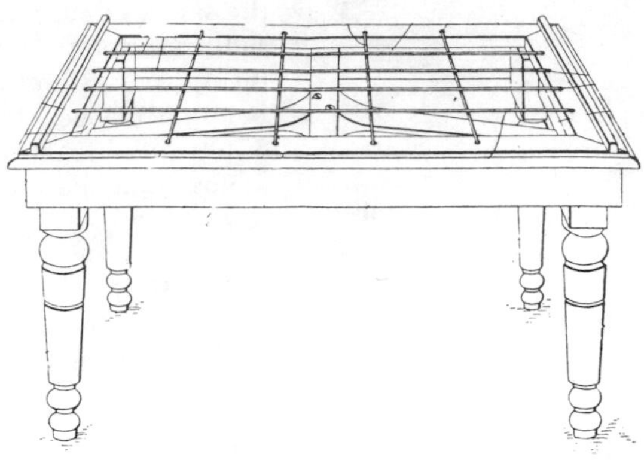

20958/1904
Charles Edwin Mitchem *Merchant*
Harvard, Illinois, USA

Pugh's Method of Preventing and Curing Consumption and other Kindred Wasting Diseases follows the lines laid down by the highest medical and most competent physiological authorities: the best known treatment is the inhalation of pure dry air and the employment of medicines has not been, and cannot be, efficacious. All scientific investigations point to the certainty that in the strata of the atmosphere some thousands of feet above the earth are to be found the ideal conditions for the treatment of such complaints. Recent experiments by leading scientists have demonstrated that patients who have been taken up in balloons to a height of eight thousand feet have shown astonishing improvements in their health, and it has also been proved that rarefied air at even six thousand feet has a totally different effect to mountain air of much higher altitudes. This apparatus accordingly provides a means of providing such pure, life-giving air, free from noxious microbes, from the upper atmosphere to supply the wants of sufferers on the ground. It comes through a tube supported by hydrogen balloons of aluminium at quarter-mile intervals, the air tube being of tempered aluminium sections, corrugated and soldered together and terminating in a lofty-ceilinged building or chamber in which the patients are to be treated. One or two thousand patients at a time may be treated in this way.

26500/1904
John Pugh
335A George Street, Sydney, New South Wales

Peterseim's Improved Device for Growing Seeds or Seedlings is made in the form of a porous hollow figure which is filled with water to nourish the seeds or seedlings. These are planted in appropriate grooves on the surface and when they grow, provide a lively effect.

23296/1904
Fritz Peterseim *Florist*
5 Dalbergsweg, Erfurt, Germany

(J E Evans-Jackson & Co)

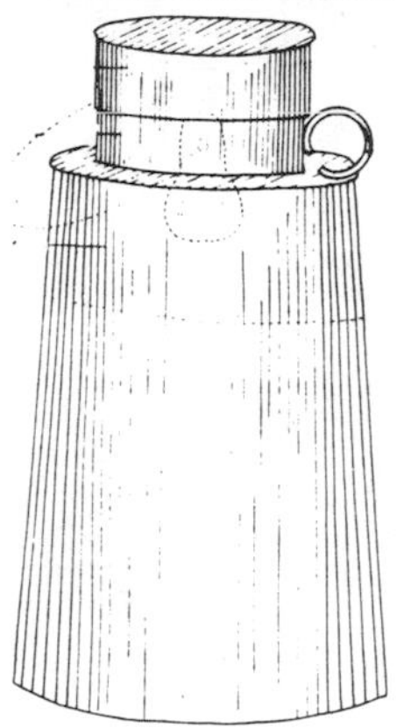

Grice's Improved Tea Can enables the milk for the tea to be carried in a separate receptacle mounted on the lid of the workman's breakfast can, so that it can be added to the tea after brewing.

28011/1904
Eliza Grice *Farmer's Wife*
Waste Farm, Kingswood, nr Frodsham, Cheshire, England

(P: Cassel & Co; C: Self)

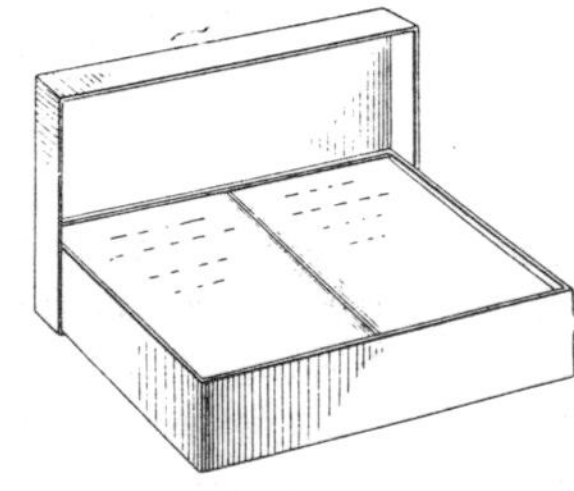

Tremaine's Improved Coffee Package is inexpensive, simple, strong, and convenient in use. It consists in a closed bag containing a measured amount of coffee which may be unwrapped and placed directly into water to infuse.

24151/1904
William Burton Tremaine *Vice President of the Aeolian Weber Piano & Pianola Company*
362 Fifth Avenue, New York, USA

(C Barnard Burdon)

Mausshardt's Outlet Swinging and Fixed Escape Galleries and Roller-floors, for Theatres, Public Halls, and the like is a constructional apparatus for removing the audience in the case of fire or other danger, a large part of the operative momentum employed being supplied by the weight of the persons thus escaping. Recent terrible fires at theatres with their consequent dreadful loss of life conclusively prove that no form of construction hitherto known satisfactorily provides for the safe and prompt emptying of a theatre in case of panic. The present invention remedies this regettable defect.

Supposing a panic to have broken out, the sliding doors are opened. The public, fleeing from the danger, pass through these doors into the galleries, which begin to swing away downward under the weight. The motion thus derived causes the pit to be moved bodily from the theatre, by means of a rack and pinion mechanism.

Fixed galleries are provided to protect those members of the public who may still continue to rush through the sliding doors from the heat of the fire, and to prevent them from being precipitated into the open space.

Brakes, such as dashpots, are provided to prevent too rapid a descent of the galleries, and they are furnished with cushioning devices such as springs.

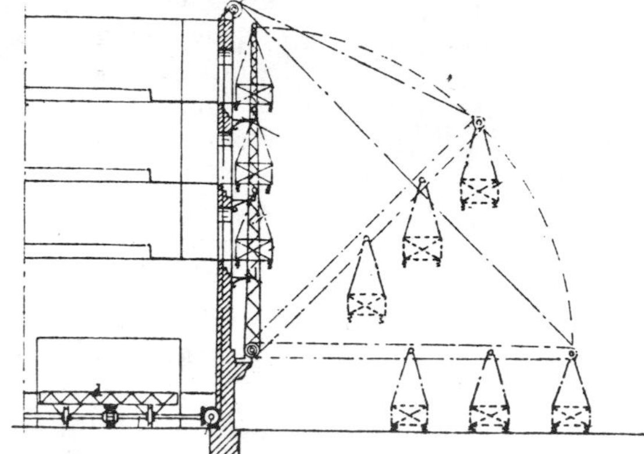

16357/1904
Michael Mausshardt *Brush Manufacturer*
Billigheim, Germany

(Marks & Clerk)

Noonan's Power Generating Gravity Action Motor makes use of falling weights to turn a flywheel which raises other weights to fall in their turn, the whole providing a means for sustaining energy undiminished.

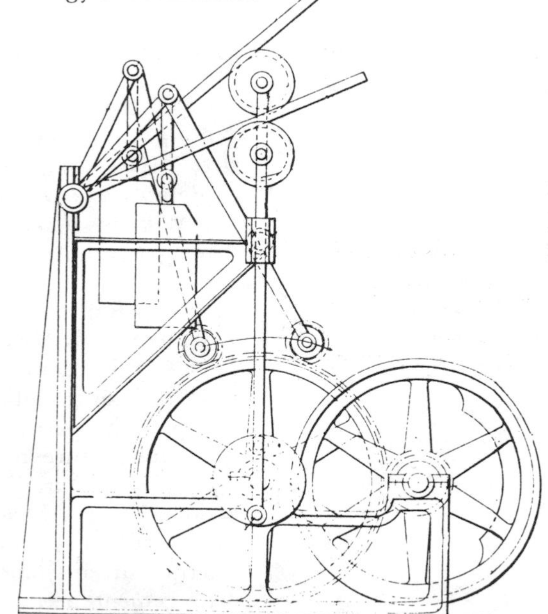

26163/1904
Henry John Noonan *Upholsterer*
Glen Holme, 54 Worting Road, Basingstoke, Hampshire, England

(Benjamin T King)

1904

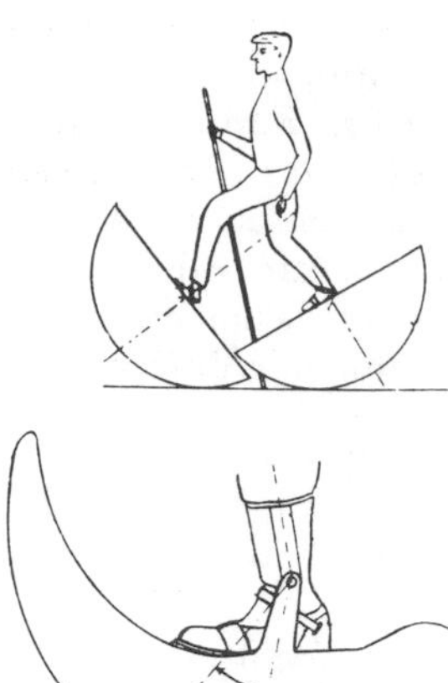

Michael's Apparatus for Increasing the Length of Stride in Walking is intended to serve for quick forward movement and, if necessary, enables obstacles to be surmounted without its being necessary for the user to leave the apparatus. It is also to be used in places where the cycle fails, on sandy or rough ground, as the flat curvature of the runners in combination with a sufficient breadth prevents excessive sinking.

The simple version of the apparatus is attached directly to the foot, but the rolling may be increased by the use of cranks, hand-levers etc.

27253/1904
Robert Michael *Engineer*
22 Coburgerstrasse, Leipzig-Raschwitz, Germany

(W H Beck)

Brown and Williams's Combination Fork and Rake provides a single implement which by means of a simple adjustment may be made to serve as either one or the other. It is especially useful for farmers, gardeners and stablemen and can be used for first raking hay or straw and then as a pitchfork for throwing up wheat or hay heaps.

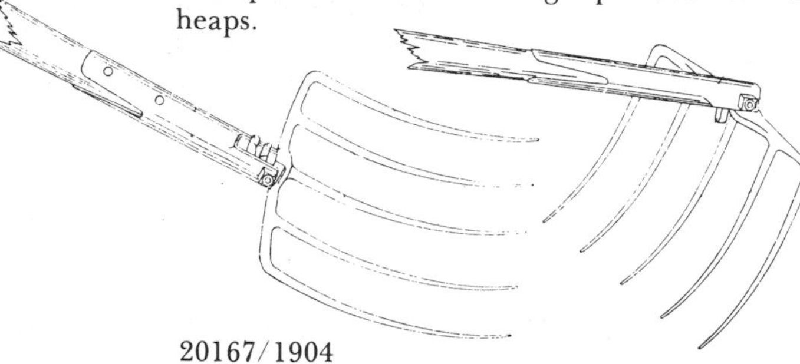

20167/1904
Frederick Collard Brown *Farmer*
Carl Williams *Blacksmith*
Penong, South Australia

Smith's Improved Bath for Therapeutic Treatment by Means of Electric and other Forms of Radiant Energy comprises a couch, curtains, and adjustable lifting reflecting canopy. The patient reclines on netting which spans an aperture in the couch, which is made of insulating material, and is subjected to the action of X-rays, heat and light rays, Finsen light rays, radium emanations or high-frequency electric currents, all under perfect control; the arrangement permits the intensity with which the various forms of radiant energy are permitted to act upon the bather to be varied with great nicety from nil to a maximum, whilst he remains uninterruptedly enclosed within a warm and luminous air space. Consequently any local applications of radiant energy can be carried on whilst the patient is rapidly throwing off the products of tissue metabolism in the heated atmosphere whereby he is surrounded. It is beneficial in the treatment of neoplasms and other disorders. In order to provide against the danger of the reflecting canopy, when raised, falling upon the bather in consequence of the suspension cords being broken, an additional arrangement of "emergency" wire cords is provided.

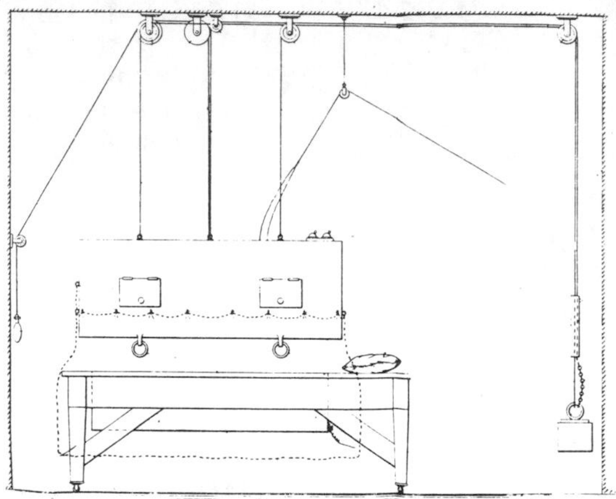

20536/1904
William Johnson Smith *Doctor of Medicine*
The Bournemouth Hydropathic, Bournemouth, Dorset, England

(A M & WM Clark)

Maxim's Improved Device for Producing Illusory Effects is a rotatable spherical structure with a paraboloidal floor and suitably hung mirrors. Outside, it may be painted to represent the globe. Admittance is gained by fixed stairs and platforms; the whole structure may be made to revolve and, because the occupants cannot see out, some very curious effects may be obtained; when persons enter the hollow sphere they will not be able to tell whether it is revolving or standing still and by reason of the parabolic floor, persons near the outer edge would, to the persons standing near the centre, appear to be walking with their heads directed inward. The throwing of a ball from the centre outward and *vice versa* will move in an unexpected direction that will be very puzzling; a person mounted on roller skates will be able to perform some very complicated and amazing evolutions.

The interior surface of the sphere might be decorated with pictures or signs of the zodiac. The mystery of the contrivance might be enhanced by keeping from visitors the knowledge that the sphere revolved and arranging for them to enter while it was stationary and then setting it in motion and, by some pretended magical influence, producing the phenomena and illusions which would cause the parabolic floor to appear level to everyone standing on it and half the people to appear standing on their heads.

A visit to this contrivance might be termed "A voyage to the interior of the earth".

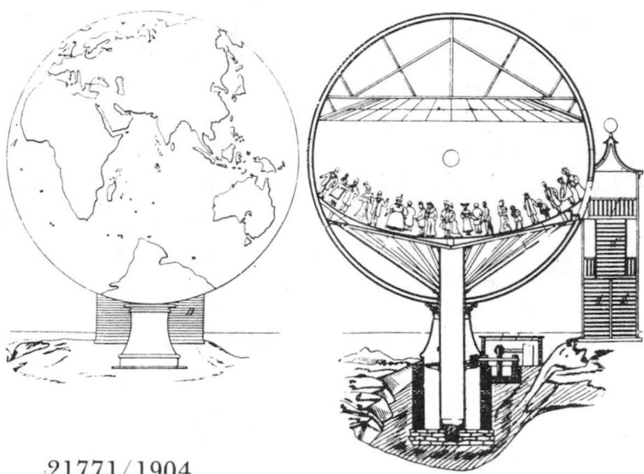

21771/1904
Sir Hiram Stevens Maxim *Civil, Mechanical & Electrical Engineer*
Thurlow Lodge, West Norwood, Surrey
England

(Haseltine, Lake & Co)

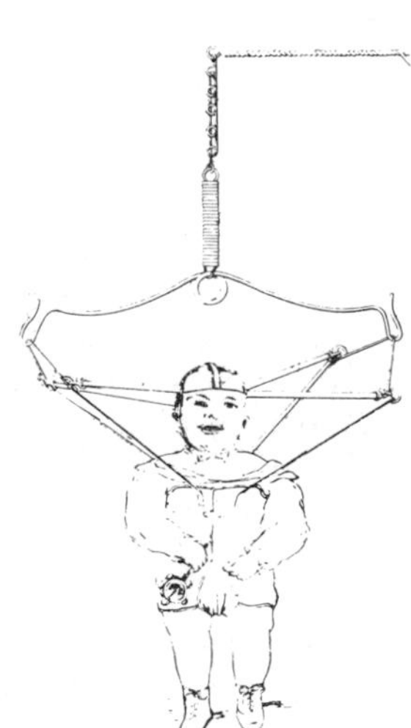

Latshaw's Improved Baby-exerciser supports the infant elastically so that it may use its legs freely in swinging or jumping, thus combining the benefits of exercise and amusement without requiring close watching or attendance by the mother or nurse. It is easily adjusted to allow for the child's growth and, furthermore, provides support for the head of a child having a weakened spine.

25981/1904
Charles Edward Latshaw *Commercial Traveller*
Lincoln, Nebraska, USA

(J G Lorrain)

Fischer's Improved Motor Car has been designed since ordinary motor cars cannot pass through doors of normal width or go up staircases. Special premises, on the ground floor of dwelling houses, are therefore required for the purpose of storing them when not in actual use. Two-wheeled motor cars, otherwise known as motor bicycles, have the advantage that they can pass through doors of ordinary width and up staircases, but their use is not unattended with danger.

Now, this invention has for its object to provide a four-wheeled motor car which shall be able to pass through doors of ordinary width and up staircases with such ease that even persons residing on the upper floors of ordinary dwelling houses will be able to keep such cars without the necessity of providing special storage space on the ground floor.

This is achieved by providing a frame of very small width and very small wheels; to increase stability the driver's seat slopes and is at a height of not more than 45cm from the ground.

A motor car constructed with a narrow frame according to this invention may be adapted to seat more than one person by placing seats behind one another.

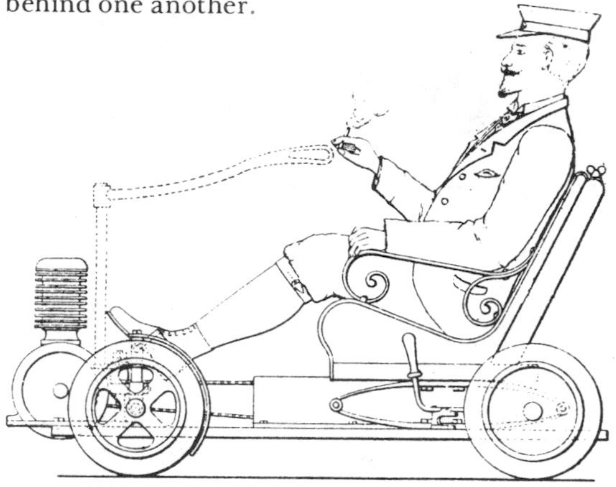

21201/1904
Martin Fischer *Engineer*
60 Gloriastrasse, Zürich V, Switzerland

(F Wise Howorth)

Fergusson's Improved Hat is particularly adapted for automobiling and touring and can be manipulated to protect the head and hair and also the face of the wearer, or arranged to present a highly pleasing and stylish appearance. Unlike the veils and other unbecoming headgear normally used, this combination hat is not only capable of being arranged to effectually protect the hair of the wearer from dust and dirt and the face from the cutting effects of the wind, but can also be arranged so as to lose all appearance as an automobile hat and present a stylish and ornamental appearance so that the wearer could attend a social function while *en route* on an automobile trip without changing her hat.

27898/1904
Constance Tucker Fergusson *Inventor*
420 West 118th Street, Manhattan, New York, USA

(J E Evans-Jackson & Co)

1904

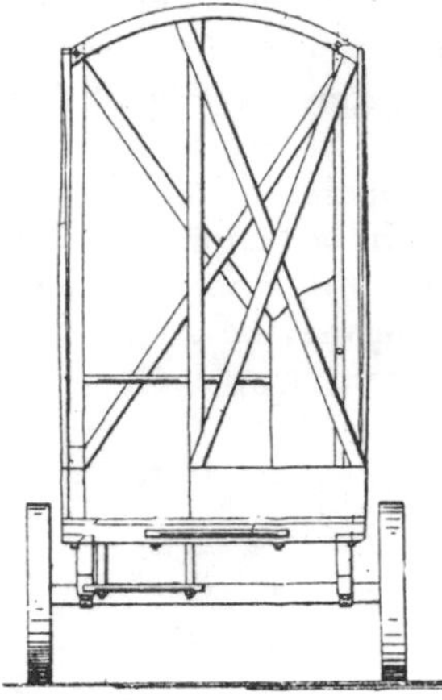

Shanly's Improved Sanitary Bathing Machine has a wooden framework on wheels covered with canvas, and may be fitted with toilet arrangements for use with fresh water. The use of canvas for sides and roof gives improved ventilation, and owing to the extreme lightness of construction the machine can be easily drawn down the beach or pushed back by the occupants by the handle.

It can also be easily taken to pieces and stored away when not required for use.

22865/1904
Michael William Shanly *Refreshment & Garden Chair Caterer*
33 King Henry's Road, South Hampstead, London NW, England

(Browne & Co)

Seligstein's Improved Self-opening and Closing Umbrella makes use of fluid or gas pressure to open it and a spring to close it when the pressure of the fluid admitted to the piston from the reservoir is released.

The holder for the fluid or gas should be large enough to contain sufficient to enable the umbrella to be opened repeatedly before it is exhausted.

23625/1904
Abraham Seligstein *Merchant*
6 Rückerstrasse, Munich, Germany

(W P Thompson & Co)

The Whirlpool Amusement Company's Improved Recreational Railway is designed to give the occupants of the cars a sensation similar to that experienced by those of a ship caught by a whirlpool. An enclosed spiral track is made to run round inside a building and the effect of being in a whirlpool is heightened by pumping water along the track and by suitable decoration. At the bottom of the whirlpool the cars pass through a tunnel, whose walls depict corals, rocks and other objects found at the bottom of the sea. The surroundings, accessories and scenery are all arranged so as to create the real impression of being sucked into a natural whirlpool. It might appear that this would be anything but pleasing, but the passengers are conscious that all danger is eliminated and therefore greatly enjoy the trip.

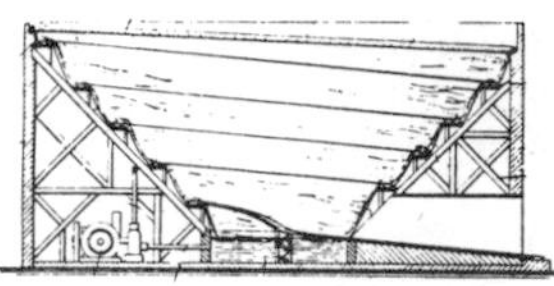

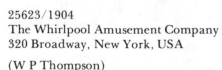

25623/1904
The Whirlpool Amusement Company
320 Broadway, New York, USA

(W P Thompson)

Rosenfeld's Illusion Apparatus for Exhibition Purposes enables the illusionist to effect the enigmatical appearance of a person inside a glass tank, the interior of which is visible from all sides and which preferably contains water.

The illusion is effected by the person lying concealed on an intermediate or false bottom underneath the vessel. After the release of its catches and powerful springs it rises to the height of the base of the vessel. The illusion is completed by the provision of a cover adapted to be retained under tension over the concealed person so that when released it springs aside to permit the person to rise suddenly into view and is now able to rise from the water. A rubber tube passes through the hollow part of the vessel, fitted with a mouthpiece, for breathing.

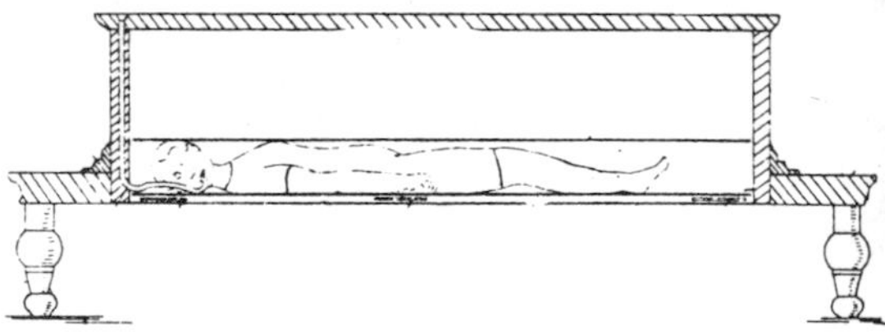

19063/1904
Carl Rosenfeld *Director*
52 Behrenstrasse, Berlin, Germany

(Marks & Clerk)

Winchcombe's Improved Cup, Saucer, or similar Vessel has printed on it divisions containing signs or words by which characters or fortunes may be told, to serve as a pastime at the tea table and in other situations, whereby a vast amount of amusement will be afforded.

The tea leaves or other sediment remaining in the vessel are agitated so that they settle in an accidental and variable manner within the various spaces and the signs thus indicated may be read off in any fanciful manner portending good and ill luck, various functions such as weddings and christenings, good and ill news, money, love and the like.

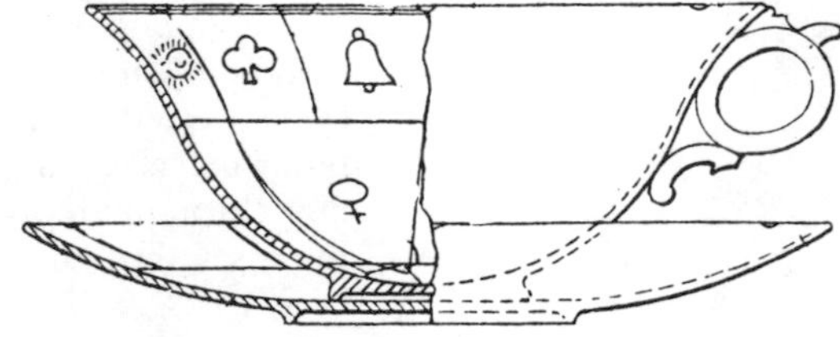

23607/1904
Ellen Harriette Winchcombe *Toilet Specialist t/a* Mrs Neville Ross, 29 Margaretta Terrace, Oakley Street, Chelsea, London, England

(White & Woodington)

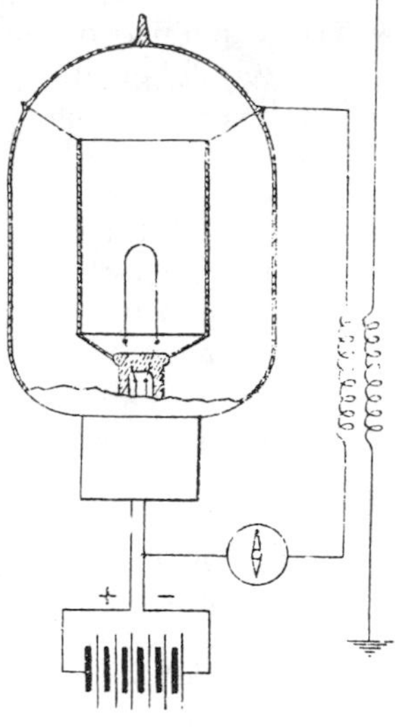

Fleming's Improved Instrument for Detecting and Measuring Alternating Electric Currents consists in a highly incandescent conductor and a cold conductor enclosed in a highly exhausted glass vessel. Negative electricity can pass *from* the hot conductor to the cold one, but not *vice versa,* through the vacuum.

24850/1904
John Ambrose Fleming *Doctor of Science*
University College, Gower Street, London, England

Wynn's Improved Ventilated Water Closet has as its object the avoidance of any smell in connection with the use of water closets by ensuring thorough and direct ventilation of and through the pan or basin. Accordingly, the seat is spaced from the rim of the pan and a current of air is induced by a gas burner, electric fan etc. To avoid any draught about the person of the user, a depending rim or flange is provided to deflect and turn the incoming current of air downwards towards the bottom of the basin.

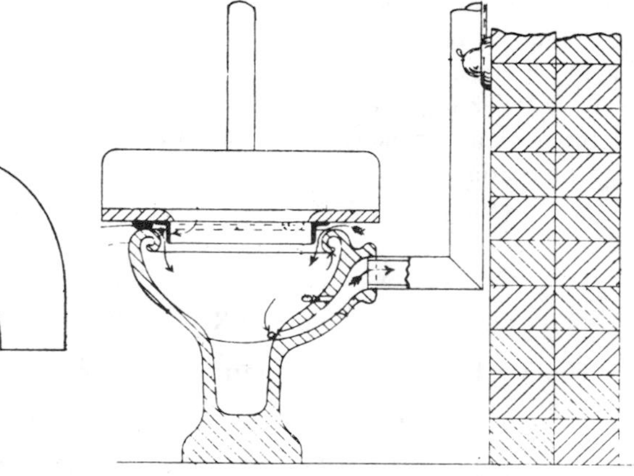

27817/1904
Arthur Ernest Wynn *Engineer*
Rathgowrie, Hereford Road, Harrogate, Yorkshire, England

(J Clark Jefferson)

Hoch and Hannemann's Improved School Desk prevents dust rising in school rooms during lessons as well as when cleaning. It is provided with a slatted footboard with a cloth, saturated in disinfectant, beneath. Dirt from scholars' feet falls on to the cloth which may be wound on a roller and removed.

24944/1904
August Hoch *School-Director*
Carl Hannemann *Master-Joiner*
Schloppe, Prussia

(F Bosshardt & Co)

Crosse's Improved Cannon for Acrobatic Performances is for use in music halls for projecting a performer or an inanimate object through the air.

The performer projected may be caught by means of a trapeze, net or mattress, or by a second gymnast.

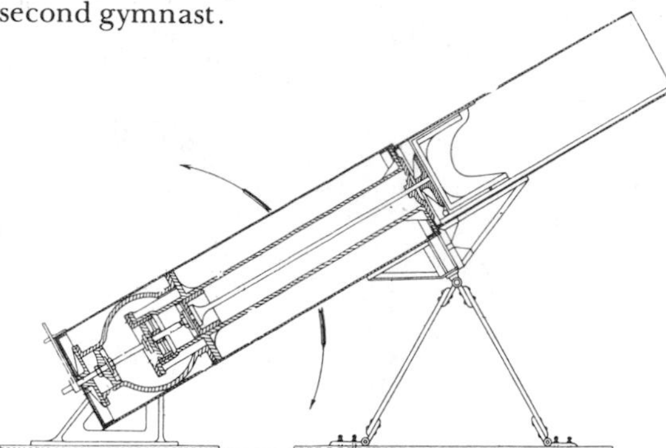

24034/1904
Alice Josephine Crosse
81 Avenue Parmentier, Paris, France

(Charles Bauer, Imrie & Co)

Rosen's Improved Construction for Theatres, Hospitals and similar Buildings provides inclined spiral gangways instead of staircases so that crushes are prevented on alarm of fire. By this construction not only the dangers connected with stairs are overcome, but the gangways offer the further advantage that one is no longer obliged to undergo the tiresome climbing of steps, which is particularly valuable in hospitals.

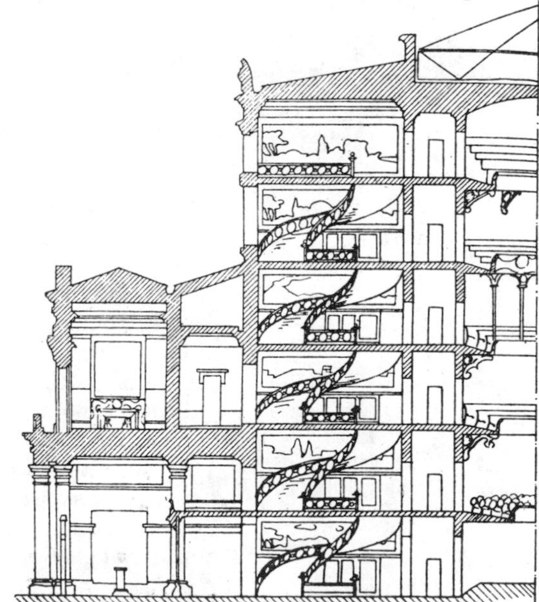

21280/1904
Michel Rosen *Manufacturer*
1 Rue Lemercier, Paris XVII, France

(Ferdinand Nusch)

1905

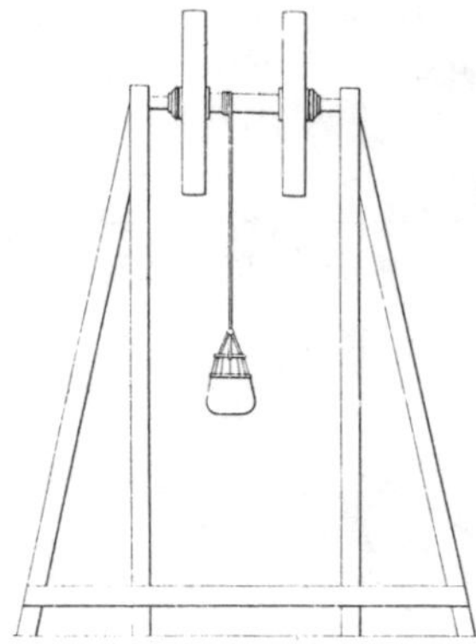

Wood's Apparatus for Recreative Purposes provides pleasure of a novel and exhilarating character. It consists in a car attached to a rope, whose other end is attached to a revolvable bar furnished with flywheels and mounted on a support frame.

The passengers enter the receptacle, which is wound to its uppermost position and then liberated. When the rope is completely unwound, the momentum of the flywheels causes the receptacle to be carried nearly to the top, and it oscillates vertically until the momentum is exhausted.

To provide against mishap, netting or canvas is slung underneath the air-chute.

The apparatus may also be constructed as a toy.

70/1905
Ebenezer Wood *Patentee*
39 Catford Hill, Catford, Kent, England

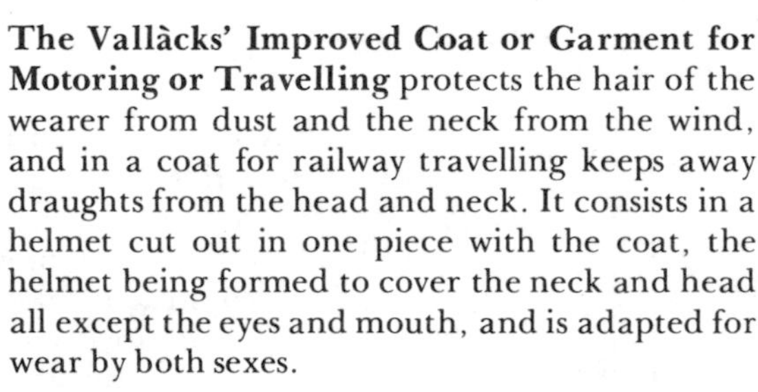

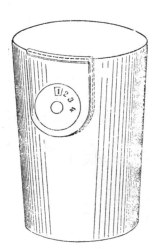

The Vallàcks' Improved Coat or Garment for Motoring or Travelling protects the hair of the wearer from dust and the neck from the wind, and in a coat for railway travelling keeps away draughts from the head and neck. It consists in a helmet cut out in one piece with the coat, the helmet being formed to cover the neck and head all except the eyes and mouth, and is adapted for wear by both sexes.

820/1905
Eliza Jane Vallàck
William Edward Vallàck *Costumiers*
21 Stamford New Road, Altrincham, Cheshire, England

(H B Barlow & Gillett)

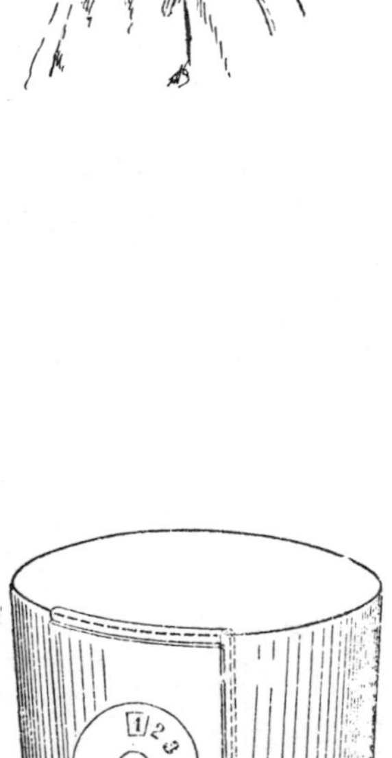

Wippermann's Improved Mouth-protector and Registering Device applicable to Drinking Vessels relates to that class of protectors for the mouth against infectious diseases conveyed by drinking out of glasses of any kind, especially in public places, and its main object is to combine the same with a device for registering and indicating the number of glasses of liquid served to, or consumed by, the drinker in question. This latter aim is achieved by providing it with a rotatable disc with an opening which may be brought over a numeral corresponding to the number of glasses consumed.

870/1905
Paul Georg Wippermann *Associate Engineer*
Aschen, Kingdom of Prussia, Empire of Germany

(Haseltine, Lake & Co)

Taylor's Improved Golf Tee is an object of the portable type that will be readily followed, by the eyes of those interested in the game, when in flight, and when it has to come to rest; but which when in use as a tee or support for the ball, will not present any feature likely to distract the attention of the player from his stroke or interfere with his distinct view of the ball.

It consists in a hollow dome-shaped body of gutta-percha or other suitable composition, provided with a recess or aperture on which the ball rests during the aim.

Attached to the hollow dome body is an attachment designed to betray the flight and position of the tee should it be struck and carried away. This attachment consists in a cord, or strands and tassell, which for greater conspicuousness may be of red or other distinctive colour or appearance.

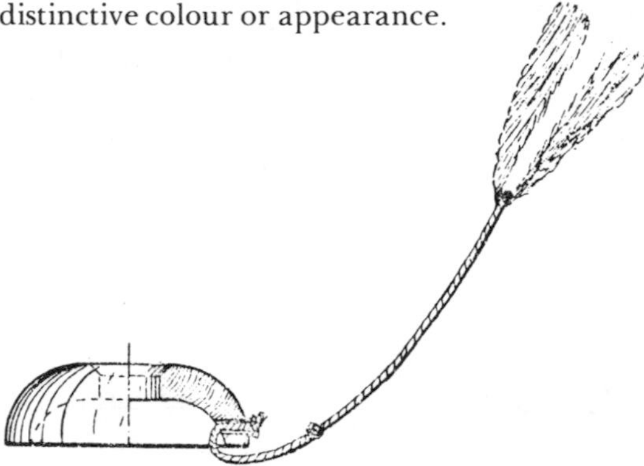

826/1905
Edward Herbert Taylor *Clerk in Holy Orders, Rector of Plaxtol*
Plaxtol Rectory, Sevenoaks, Kent, England

(William Brookes & Son)

Weiss's Dental Syringe for the Introduction of Cocoa, Butter and like Substances into cavities in the teeth and other parts of the body is intended for injecting these and similar solid or half solid waxy substances which are hard in ordinary temperature but melt in blood heat. The bougie substance is placed in the reservoir of the syringe and a piston forces the mass through the very thin nozzle into the cavern of the tooth in a very fine form, premature melting having been avoided. The end of the nozzle is roughened for the reception of cotton wool.

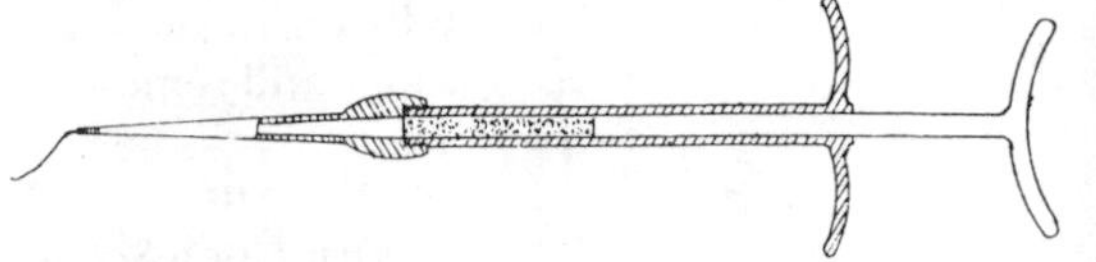

10994/1905
Richard Weiss *Wholesale Chemist & Import Agent*
27A St Mary-at-Hill, London, England

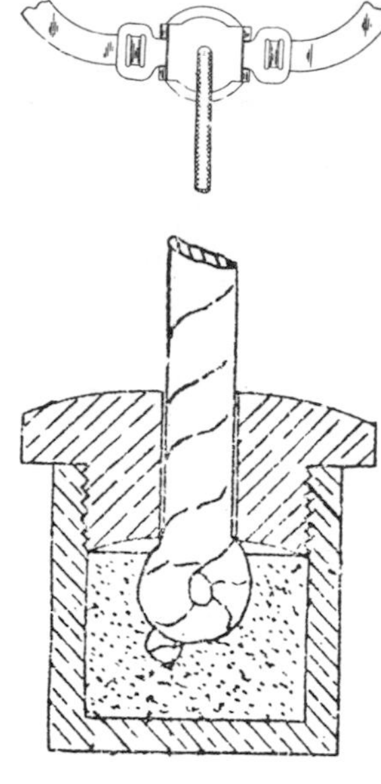

Richmond's Therapeutic Apparatus is a thermo-electric appliance devised for the purpose of overcoming certain local or general morbid conditions in the body of a patient. It consists in two metallic bodies connected by a conductor. One is a shell containing metal wires, powdered plumbago, sulphur or the like, and adapted to be heated or cooled; the other is in contact with the part of the body to be treated. The system produces such a degree of attraction as to cause the blood of a patient to expel any positive diamagnetic gases such as hydrogen that may be in the body. This action on the body of the patient will rid the blood of any diamagnetic poisons in gaseous form by attracting them to the magnetic cell, thus leaving him invigorated and restored.

426/1905
Arthur Percy Richmond *Commercial Traveller*
"Kelvin", Enmore Road, Marrickville, nr Sydney, New South Wales, Australia

(Wheatley & Mackenzie)

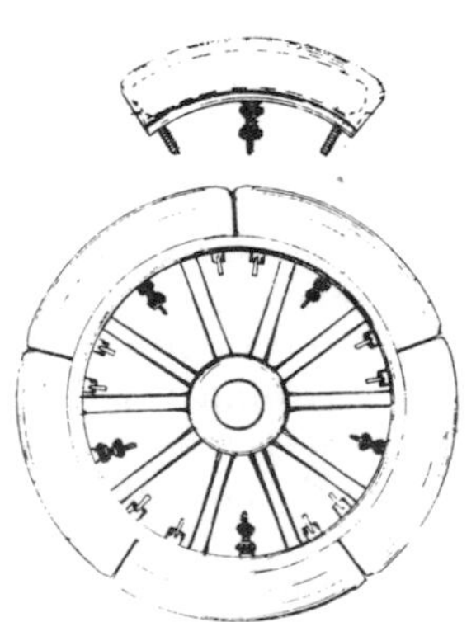

Hammond, Mason and Brown's Improved Pneumatic Tyre is formed in sections which may be bolted to the rim of the wheel, and whose ends abut on inflation. It is designed to be readily repaired without any necessity for jacking up the vehicle, and also any part of the tyre can be renewed as required, in an expeditious and economical manner.

1608/1905
James Hill Hammond *Manager*
51 Arundel Street, Leicester, England

Charles Tookey Mason &
Sam Randall Brown *Electricians*
Volta Works, Rutland Street, Leicester, England

(P: W P Thompson & Co; C: W Bestwick Maxfield)

Conze's Improved Sound Reproducing Apparatus is designed to eliminate the distinct whirring and grating noises which very prejudicially and disagreeably affect the performance of musical or lyrical pieces on the phonograph.

It consists in a stand which holds an adjustable damping device, preferably spherical or oval and made of a material such as cork or leather, in a suitable position in the sound trumpet.

2985/1905
Albert Conze *Gentleman*
41 Königin Augusta Strasse, Berlin, Germany

(Haseltine, Lake & Co)

Booth's Improved Device for use in Games and Athletic Exercises prevents injury to the private parts during games and athletic exercises, such as cricket, football, wrestling and the like, wherein such parts are liable to be injured either by a ball or playing device or by the action of an opponent accidentally or otherwise.

Such devices being only required for occasional use, such as during the game or exercise and generally where there is a multitude of onlookers, it is a desideratum that they should be applied and removed with due regard to privacy and decency.

Existing devices do not permit of this; the improved device can be lowered into usable position and adjusted by simply slipping it down the trousers without the knowledge of the bystanders.

It may consist of part of the shell of a coconut suitably shaped, whose edge may be fitted with a pneumatic tube to prevent chafing. When not in use, the edging can be deflated and the device may be slipped into the pocket.

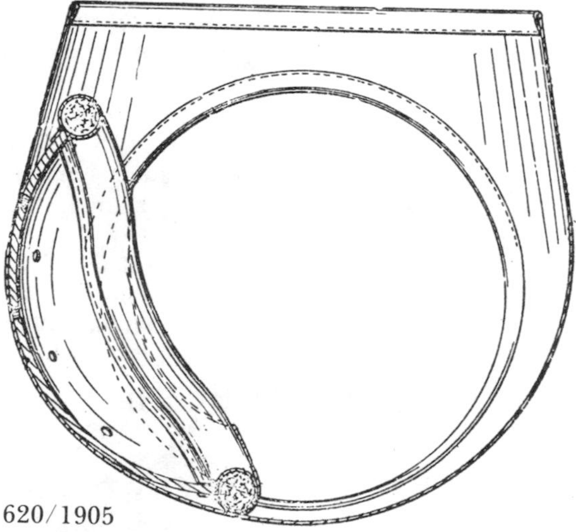

620/1905
Walter Booth *Fruiterer*
213 High Street, Deptford, London SE, England

(Day, Davies & Hunt)

Boswell's Improved Apparatus for Public Amusement consists in a wheel mounted on an inclined shaft and provided with seats adapted to carry performers who, when a certain degree of velocity is attained, purposely fall off more or less inelegantly and by their antics create some amusement among the spectators.

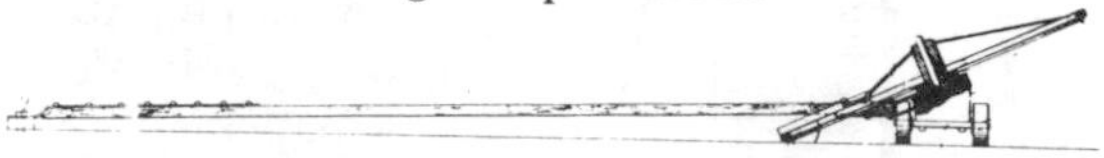

976/1905
James Clements Boswell *Amusements Contractor*
Oadby, nr Leicester, England

(Boult, Wade & Kilburn)

1905

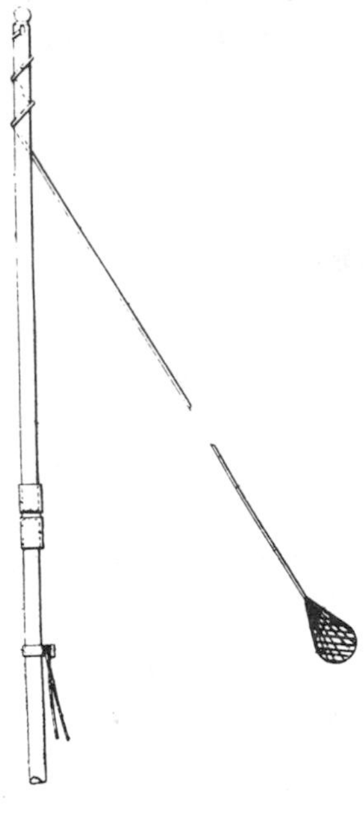

Denny's Improved Indoor Game has a xylonite ball attached to the top of a post by means of a cord. Each player has a bat, and tries to strike the ball so as to wind the cord round the post. The post may be attached to a chair; should a player in his eagerness knock against the upright rod in such a manner as to upset the balance of the chair, then that player loses a point.

4377/1905
D'arcy Denny *Gentleman*
44 Granville Park, Blackheath, London SE, England

(Rayner & Co)

Morrison's Improved Method of Manufacture of Newspapers, Books, and other Publications obviates the problems of the difficulty of selling left-hand page advertising space. The problem is that the right hand page, being directly in front of the readers, is much preferred by the advertisers, who will pay little or nothing for advertisements on the left. This difficulty can be easily overcome by printing the left-hand advertisement pages upside down, leaving the reader who has looked through the publication right way up to reverse it and go through the left hand advertisement pages which will now have become right.

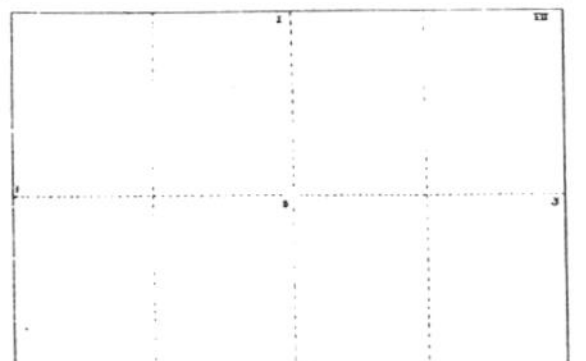

10529/1905
George Ebenezer Morrison *Barrister-at-Law*
8 King's Bench Walk, The Temple, London, England

(G G M Hardingham)

Jander's Clip for Attaching Newspapers to Doors is for the use of postmen or other messengers who deliver such matter.

It is well known that newspapers delivered by post or messenger, if a letter box with a large slot is not provided on the door, are simply thrown in front of the door.

This is very objectionable from a hygienic standpoint as microbes settle on the paper and it is highly unpleasant to feel that the newspaper read at breakfast has already been on the dirty floor or has perhaps been dirtied by being tramped upon. This disadvantage is overcome by providing, attached to the door, within easy reach, a suitable clamping device in which the newspaper is clipped by the messenger.

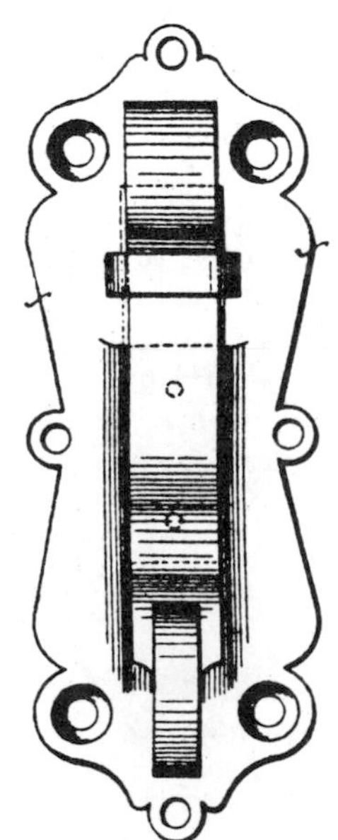

10590/1905
Friedrich Emil Jander *Merchant*
6 Burgkstrasse, Dresden, Germany

(Cruikshank & Fairweather)

Wilson's Improved Tea-pot has an internal receptacle for the tea-leaves arranged at the top and at one side of the usual opening of the pot, so that by canting the pot over on its side the tea in the receptacle can be submerged in the water.

This arrangement avoids the necessity for using special infusing devices which are more or less of a trouble and seldom properly attended to, is less ugly in appearance, and renders the lid less likely to fall off when pouring.

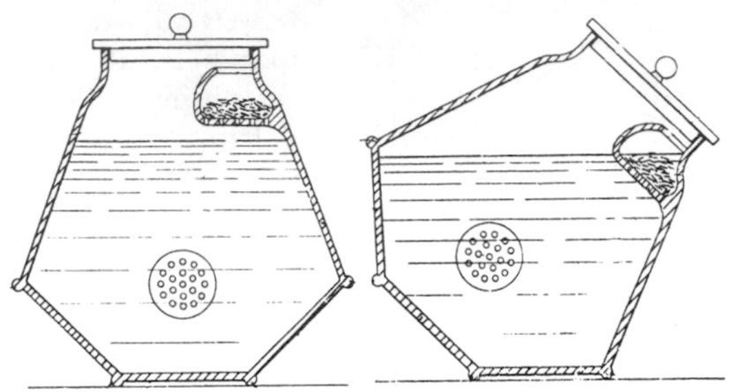

6023/1905
Arthur Edward Wilson *Stationer and Printer*
58 Cadogan Street, Glasgow, Scotland

(H D Fitzpatrick)

Ward's Improved Apron for Riders in Motor Cars is adapted to enclose the occupants from their waists down, and effectively to protect the lower portions of their bodies from wind and rain while at the same time not impeding the free movement of their feet. The apron is fastened around the dashboard, sides and back of the seat. The steering column passes through a laced slit. A window of mica or celluloid may be provided to enable the dashboard to be viewed. One great advantage of this invention is that the heat from the engines is retained, as when the parts are strapped up it is practically airtight.

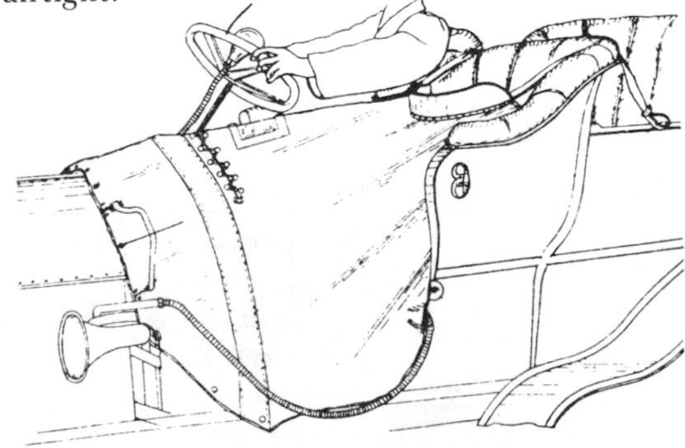

8337/1905
George Herbert Ward *Engineer*
The Motor House, Gordon Street, Southport, Lancashire, England

(Charles Coventry)

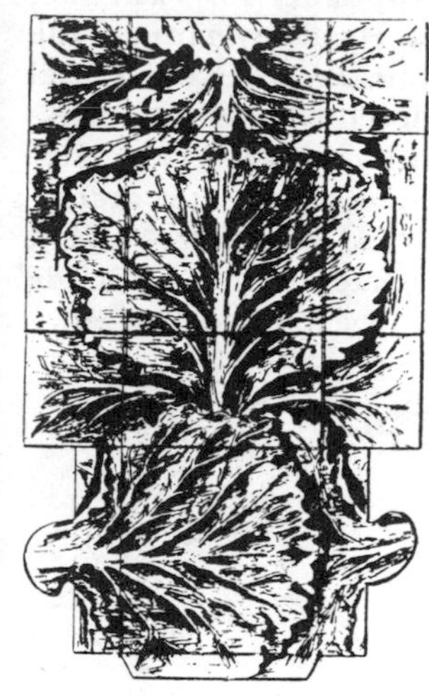

Chambers's Improved Method of Packing Dairy Produce is new in that hitherto it has been customary to pack dairy produce, such as butter, in cabbage leaves, which are, however, not always clean and at certain times of the year not easily obtainable or in good condition. This invention overcomes these drawbacks, and consists in a collapsible wrapper with its exterior printed, or embossed, or impressed, so as to represent real cabbage leaves. When it is folded into a package the positions of the leaves correspond with those of a package made of genuine cabbage leaves, and thereby give it such character.

5001/1905
William James Chambers *Art Publisher*
420 Corn Exchange Buildings, Hanging Ditch, Manchester, England

(F Bosshardt & Co)

Kaufmann's Tongue-cleaner can be made to meet all requirements for suitable treatment of the tongue of men and animals. It consists in a frame formed with a handle at one end and a bristle-covered loop at the other. The bristles are preferably longer at the front end than they are towards the handle.

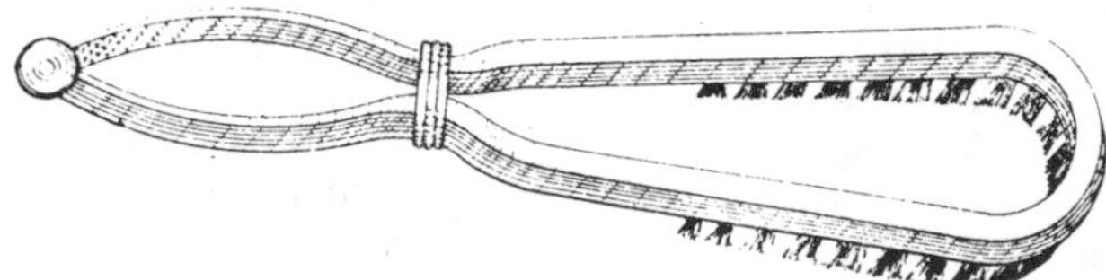

11403/1905
Richard Kaufmann *Merchant*
59 Kaiserstrasse, Rastatt, Baden, Germany

(Ferdinand Klosterman)

Watts's Improved Apparatus for Recording and Displaying Messages affords a convenient means of communication between persons who have arranged to meet at a certain place but do not arrive simultaneously.

The apparatus consists in a desk-like structure with a glazed casing in which messages are displayed. Upon inserting a coin, a piece of paper upon which a message may be written is uncovered. When the lid is afterwards closed the paper moves behind a display window so that it may be seen by the next visitor. The time may be printed automatically upon the paper.

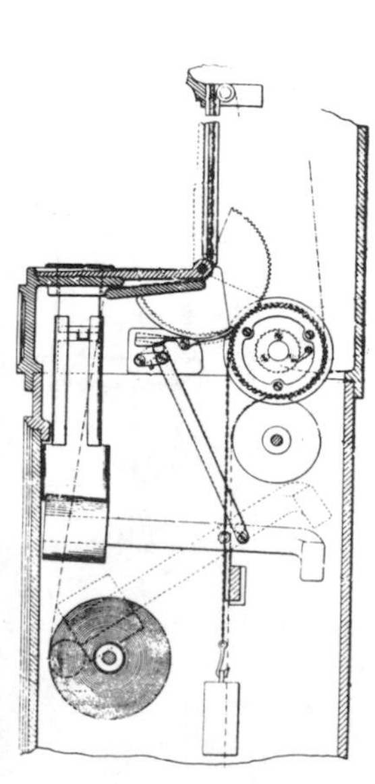

5124/1905
Alfred Monsell Sprainger Watts *Engineer*
The Vicarage, Waterbeach, Cambridge, England

(Boult, Wade & Kilburn)

Mathews's Hand Rake with Handle at the End prevents injuring tender shoots and flowers growing close to the ground, since the handle is in line with the head of the rake and therefore must be used close to the earth.

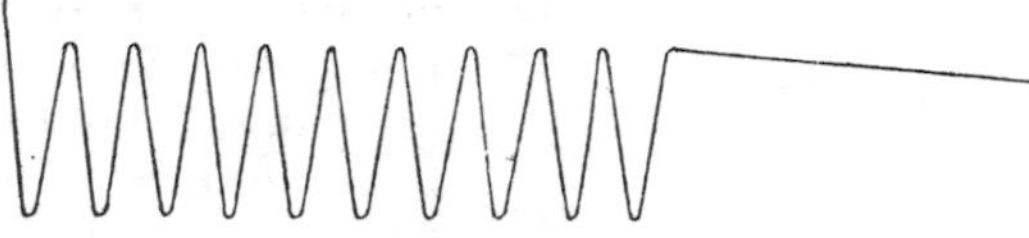

5489/1905
Francis Joseph Mathews *Machine Fitter*
52 Harcombe Road, Stoke Newington, London N, England

Macleod's New Method of Raising and Giving the Initial Impetus to Flying-machines and the like uses cylinders filled with explosives, compressed air, or compressed air and gas, and fitted to the machine in an inclined position such that, on the gas being suddenly discharged from the rear end of the cylinder, or the chemicals being fired by electricity or other means, the result will be an enormous impetus in an upward and forward direction.

If the machine is started from the surface of a body of water, extensions may be fitted to the tubes to project below the water.

1763/1905
Malcolm Campbell Macleod *Gentleman*
36 Harrington Gardens, South Kensington, London SW, England

(W D Rowlingson)

Lanchester's Improved Mechanically Propelled Road Vehicle has a heavy flywheel or "spinner", rotating rapidly, to provide motive power. The object is to provide for the storage of energy in an easily available form, and its subsequent utilisation to assist the prime mover, so that the necessity for change gear mechanism is obviated. A modification permits of the power being supplied by stationary engines or otherwise at prearranged points *en route*. This latter method is preferred when the principle is applied to a stage coach or omnibus.

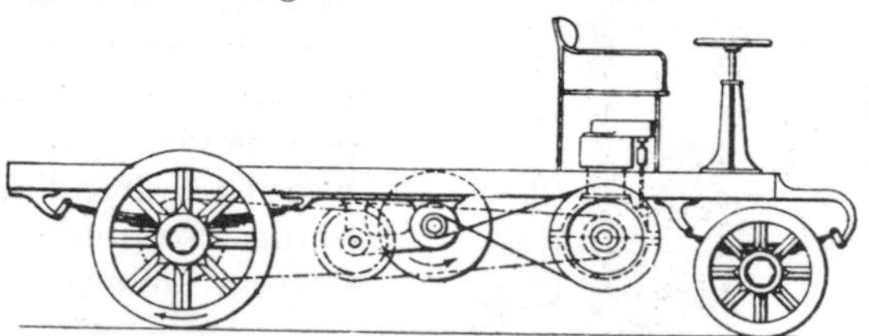

7949/1905
Frederick William Lanchester *Engineer*
53 Hagley Road, Edgbaston, Birmingham, England

(Marks & Clerk)

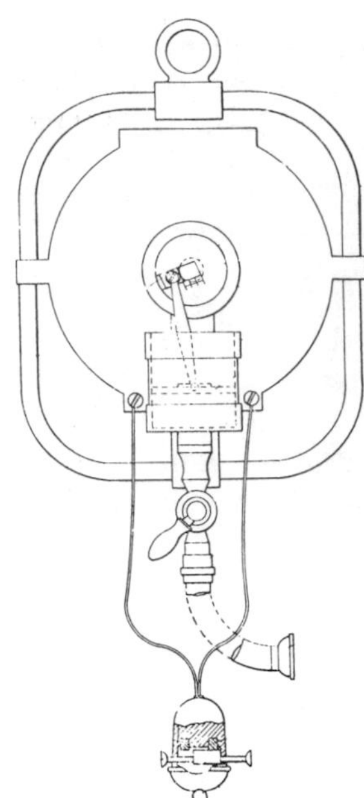

Muschik's Improved Pneumatic Massaging Appliance is designed to correct defects of beauty, by raising or filling up hollows in cheeks, necks, chests and other parts of the person, by means of a suction-instrument. For the purpose of carrying out the treatment, the tube-end is placed air-tight to the body; at each stroke of the pump, the flesh at its base is sucked into the vacuum space of the fitting, drawing by this the blood into the raised parts, which, as it carries with it the nutritive matter, tends to increase the volume of the textures.

2800/1905
Otto Emil Muschik
t/a C Bellack & Co, *Electrical Engineers*
12 Marble Arch, London W, England

(Vaughan & Son)

de Macedo's Improved Construction for Double-decked Tramcars and other similar Vehicles overcomes the problem that, as the cars are now constructed, when the conductor is on the platform he cannot see the top of the stairs and so often gives the signal to start when a passenger is on the point of stepping down; and when he is on the upper deck, he cannot see the lower platform and so often starts the car while a passenger is in the act of entering or leaving; in both of these cases serious accidents frequently happen. This invention obviates this danger by placing a mirror about two feet high by one and a half feet wide in such a position that it allows the conductor, whether on the platform or on the upper deck, to observe the movements of the passengers.

2882/1905
Joaquim Antonio de Macedo *Gentleman*
Leventhorpe Hall, Swillington, Yorkshire, England

Spencer's Improved Frying Pan prevents the grease dropping from the pan when it is hung up after use. It is provided with a pouring lip at each side and an inturned edge distal from the handle, so that, when the pan is hung up by its handle, grease collects in the cavity formed by the inturned edge. The pan can thus be put away as soon as it is used without danger of the melted fat dropping from the lower edge, and the lip also prevents dust from the back of the fireplace falling into the pan while in use.

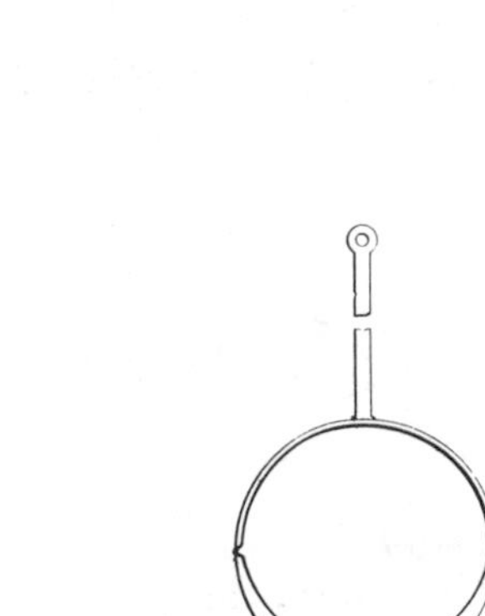

2779/1905
William John Spencer *Carpet Planner*
23 Meteor Street, Cardiff, South Wales, England

(Rayner & Co.)

Marshall's Improved Appliance for Fastening a Lady's Hat, or similar Head-covering, to the Hair, and keeping same in a requisite and sufficiently Firm Position on the Head during Wear consists in a thin band of steel, bone or other suitable material eyeletted to accommodate an elastic band and with a cut or bent edge, or pieces of hard material, to fix it inside the hat. Other pins fasten the band to the hair.

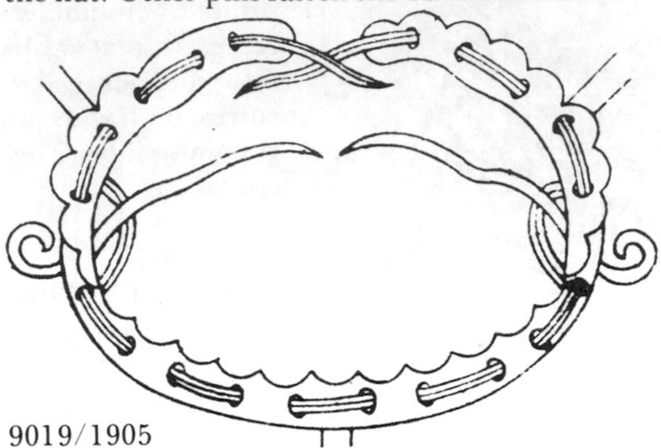

9019/1905
Alfred Marshall *Assurance Superintendent*
Ruddington, Nottingham, England

Williams's Electrical Signalling Device for Motor Cars and other Vehicles, and for other Purposes, available for Night and Day Use may also be used in ships, factories and other buildings, and for a great variety of purposes. When used in a motor car, it enables persons seated in the passenger compartment to give instructions to the driver. It consists in an electric arrangement which rings a bell and then lights a lamp behind a transparency, placed so as to be easily seen by the driver, on which are written orders or messages, such as "slower", "faster", "stop", "home", "left", "right", "ahead", "astern", and the like. The mode of operating the device is to depress the button or switch which is identified with the message to be given.

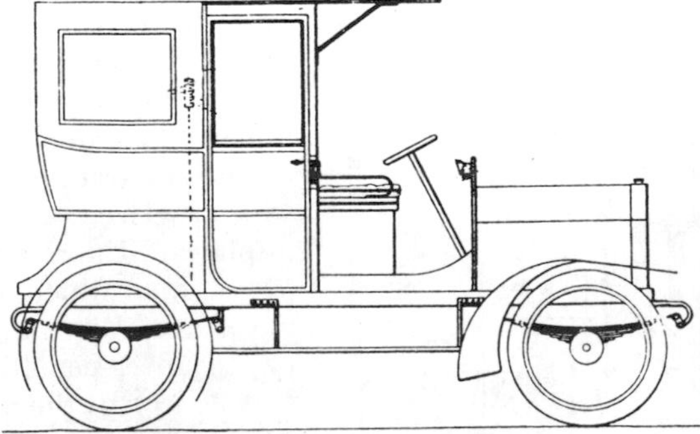

7828/1905
William Edwin Williams *Secretary to a Company*
47 Croftdown Road, Highgate, London N, England

(S S Bromhead)

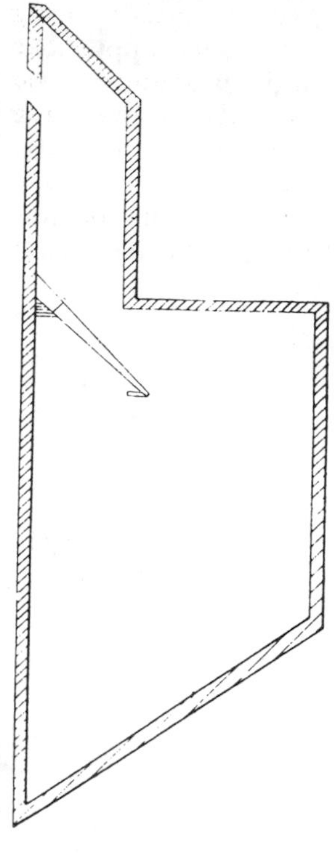

Ferrie's Improved Letter-box is constructed so as to prevent unauthorised withdrawal of its contents, being provided with a comb, with sharp springs on its prongs, placed between the letter slot and the collection chamber.

5641/1905
William Ferrie *Manager*
"Cumbrae", Merivale Road, Harrow, Middlesex, England

(Hughes & Young)

Baldauf's Improved Cycle-sledge has an adjustable connection between the driving-wheel frame and the runner frame. The centre of the faying surface lies after the handle bar, in order to drive without using the handle.

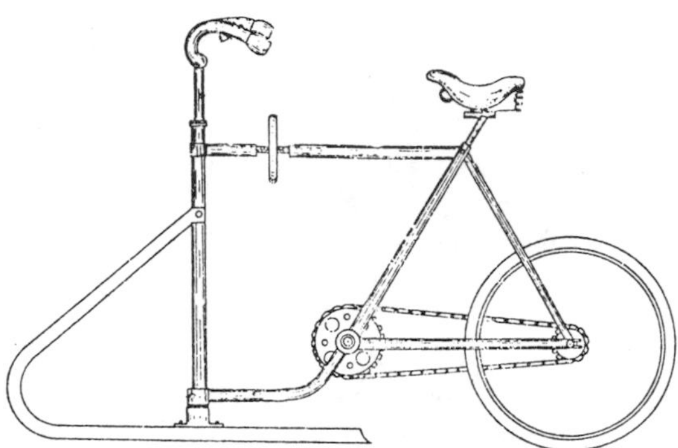

2835/1905
Josef Baldauf *Merchant*
Oberstaufen, Germany

(J A Nees)

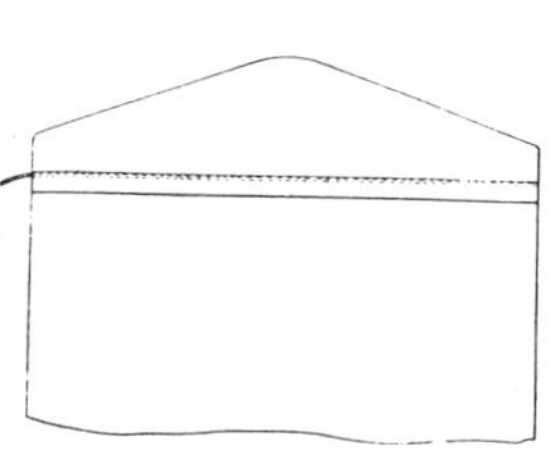

Curtis's Improved Envelope or Wrapper is furnished with a string or tape, arranged so that pulling it opens the letter or wrapper by quickly and neatly tearing it along the hinge or fold of the closing flap.

5915/1905
Thomas Curtis *Merchant*
85 Stockport Road, Levenshulme, Manchester, England

(John G Wilson & Co)

Arens's New or Improved Game has as its object to cause resilient balls to bounce either into a large central pocket or into one of a number of smaller pockets surrounding it.

4427/1905
Carl Arens *Gentleman*
16 North Audley Street, Grosvenor Square West, London, England

(Chatwin, Herschell & Co)

Francis, Butler and Amoore's Improved Device to Prevent Lying on the Back is to be worn as a belt by a person lying down, to prevent the heating of the spine. It consists in a wood or metal block with a strap to attach it to the body of the patient. The block is grooved to permit of the free passage of air.

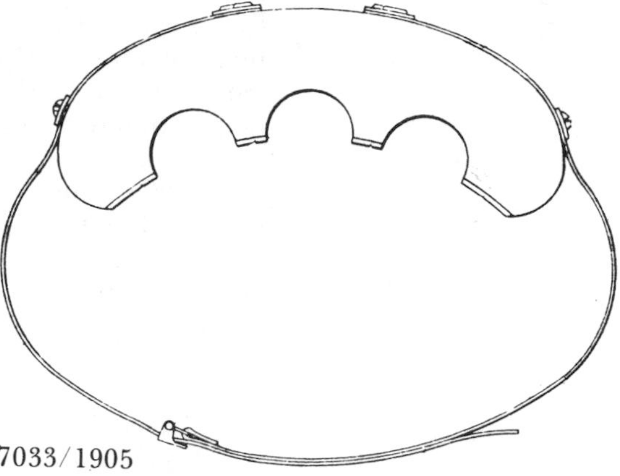

7033/1905
Arthur Francis
Ernest Frederic Butler
Henry James Amoore *Medical and Health Culture Appliance Makers*
272 Uxbridge Road, West Ealing, London W, England

(J B Fleuret)

The Camerons' Machine having the Inherent Power to Generate Motion derives its power from two weights suspended by ropes and driving one another by a system of levers and ratchet wheels. The arrangement may be multiplied to increase the power obtained.

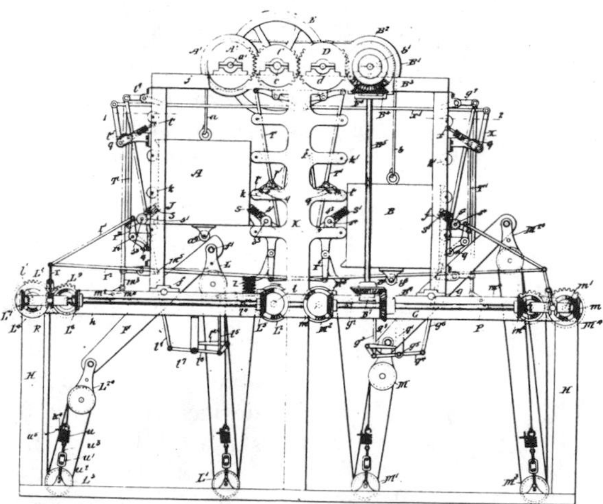

7789/1905
Alexander Cameron *Engine Fitter*
John Henderson Cameron *Power Loom Tenter*
William Cameron *Moulder*
77 Glover Street, Newtown, Perth, Scotland

(Johnsons)

1905

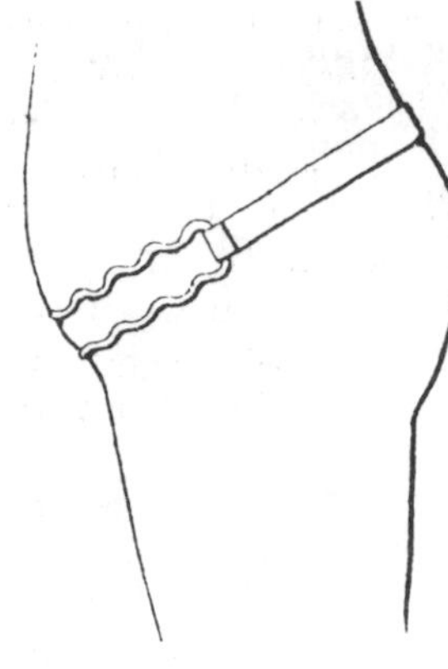

Burns's Improved Abdominal Massage Device for reducing and preventing protrusion of the abdomen co-acts with the muscles to cause them to knead or massage themselves whenever bodily movements of the wearer occur.

The device is to be secured in a fixed position on the body; projections on it slightly indent the flesh, which freely expands into the reserves between them. When any muscular or other movement takes place, the motion of the flesh under the projections is restricted by them, while its free movement in the recesses is permitted, resulting in a mechanical action similar to that produced by the fingers of an operator.

10034/1905
Sidney Herbert Burns *Manufacturer*
1133 Broadway, New York, USA

(Haseltine, Lake & Co)

Padmore's Improved Rest for Knives, Forks and Spoons comprises a holder designed to receive the aforementioned articles and keep them elevated, its primary purposes being to save embarrasment and to avoid soiling the table cloth. It is well known that many persons are uncertain what to do with their knives and forks when called upon to pass their plates for food; by the use of this invention all uncertainty in regard to their disposition is avoided and the diner placed at ease.

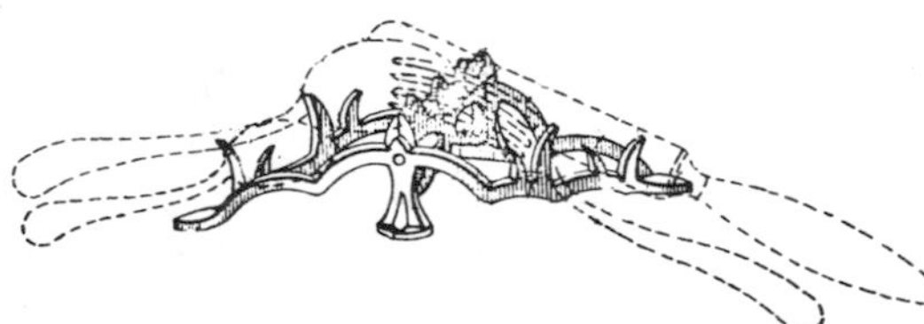

13452/1905
Arthur Meyrick Padmore *Gentleman*
Lead City, South Dakota, USA

(Marks & Clerk)

Pollock's Improved Washable Appliance for the Prevention and cure of the Condition known as "Double-Chin" overcomes the objectionable qualities of the india-rubber devices used heretofore, where the heat from the skin rapidly spoils the elastic and the same effect is produced whenever the appliance is washed. This improved device is made of xylonite or celluloid to render it conveniently washable, and is secured under the chin by ribbons.

10654/1905
Mary Fava Pollock *Surgical Appliance Manufacturer*
31 Aspinall Road, Brockley, Kent, England

Wilkins's Improved Cosey or Jacket applicable for use with Teapots and other Receptacles provides a simply effective and ornamental contrivance whereby the receptacle may be expeditiously enclosed while offering no obstacle to its ready use, and whereby the handle does not become so highly heated as to necessitate the use of a holder.

10033/1905
Frank Richard Wilkins *Commercial Traveller*
132 Chiswick High Road, London, England

(P: Haseltine, Lake & Co; C: J B Fleuret)

Macaura's Improved Means of Treating Patients by Static Galvanic or like Electrical Currents makes it possible to treat at one and the same time, from one machine or generator, as many as twenty or thirty patients, without necessitating a nurse or assistant for each patient.

The apparatus for administering electricity to a number of persons comprises a metal rod connected to a suitable machine and running round a room divided into cubicles. By this means, a great number of patients may be treated in a day, and consequently the patients' time is not unduly wasted through waiting their turn and in addition there is a saving of fees for the patient.

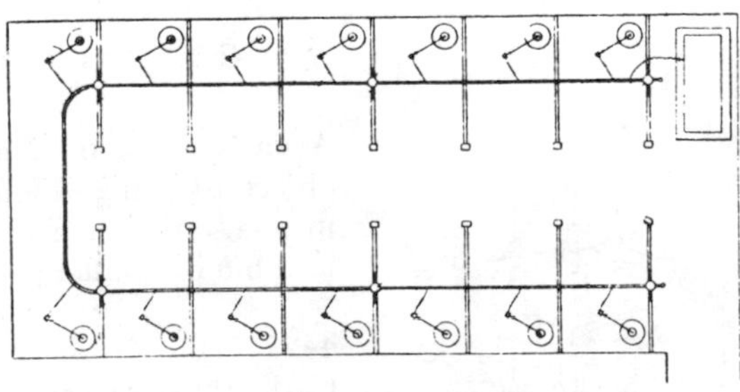

10067/1905
Gerald Joseph Macaura *MD*
4 Spring Bank, Bradford, Yorkshire, England

(John Waugh)

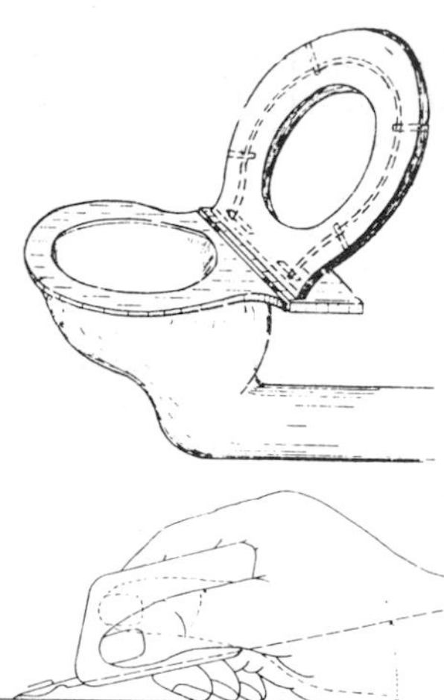

Richter's Improved Closet Seat Guard consists in a block of appropriately shaped papers which may be folded on to the seat one at a time and discarded after use.

11734/1905
Alfred Richter *Merchant*
1 Osterstrasse 78, Hannover, Germany

(J A Nees)

Breymann-Schwertenberg's Improved Holder for Pens, Pencils and the like recognises the fact that many people nowadays are obliged to write until they are very tired and if in consequence of this over-exertion "writer's cramp" is caused, then such people are hindered in their occupation or even their very livelihood is threatened. The provision of a typewriting machine is not possible for everyone; moreover the use of it requires time and practice.

The remedies prescribed against writer's cramp, such as the use of finger-holders, or of cork or rubber balls or lumps of gypsum upon the holders, are only palliatives which can never succeed in restoring the legibility or correctness of the writing: consequently there is a need for a pen holder such as this, which prevents the becoming tired through the natural and convenient position of the fingers and hand and thus greatly increases the capacity for working. It consists in a block to hold the writing instrument, with a smooth cover to slide easily over the paper. The block may be hollow to receive weights, and the weighting of the holder can take place in proportion to the tiredness or the occurring tension or cramp. It can be used for writing with ink, Indian ink, painting brushes, styluses or pencils.

10300/1905
Gustave de Breymann-Schwertenberg
Diplomaed Engineer
37 Joseph Korut, Budapest, Hungary

(W P Thompson & Co)

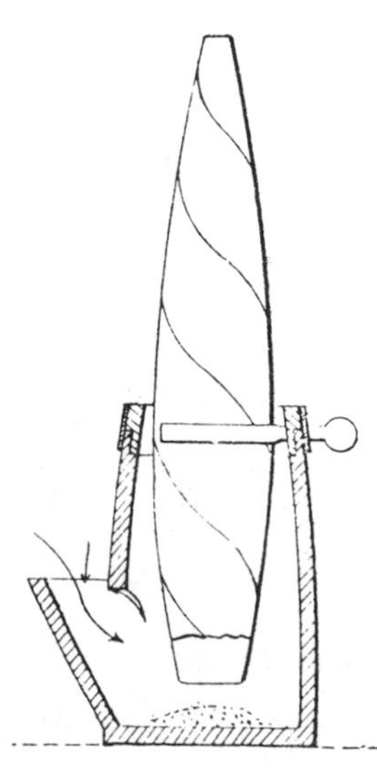

Rhodes's Improved Ash and Wind Guard for Cigars affords comfort in smoking in high winds, or when motoring or driving, and allows the cigar to be put down on a card, billiard or other table without risk of the ashes falling on the table, carpet or clothing, and without the risk of the burning end of the cigar coming in contact with any object on which the cigar is placed.

3853/1905
John Rhodes *Manager*
29 King Street, Fenton, Staffordshire, England

(John G Wilson & Co)

Mujata's Improved Amusement Apparatus consists of a dancing platform revolubly mounted on a wheeled truck, which is adapted to run on a railway. There is a guard-rail round the platform, and it is provided with a piano and a changing-room. The truck is driven by an electric motor, which picks up its power from underground or an overhead trolley wire; the same power may be used to light decorative lamps on the truck.

The platform, which may be four or more yards square, is moved slowly along while it is employed for dancing, and may be turned slowly around from time to time, thus serving two purposes: first, the amusement of the dancers themselves by moving and turning while dancing and, second, the amusement of the spectators who are not participating. The roof upon the platform is very desirable to protect the dancers from the heat of the sun or from rain in case of sudden showers, and the effectiveness of the apparatus may be greatly increased by extending an ornamental tower above the roof and decorating the same with electric lights, artificial flowers, and other appropriate objects.

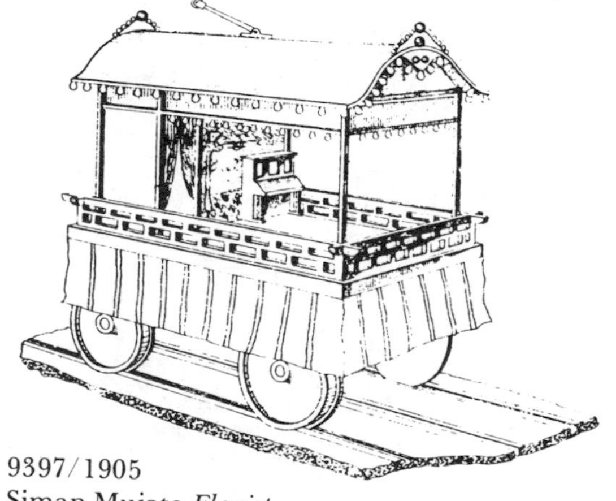

9397/1905
Siman Mujata *Florist*
191 Valley Road, West Orange, New Jersey, USA

(Thos S Crane)

Fletcher's Ladies' Veil Retainer consists in a length of india-rubber, bearing at each end an eyelet provided with a blunt point. It provides an easy and quick method of holding a lady's veil in position, fastened under her chin, over her head, or to her hat and covering the face, and of as easily removing it to any other position over the head or face, instead of twisting the veil into a tight and disfiguring knot under the chin.

12340/1905
Louise Fletcher *Gentlewoman*
8 Euston Square, London NW, England

1905

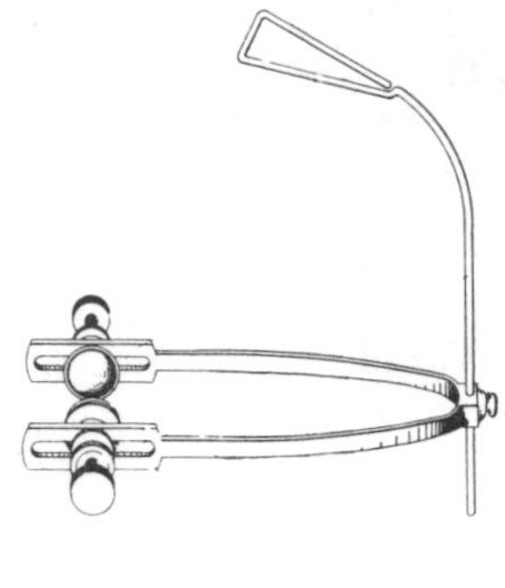

Bhisey's New Instrument for Curing or Alleviating Headache applies pressure to the temples by means of pads which work in screw threads in a frame which fits round the back of the head. The pads may be pneumatic, and saturated with a medicament, continuously supplied by capillary action from a reservoir. In operation the device is arranged upon the head of the patient, with the semi-circular frame in a horizontal position and the position of the pads adjusted to press on the temples. There may also be a third pad to act on the nerves in the back of the head.

10662/1905
Shanker Abaji Bhisey *Engineer*
323 Essex Road, Islington, London, England

(Hughes Son & Thornton)

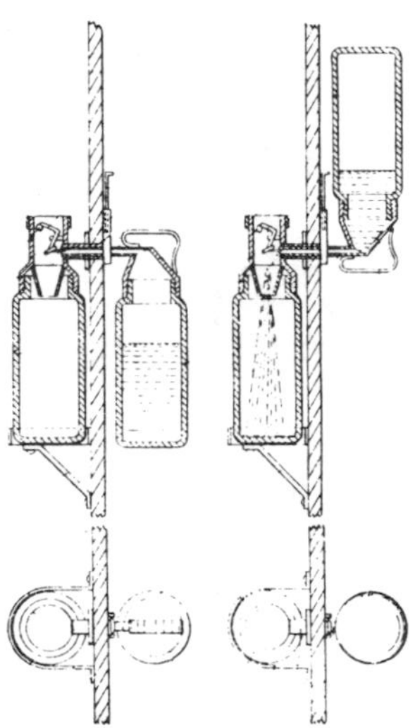

Graaff's Improved Means for Receiving Milk in Houses obviates the inconveniences in supplying milk to the household by the ordinary method, as, if nobody is at home to receive the milk, it can be easily misappropriated when accessible to unauthorised persons. Accordingly, a receiving vessel is fastened inside the house door; the milkman pushes the spout of the delivery vessel through a hole in the door, where a pin on the collection vessel opens the delivery valve. He then inverts the delivery vessel and the contents run into the collection vessel by gravity. The hole in the door is closed by a slide when the can is removed.

7014/1905
Wilhelm Graaff *Merchant*
10/11 Potsdamerstrasse, Berlin, Germany

(Philip M Justice)

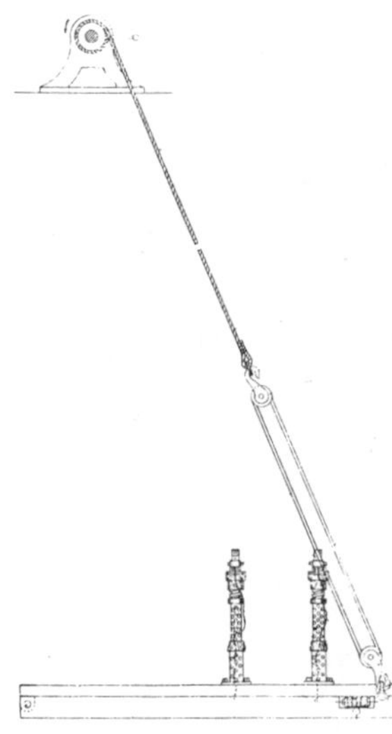

Wulff's Improved Apparatus for Throwing Animals into the Air for Exhibition Purposes relates to equipment for throwing animals into the air to take a "death leap", or salto-mortale.

It is an improvement on the apparatus already described in 8713/1904, having means for effectively securing the suspension rings of the animal on the flat hooks of the standards until the moment at which the animal is thrown off the movable plate. Furthermore, power devices are substituted for the springs previously described.

8894/1905
Eduard Wulff *Circus Manager*
10 Rue du Moniteur, Brussels, Belgium

(Boult, Wade & Kilburn)

Ratcliffe's Improved Towing Arrangement for Cycles and Motor Road Vehicles is attached to a vehicle and enables a gentleman rider to render assistance and considerable relief to ladies or other companions riding a cycle when travelling over heavy rough roads or when going uphill.

The gentleman rider in front throws the loop to the lady who then drops back until the coil has run out and becomes taut. When it is no longer required, the lady simply lets go of the loop, which will then automatically spring forward and wind itself up in its case, which by preference is placed underneath the saddle of the leading machine, until again required.

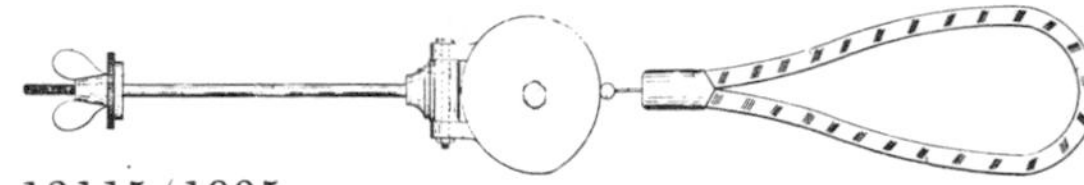

13115/1905
Richard Ratcliffe *Foreman Moulder*
55 Gilmore Road, Lewisham, London, England

(S S Bromhead)

Héroult's Improved Method of Effecting Marine Propulsion In all known methods of navigation the floating vessel is held *in equilibrio* by the displacement of a quantity of water about equal in weight thereto. When the floating body is moved it experiences a resistance due to the displacement of the surrounding masses of water and to the friction on the surfaces; this invention renders the apparatus for marine propulsion independent of these two forms of resistance to forward motion, which serve to limit the speed of other marine motors.

It depends on the principle that the work needed to lift a vessel from the water approaches zero when the number of lifting impulses increases indefinitely.

Accordingly, the vessel is lifted by means of a paddle wheel at each corner, the reactions of whose blades perform the dual offices of lifting and propulsion. Moreover, the arms upon which the wheels are mounted are hinged and spring-loaded so that they can ride the waves without the vessel partaking of their motion.

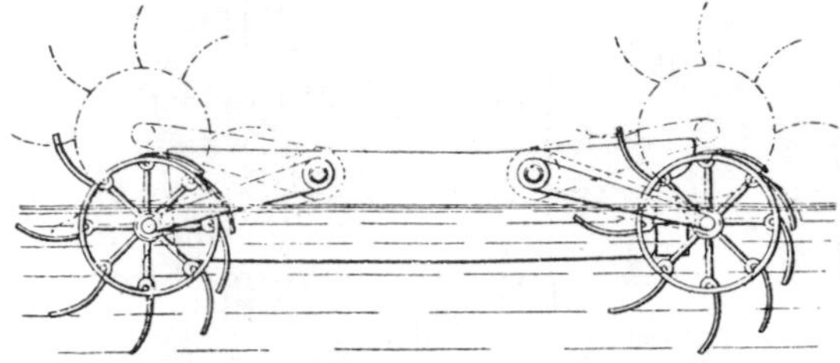

12600/1905
Paul Louis Toussaint Héroult *Merchant*
La Praz, Savoy, France

(Abel & Imray)

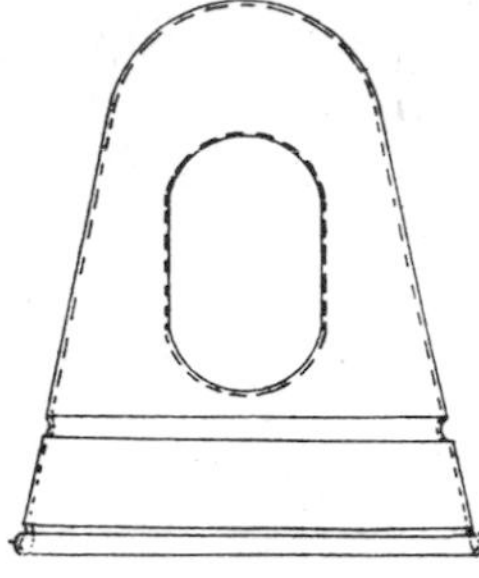

Nesbitt's Improved Thimble has one or more holes in it, and is provided with an inner lining which also has one or more holes in it and may be made to revolve. The idea of the hole or holes is that a miniature portrait, device, motto, name, date, advertisement, or any other matter may be seen through them.

15963/1905
Albert Nesbitt *(No profession stated)*
32 Stamford Hill Mansions, Stamford Hill, Middlesex, England
formerly of Dormers Well Farm, Southall, Middlesex, England

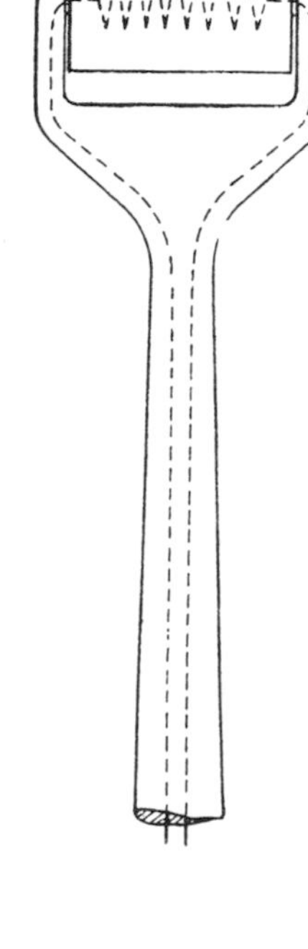

Muschik's Appliance for Face Massage is a heated roller which is applied to the skin for the withdrawal of impurities and the removal of freckles and wrinkles and restoring the natural chemical tension of any flabby and degenerate skin. Either the face or the roller are covered with absorbent material and the roller may be heated by electricity, burning carbon, hot water etc. The device may be controlled by a rheostat.

13825/1905
Otto Emil Muschik *Toilet Specialist*
12 Marble Arch, London W, England

(Chatwin, Herschell & Co)

Davenport's Artificial Breathing Apparatus restores animation in cases of supposed drowning, suffocation, trance, syncope etc. The body is placed in an airtight container and connected to a valveless pump so that the pressure variations in the chamber when the pump is operated promote respiratory movements.

In cases of drowning, water is also expulsed, as are the foul airs and gases in cases of asphyxiated persons.

An apparatus of this description for artificially restoring respiration will be found of incalculable advantage, as the body, after being reverently placed into the case, need not be handled again until animation is restored or death has been finally certified to have taken place.

9538/1905
William Davenport *Gentleman*
156 Stamford Street, London, England

(H W Denton Ingham)

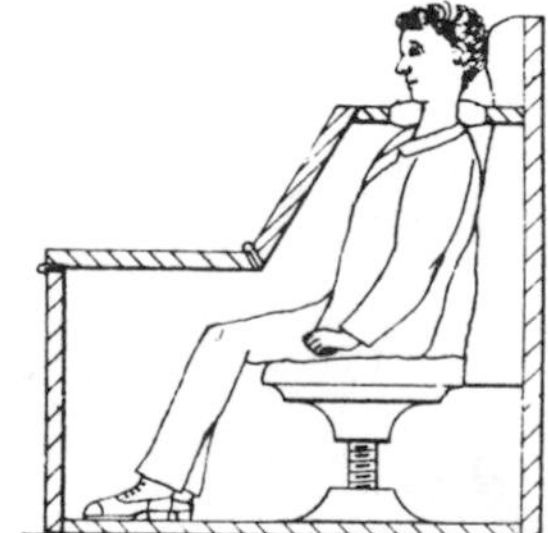

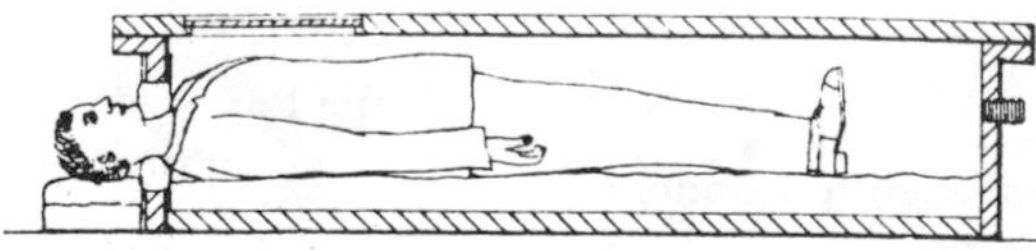

Carysfort-Loch's Improved Marking Instrument, be it pen, pencil, brush or other, is provided with an electric lamp to enable it to be used in the dark, or in a badly lighted place.

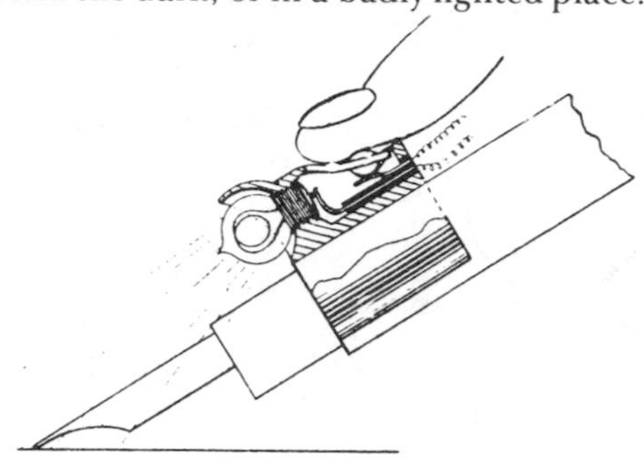

11070/1905
Violet Frances Carysfort-Loch *Married Woman*
Trivandrum, Trivancore, South India

(R Core Gardner)

Fleming's Improved Means for Generating and Using Hydrocarbon Vapours for Heating and Lighting is portable and uses very little fuel, and is designed so that the vapouriser does not get choked up with residue.

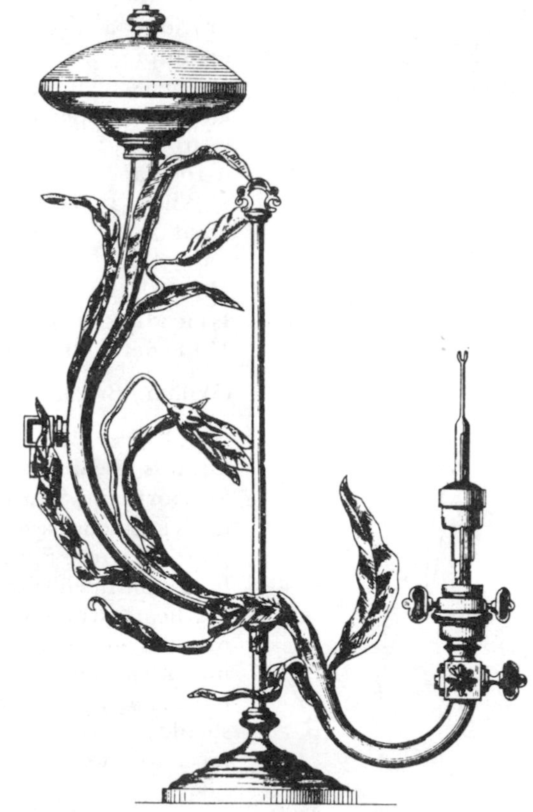

10616/1905
Josephine Marie-Louise Fleming, *née* Imbert
Chateau des Ormes sur Voulzie, France

(Imrie & Co)

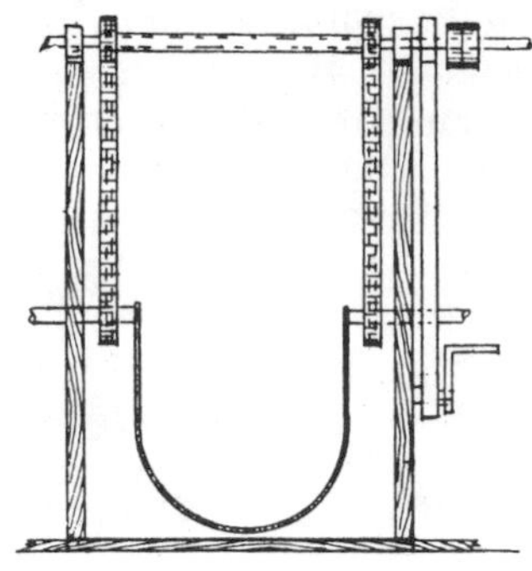

Ford's Improved Recreation Apparatus is designed for use in showgrounds, parks, schools and other public places and consists in a skipping rope which is swung mechanically, or by another hand than the skipper's.

Several such ropes may be mounted on one frame, and it is even possible to have two ropes swinging in opposite directions to allow of cross or French skipping. A safety device is incorporated to ensure that the rope is easily stopped by the skipper's feet and the mechanical arms swinging the rope are pliant or jacketted with india-rubber so that, though strong, they will not injure the skipper.

11098/1905
William Ford *Engineer and Millwright*
57 Duke Street, Liverpool, England

(J A Coubrough)

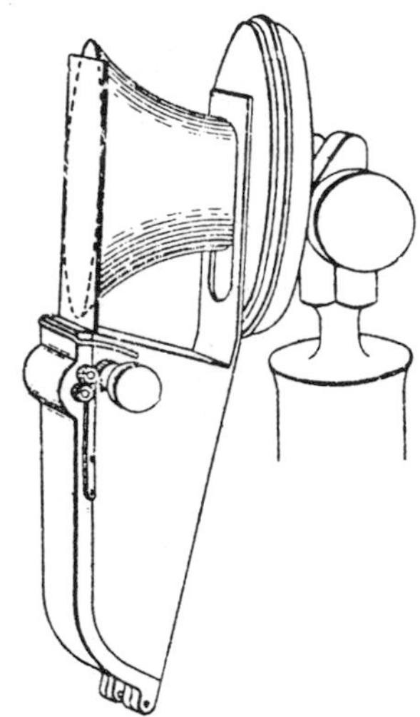

Slocum's Improved Sanitary Attachment for Telephones obviates the well-established fact that contagious diseases — such as diphtheria, scarlet-fever, consumption etc — are often transmitted by mouthpieces of telephones and other sound-conveying instruments, especially those used in public places. This invention provides a novel device for protecting the mouthpiece from contact with the lips of the person using it. It consists in a supply of sheets of antiseptic paper which are moved one at a time in front of the mouthpiece which forms a guide to the paper.

When a sheet is soiled, it is withdrawn by hand.

12341/1905
Isaac Michael Slocum *Mill Agent*
164 Federal Street, Boston, Massachusetts, USA

(William Brookes & Son)

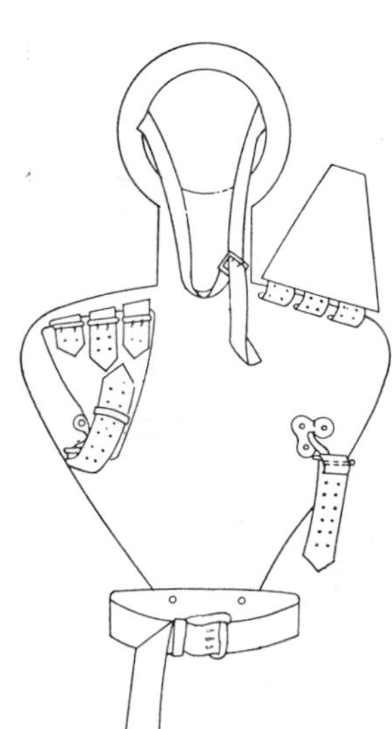

Grant's Head to Waist Back Board with Metal Supports and Strappings is to be worn only over one's garments several hours a day and insures a perfectly erect figure. It is made of wood, and has a hole lined with chamois leather to be soft to the head, straps to attach it to the body, and metal arm rests to prevent the usual cutting under the arms and painful discomfort. It is particularly suggested that this backboard should be used to support the back and keep erect the heads of children during their study hours.

13550/1905
Eleanor Grant
Cromwell Road, South Kensington, London SW, England

Brink's Arrangement for Fastening Together and Hoisting Benches, Desks, Beds and Other like Furniture is an improvement on the known arrangements for uniting all the school benches of a classroom into one suspendable whole in such a manner that by the intermediary of a windlass they may be raised from the floor, which are not very easy to apply in practice because, with such arrangements, at least three walls of the room are occupied by the lifting tackle, the gangways are blocked by connecting beams and lattice work, and the free moving of doors and windows is hindered.

All these inconveniences are diminished by means of this patented arrangement, whereby one wall only is interfered with and the gangways are left free by securing the pieces upon one or more longitudinal girders, arranged one behind the other and connected at one end directly, and at the other by a geared hoisting arrangement.

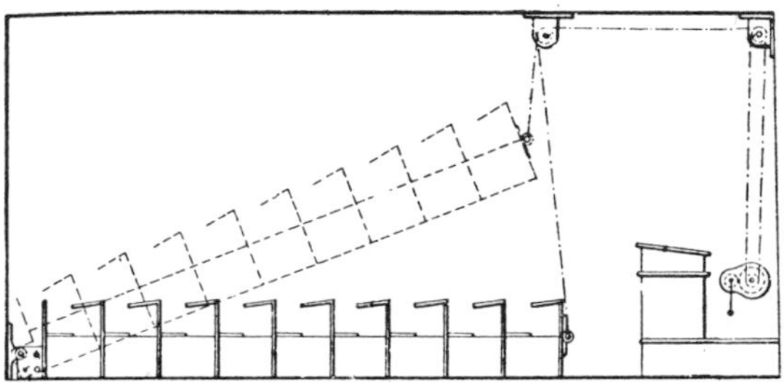

12467/1905
Carl Theodor Heinrich Brink *Manufacturer*
Kohlenstrasse 127½, Wahlershausen, nr Kassel, Germany

(Witold Baronowski)

Poitel's Wind-guard for Cigars or Cigarettes prevents too rapid combustion, and retains the ash when the smoker is travelling rapidly in the open as, for instance, in a motor-car or on a boat.

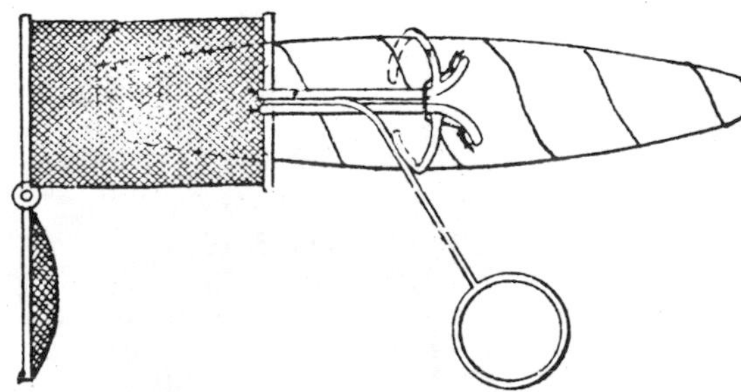

13887/1905
Emile Poitel *Merchant*
6 Rue Fontaine au Roi, Paris, France

(A F Spooner)

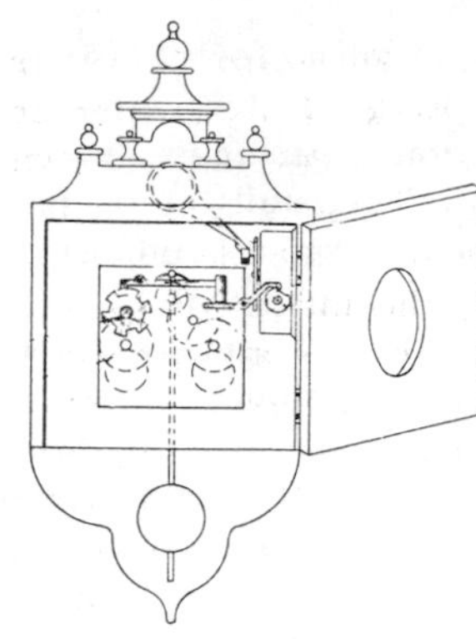

Wetzel's Improved Clock eliminates the noisy striking works of the usual clock and replaces them with a device which is less noisy, takes less time and can equally well be heard. A phonograph, gramophone, or the like, is combined with the striking works and the usual bell or gong is done away with. A trip mechanism is connected to the gramophone, and a gramophone-plate, which reproduces a human voice, calls out, according to the time of day: "half past twelve", "one o'clock" and so on.

The sound-funnel or trumpet may be made to project out of the case of the clock at any suitable place.

12604/1905
Paul Wetzel *Merchant*
30 Markgrafenstrasse, Berlin, Germany

(Haseltine, Lake & Co)

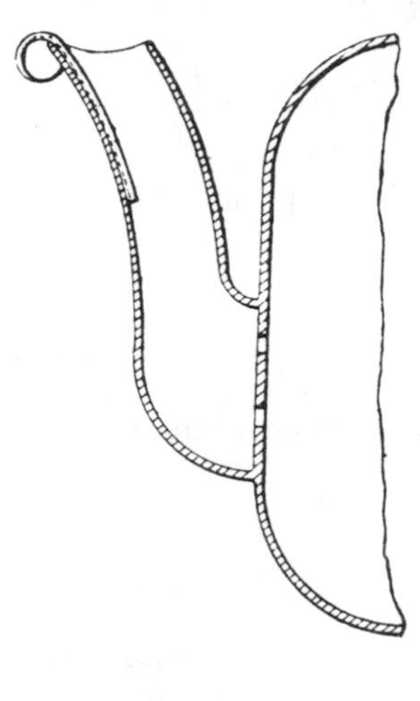

Cox's Dripless Spout for Teapots uses capillary action to hold the drop of liquid which usually remains after pouring, and so save it from falling or running down the outside of the lip or spout.

17779/1905
William Cox *Scientific Mechanician*
Tyler Hill, Canterbury, Kent, England

(Marks & Clerk)

Kennedy's Signalling Device for Indicating Alteration of Speed of Self-propelled Vehicles is applicable to motor cars, especially those travelling on ordinary roads. When such vehicles are travelling on more or less crowded thoroughfares, or rounding corners and so forth, it is frequently necessary to reduce speed suddenly. In such cases there is often a danger that the following vehicle will not notice the alteration in time, and will run into its predecessor unless the driver has given a signal of some kind. In horse-propelled vehicles the driver usually signals by using the whip or hand, but with motor cars both hands are required for operating and steering the car, and to be continually raising the hand is inconvenient and sometimes dangerous.

This invention provides a signal connected with the brake of the car in such a manner that the operation of the brake automatically displays the signal, which consists in a disc or small lamp on the end of a pivoted lever operated by a cord and pulley.

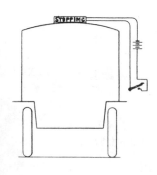

7601/1905
Murray Kennedy *Shipbroker*
Montreal, Canada, *temporarily at* The Cannon Street Hotel, London, England

(W P Thompson & Co)

Fey and Weisshaar's Improved Automatic Vermin Trap is adapted to be reset by the animal caught. It is arranged as a cage with a pivoted door and a mirror at its further end. The animal enters the cage, treads on the rocking bridge, thereby disengaging the door, and falls into the water vessel. The bridge then returns to its former closed position and the animal's attempts to escape serve to ensure that it is kept shut.

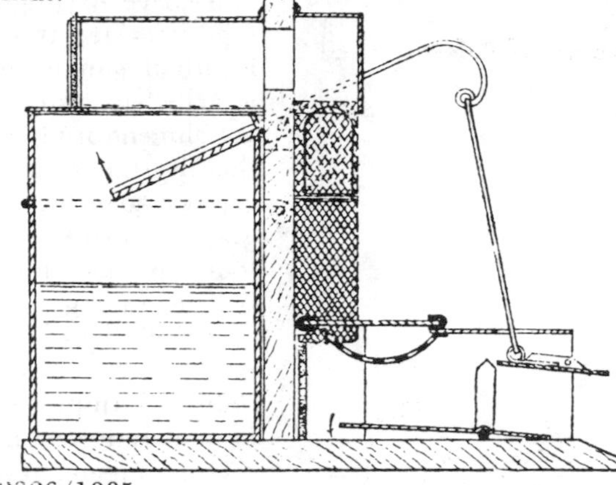

2326/1905
Cárl Fey *Master Tinsmith*
Kreis Eschwege, Kirschhosbach, Germany

Emil Weisshaar *Commission Agent*
Eschwege a/Weira, Germany

(Haseltine, Lake & Co)

Anschütz-Kaempfe's Improved Gyroscopic Apparatus is particularly for use on board ships, as a substitute for, or as a means for corroborating or correcting, the magnetic compass.

It consists in a motor-driven gyroscope, so mounted that it is not affected by the earth's rotation, but by changes in the ship's course only.

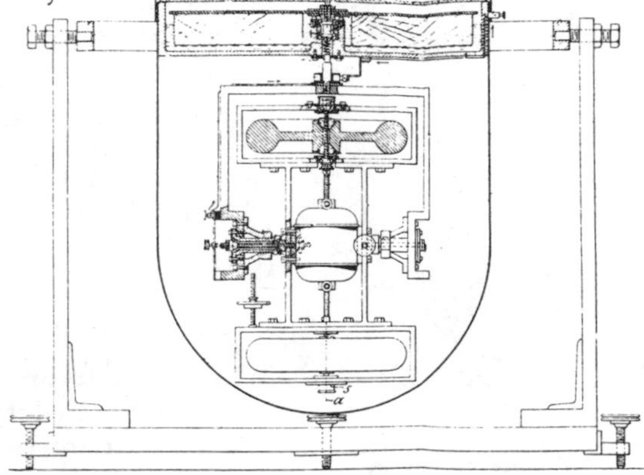

6359/1905
Hermann Anschütz-Kaempfe *Doctor of Philosophy*
13-14 Markt Platz, Kiel, Germany

(Boult, Wade & Kilburn)

1905

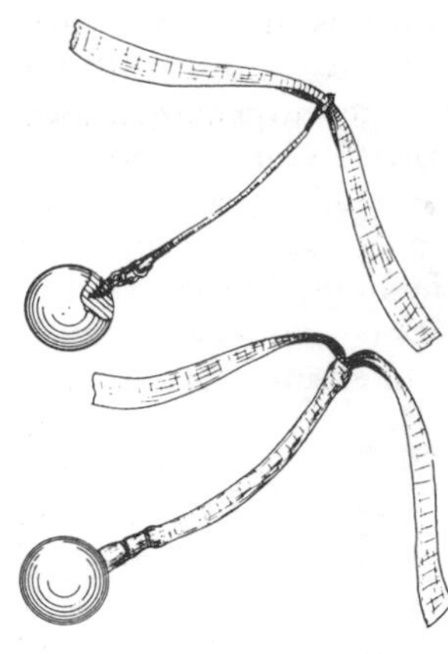

Ball's Improved Practice Apparatus has reference more particularly to games such as golf, in which the ball is projected for a considerable distance in the course of play and is liable to be lost. This invention is particularly suitable for the use of beginners practising on an ordinary lawn or field when a golf course is not convenient as it shortens the flight of the balls as well as enabling them to be found more easily.

A piece of bright-coloured (preferably red) tape is passed through a loop at the end of a length of window-blind cord, whose other end is attached to an eye screwed into the ball. The attachment of tape is effected in such a manner that it stands clear of the ball and remains exposed to view if it happens to fall in a tuft of grass or hollow, cannot with ordinary care be fouled by the club, and will not curl round the ball on landing.

A ball fitted with one tape can be driven rather more than one third of the distance of an ordinary drive; to further shorten the flight, two or more tapes can be passed through the cord, so offering greater resistance to the air. If desired two players can practise together, one driving to the other.

14452/1905
Frederick James Ball *County Inspector, Royal Irish Constabulary*
Carrick-on-Shannon, Ireland

(Haseltine, Lake & Co)

Arens's Electrical Appliance for Massage provides a simple apparatus for utilising electricity in combination with massage. It consists in a number of finger-stalls of rubber etc to which are attached electrodes connected to an insulated wrist-ring, or to a hand-strap, connected to a battery. These finger-stalls are covered with linen, chamois leather etc, which may be moistened for the purpose of saturation to increase their conductibility when applied.

Another form of the invention consists of gloves of rubber with electrodes in the palms.

Gold-leaf or foil electrodes are preferable to thin metal plates and, covered with linen and saturated when in use, this apparatus is especially suitable for medical rubbing.

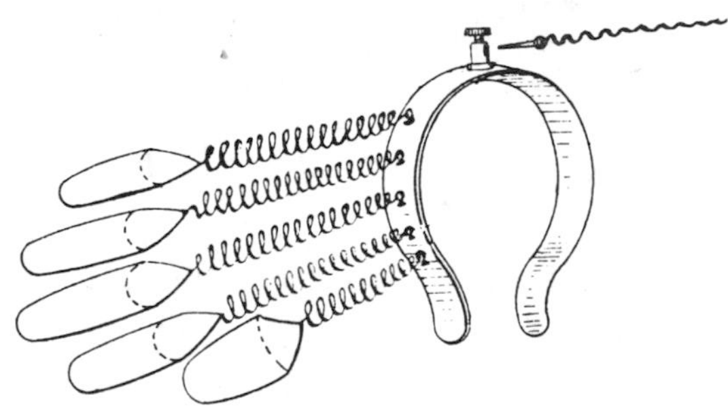

15985/1905
Carl Arens *(Gentleman) Inventor*
16 North Audley Street, Grosvenor Square, London W, England

(Chatwin, Herschell & Co)

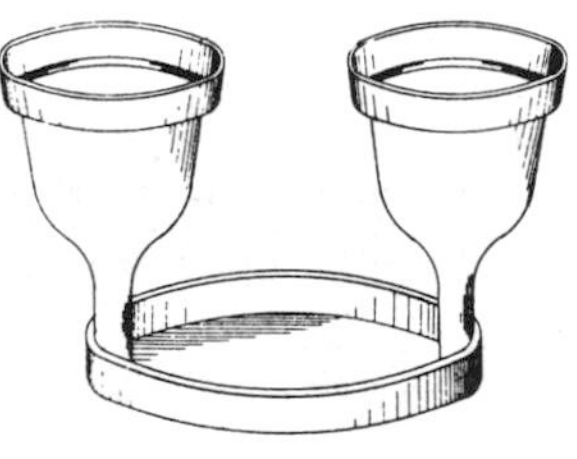

Young's Improved Egg Cup has an enlarged top, sometimes with a surrounding trough, to catch any overflow which might otherwise escape down and soil the outer side of the cup, and to receive the removed broken parts of an egg shell taken from a boiled egg. The base may be found to serve as a salt-cellar. Moreover it is possible to fit more than one cup to the same base.

16216/1905
Alexander Myatt Young *Salesman*
78 Manor Lane, London SE, England

(C H Burgess)

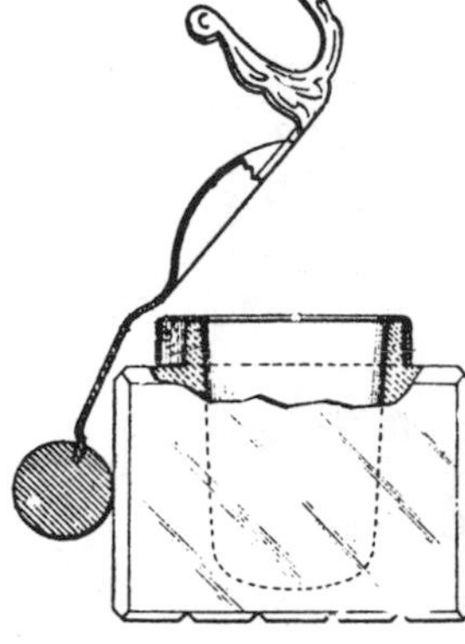

Whitehill's Improved Inkpot Lid has a pen rack incorporated with it so that when the pen is taken up a counterweight causes the lid to open.

17520/1905
Thomas Wright Whitehill *Tool Maker*
150 Fentham Road, Birchfields, Birmingham, England

(George T Fuery)

Pillsbury's Improved Device for Preventing Train Robberies consists essentially in a system for spraying the robbers with hot water or steam. It is obvious that in the event an attacking party should receive steam full in the face, he would be blinded thereby and compelled to retire.

The invention comprises a series of pipes and nozzles connected with the boiler of the locomotive engine whereby a single movement either by the hand or foot of the engineer will cause steam to be sent through the nozzles in different directions.

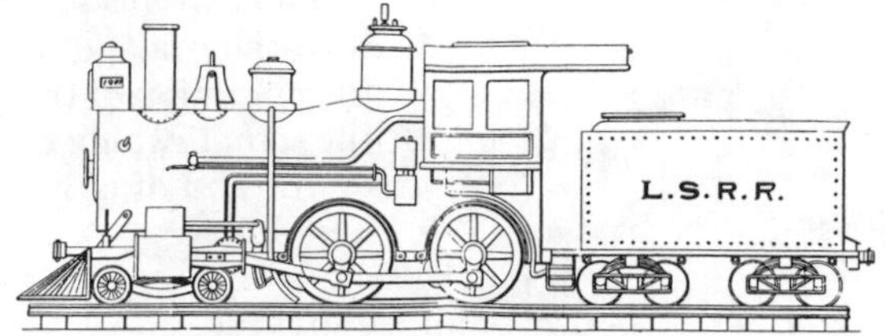

14039/1905
John Colby Pillsbury *Travelling Salesman*
Whitefield, New Hampshire, USA

(Boult, Wade & Kilburn)

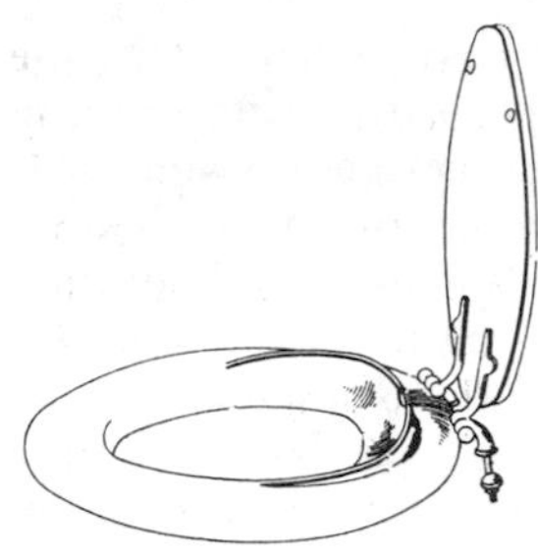

Muckenhirn's Improved Water Closet and Seat has for its object an improved bowl and seat construction whereby the user is forced to assume a position that relieves from pressure those muscles or muscular parts that are especially called into action at the time the structure is used. The special formation of the seat also forces the user to assume the position that will overcome any liability to soil the seat.

To this end, the closet is designed with a basin which is made higher at the front than at the back, and a seat made with a concave surface at the rear extending up to a ridge with a notch or recess in the middle.

16851/1905
Charles H Muckenhirn *Manufacturer*
Salem, Massachusetts, USA

(W P Thompson & Co)

Ball's Improved Golf-bag is able to assume an erect attitude. To this end it is provided with two supporting legs which normally lie close to the bag but can be moved outwards to form a stand by depressing a thumb piece.

If the bag is made of limp material, it may be furnished with a stiffening rod, secured along its side.

This avoids the well-known inconvenience incurred in stooping to place the bag on the ground before playing and lifting it again afterwards in cases where a caddy is not employed.

17450/1905
Frederick James Ball *Royal Irish Constabulary*
Carrick-on-Shannon, Ireland

(Haseltine, Lake & Co)

Devrient's Improved Means of Making Artificial Eyelashes involves cutting them out from strips of black silk fabric having two selvedges united by cross-threads; two longitudinal threads retain the material in proper condition while it is being cut into shape. The selvedges are of different coloured silk, in blue and flesh tints, specially designed for this purpose. For fixing the eyelashes, the selvedge is bent at right angles to the shape of natural long eyelashes and coated with an innocuous adhesive for attachment to the wearer's eyelids.

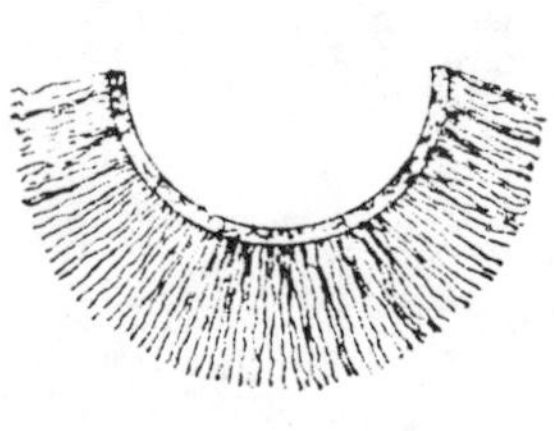

18652/1905
Catherine Goldsbrough Devrient *Operatic Artiste*
20 Abbey Gardens, Abbey Road, London NW, England

(Hughes & Young)

Bullock's Swimming Devices are attached to the arms and legs of a swimmer, and facilitate his progress by affording him an enlarged area with which to push himself forward in the water. This object is attained by fastening to the limbs light frames with movable vanes which offer little resistance on the forward or idle stroke, but afford great assistance on the working stroke. Instead of being used merely as paddles, the effort of the feet is utilised to impart their considerable muscular power to propel the body forward by connecting them to the leg vane frames through parallel motion links. If however, the feet should become tired the instep action may be disconnected and the leg movement alone used.

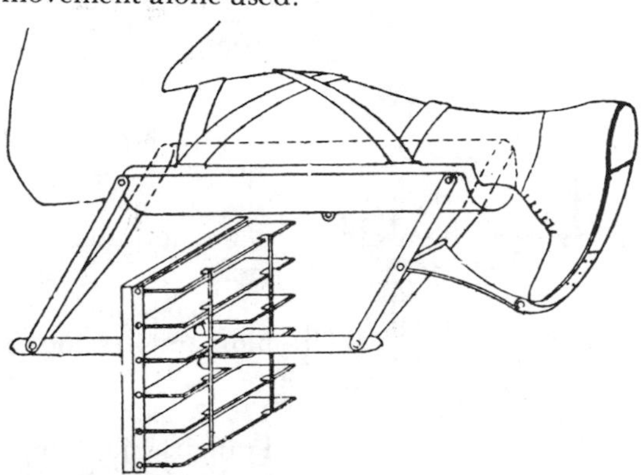

13851/1905
Amasa Marion Bullock *Insurance Agent*
853 Hamilton Street, Vancouver, British Columbia, Canada

(Rowland Brittain)

Senger's Ball-catch Game may be played by any number of persons. Half of them, "the bearers", are provided with belts with hooks at the back, and are chased by the other players, the "catchers", who, using a "catching-stick" endeavour to pass a ring attached to a handle over the hook.

The bearers, by effecting rapid turning movements, must try not to be caught but must try to remain face to face with the catchers, whereby the game becomes interesting and exciting.

16268/1905
Karl Senger *Engineer*
Weipert, Bohemia, Austria

(P Follin)

1905

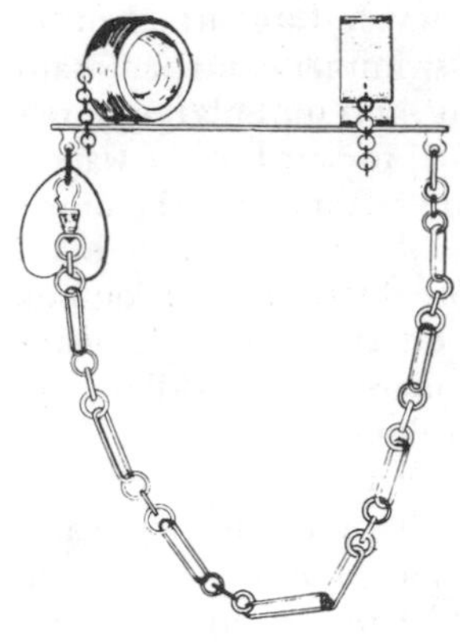

Procter's Improved Means for Carrying Umbrellas, Parasols, Walking Sticks or the like enables the user to suspend the accessory from her wrist or arm, thus making it possible for such articles to be carried with greater freedom and convenience and with less liability to misplacement and consequent loss from forgetfulness or other causes than at present obtains. The appliance is preferably made light and elegant in character so as to permit convenient use of the umbrella with the same attached thereto.

12148/1905
Annie Louise Procter *Gentlewoman*
62 Barton Arcade, Manchester, England

(William Gadd)

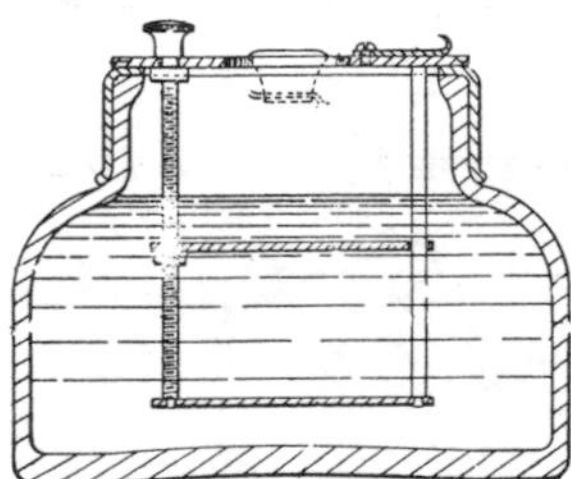

Mingay's Improved Inkstand contains means for regulating the depth to which the pen can be dipped, namely a false bottom or regulating plate, adjusted by means of a screw. By this means it is possible to prevent the pen being charged with too much ink, thereby causing blots and in some cases soiling the fingers.

17517/1905
Frank Healey Mingay *Bank Teller*
Berfield, Bridge of Weir, Renfrewshire, Scotland

(H D Fitzpatrick)

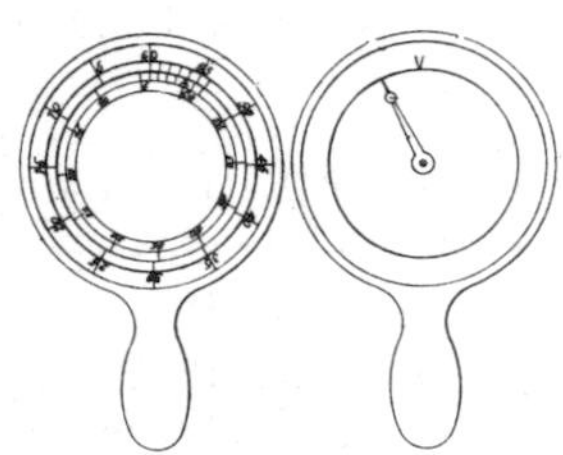

Cameron's Improved Game Apparatus is also for use in training persons in judging angles, consisting of a rotating pointer and a concealed vernier scale turned by the pointer, indicating accurately the position to which it has been turned. Any number of players may take part, arousing competition in the skill with which they estimate the position of the pointer, and a large model of the device is of particular use in places of entertainment or public schools.

15730/1905
George Cameron *Engineer*
3 Bignell Road, Plumstead, Kent, England

(Rayner & Co)

Stout's Improved Buoyancy Motor uses atmospheric air as the actuating force, in such manner that the sum of activity of the movable parts is constantly equal to this initial actuating force.

In carrying the invention into effect, the buoyancy or ascending power of submerged floats is utilised for imparting rotary motion to a shaft, from which connection may be made to an object or machine to be driven thereby.

The motor mechanism consists in a number of cylindrical vessels, floats, fitted with pistons arranged on chains and immersed in water. The vessels are interconnected by means of flexible tubes, and are made to increase their buoyancy when they are at their lower position in the water. The action of the floats is uniform and steady, as is the motion of the shaft operated thereby; as each float rises to the surface of the water the same is caused to descend again beneath the water, so that the operation is thus rendered continuous. Previous to the passing of each float into the water, the air is caused to be displaced or almost immediately exhausted therefrom, for the entrance of water thereinto, and in this way the float is brought substantially into equilibrium with the full body of water, thereby rendering it non-resistant to the action of the ascending floats in pulling the same downwards through the water. In this way, also, the displacement of the water caused by each descending float gives increased buoyancy or power to each ascending float, and thus is the operation of the motor rendered effective and reliable.

A plurality of such motors may be arranged to work together.

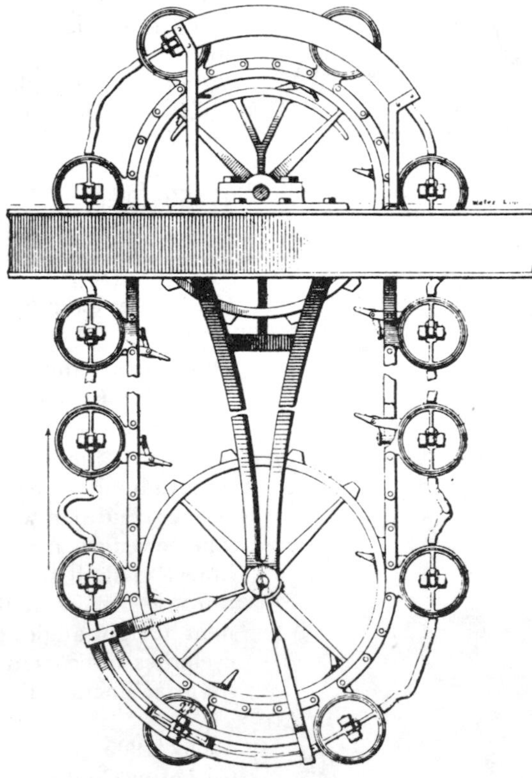

15746/1905
Cornelius Stout *Manufacturer*
Pomona, California, USA

(A M & WM Clark)

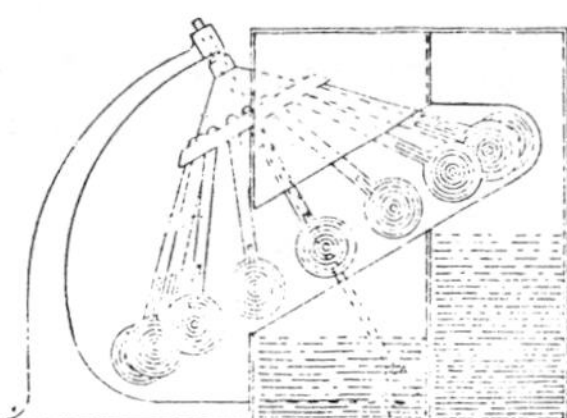

Heywood's Machine for Obtaining Power or Motion by the Expansion or Contraction of Mercury, Air, Spirits, Metals and the like utilises the temperature of air in ordinary rooms, the temperature of air which stands in a vessel partly filled with water, the temperature of water, the temperature of the sun's rays and temperature of the shade from the sun, and other heat such as a fire.

It consists of a wheel whose spokes are separate thermometers arranged so that there is a temperature difference between one side and the other. This disturbs the balance and sets up a rotating or oscillating motion because as the mercury or other expanding agent in each tube moves with the change of temperature the centre of gravity moves nearer to the shaft or fulcrum, causing the cooler side of the wheel to overbalance and therefore move.

In another configuration, the thermometers are arranged in a conical, rather than a radial, fashion.

In another embodiment, the expansion and contraction of air in elastic vessels may be used to create the imbalance.

17600/1905
John James Heywood *Draughtsman*
19 Deardengate, Haslingden, Lancashire, England

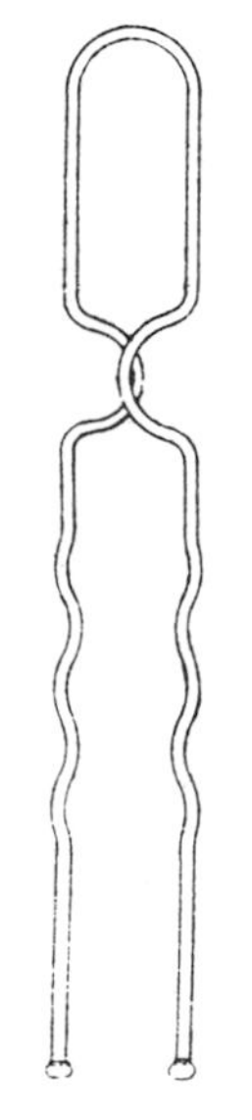

Dick's New Hair Pin is designed to obviate the inconvenience and annoyance now so frequently experienced owing to hair pins working loose, falling out, and being lost. This is obviated by providing an inwardly curved portion on one limb which enters, when the limbs are compressed, a slot on the other limb.

18108/1905
James Alexander Dick *Merchant*
Holmbank, Polmont Station, Stirlingshire, Scotland

(Elt & Co)

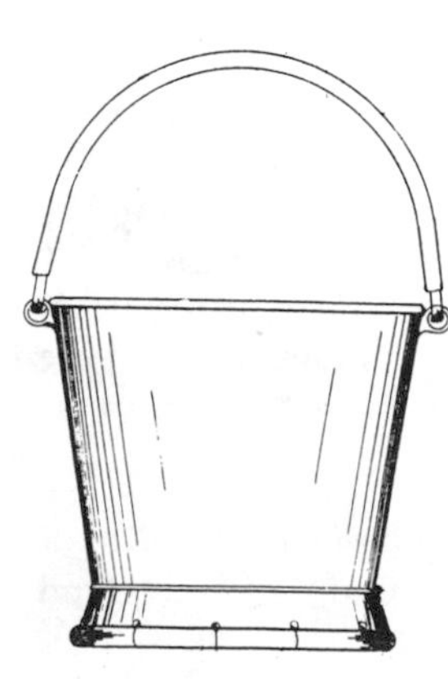

Norledge's Improved Pail for use in Apartments for the Sick has its handle covered with rubber or other resilient material so that when it falls on the rim it will not make much or any annoying noise. A rubber ring attached to its foot also prevents it from making more than a deadened sound when stood down.

18654/1905
Sarah Ann Norledge *Matron of Nursing Home*
Coolavin, Hawkwood Road, Boscombe, Hampshire, England

(Hughes & Young)

Villacampa y Villacampa's Improved Ventilating Apparatus for the Interior of Foot-wear comprises an air suction and compression device in the form of a closed box in the heel. This acts as an air-pump put into action by the natural impulsion of the foot, so that the motion of walking causes air to be distributed through a perforated tube extending along one side of the boot.

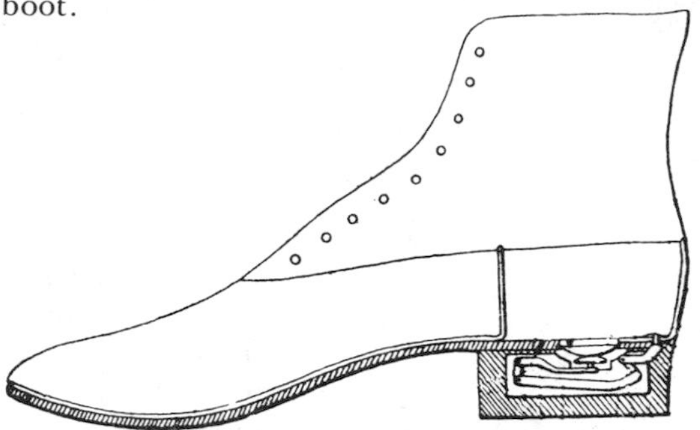

15923/1905
Miguel Villacampa y Villacampa *Mechanical Engineer*
3929 Rue Rivadavia, Buenos Aires, Argentine Republic

(Herbert Haddan & Co)

Brude's Improved Life-boat affords shelter to its occupants and may be equipped for a long voyage. The hull is in the shape of an egg, and reached through man-holes with watertight covers fore and aft. There is a seat round the inside of the hull with storage space beneath. Windows are located at such a height above the seat that they can be used as rowlocks when rowing is necessary, in calm weather or as exercise for the occupants. There is a mast socket and a drop keel. The rudder has two tillers; in foul weather the inner one is used and the steersman stands in a conning-tower; in fair weather the outside tiller may be attached. The boat is fully equipped with water closet, pumps and other necessary appliances.

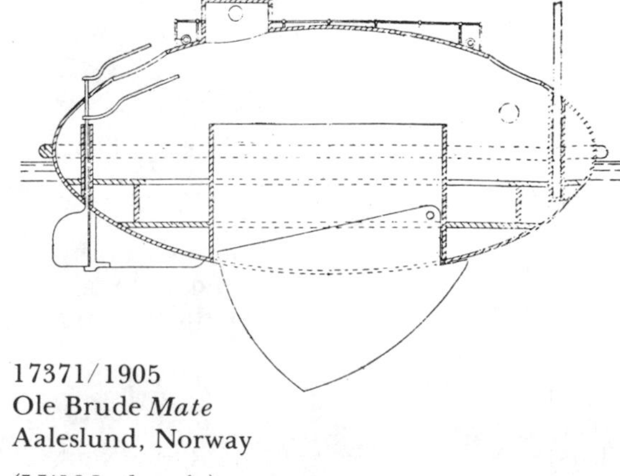

17371/1905
Ole Brude *Mate*
Aaleslund, Norway

(J W Mackenzie)

1905

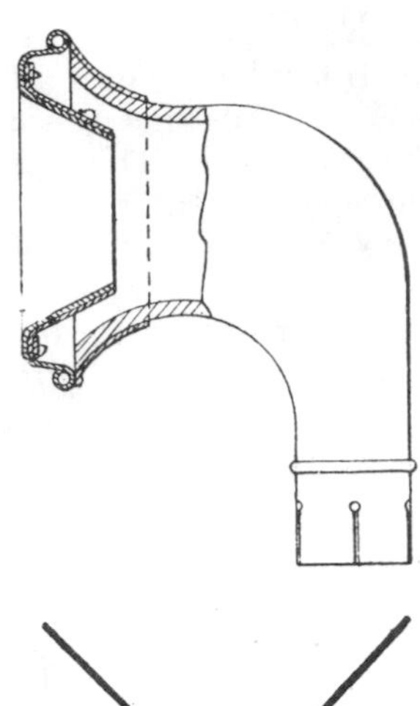

Savage's Improved Sterilizer Mouthpiece for Telephones comprises an annular insert, spaced from the mouthpiece proper, so that it may receive on its exterior surface a suitable paste-like sterilizer or an absorbent material for a liquid antiseptic.

13597/1905
Orrin Henry Savage *Manufacturer*
263 West Twenty-third Street, New York, USA

(Herbert Haddan & Co)

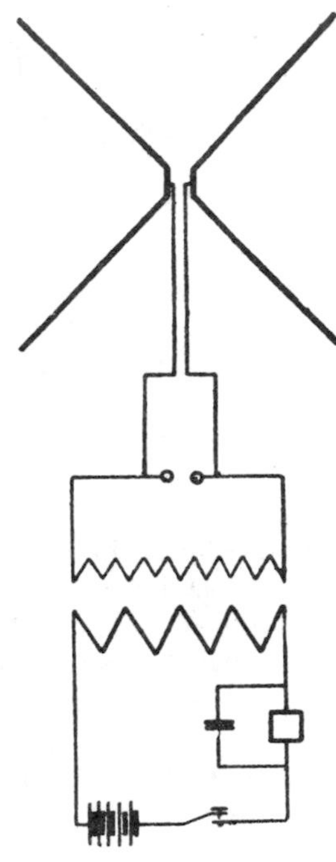

Rota's Apparatus for Preventing the Formation and Falling of Hail is based on the meteorological utilisation of electric waves.

It consists in a complete device which may be mounted in the open field, and in places usually devastated by hail, preferably on level ground with no high tree at less than 50 metres distance. The drawings show firstly a representation of the whole of the circuits and apparatus (which consists of an electro-generating system, a transforming system, line wires and accessory parts) and secondly its installation in a field.

The generating system is protected by a hut from which radiate four line wires at right angles, carried by poles of seasoned timber. The direction of one straight pair of lines corresponds with the usual axis of translation of local storms. The lines are connected in two adjacent pairs, and the pairs are connected to the two ends of the secondary winding of a Ruhmkorff coil. The electric waves generated by the coil run along the lines and radiate in the space, influencing the electric state of the clouds and preventing the formation of hail.

In order to render the action of the apparatus effective it must be actuated a little before the storm becomes dangerous, and it is convenient for it to continue to work until after the storm has abated.

24688/1905
Ettore Rota *Lieutenant in the Italian Navy*
25 Via Vittorio Emanuele II, Cassale, Monferrato, Alexandria, Piedmont, Italy

(Harris & Mills)

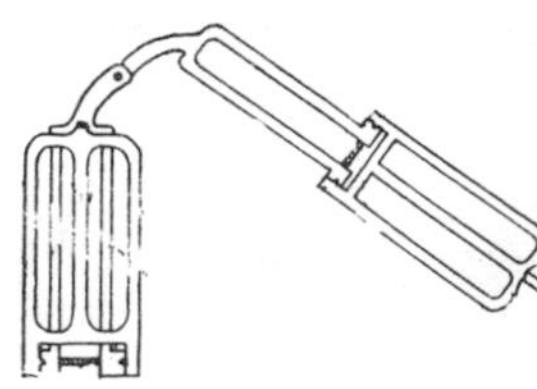

Schmidt's Improved Moustache Clip is intended to give shape to and hold up the moustache during the toilet and shaving.

17929/1905
Georg Schmidt *Perfumer*
2 Escherstrasse, Hannover, Germany

(R J Urquhart)

Goldin's Improved Stage Illusions give the appearance of a lady being fired from a cannon into a series of locked and bound nested boxes, or of a lady disappearing from a scale pan after she has been enveloped in a cloak and a pistol fired, the whole being effected in full view of the audience.

In the case of the first illusion, the effect is achieved by providing the cannon with a concealed means of egress, through which the lady descends after effecting an explosion by firing a revolver attached inside the cannon's mouth. The trunks are now conveyed from their previous position, such as the roof, and land on the stage, at which point a double dressed to resemble the lady first fired from the cannon is bodily inserted into them through another trap door, the trunks are unbound, and the artiste steps on to the stage.

The scale pan and cloak illusion is achieved by means of a plank bridge placed under the weighted scale pan, which is provided with a concealed trap door through which the lady climbs while the pistol is fired elsewhere in the theatre to distract the attention of the audience.

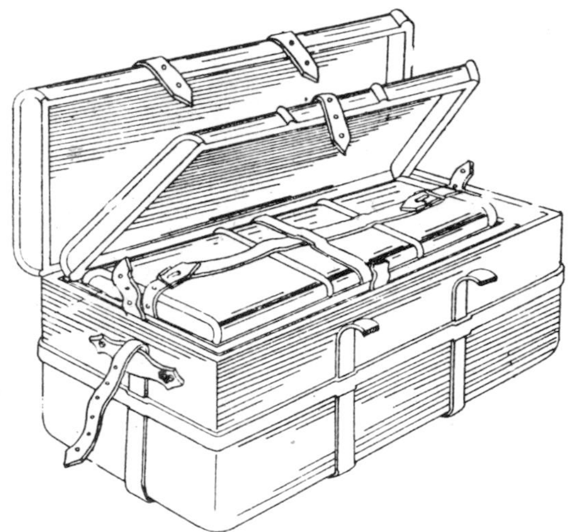

26307 & 26308/1905
Horace Goldin *Illusionist*
17 Torrington Square, London WC, England

(Lewis William Gould)

Macdonald's Apparatus for Enabling the Pressure of Internal Combustion Engines to be Utilised for Sounding a Warning Device connects the horn of an automobile, or the siren of a ship, to an explosion chamber of the engine.

26971/1905
Alexander Macdonald *Gentleman*
72 Holland Park Avenue, London, W, England

(Lewis William Gould)

Goad's Ladies' Garment Suitable for Motor Car Travelling is designed to give suitable protection against cold, wet and dust whilst affording sufficient freedom for the feet and hands. It is preferably double-breasted, and with a neck-flap. The neck may be fitted with a collar and detachable hood. The sleeves terminate in gloves or mittens, but with a slit to allow the hands out if necessary. The bottom of the garment is made large enough to enclose the ordinary skirts, and fitted with gaiters, which may be formed as boots, preferably made with stiffer material than the garment, or provided with whalebone so as to cause the lower end of the garment to hang down and resemble a skirt.

26057/1905
Gertrude Goad *Married Woman*
29 Conduit Street, London W, England

(Jensen & Son)

Paparella's Improved Protector for Trains comprises a small pilot motor which precedes the main engine, to which it is connected by a cable several hundred metres long. The pilot motor carries a number of contacts, any one of which, if closed, signals to the driver by means of a lamp or bell that it has met an obstacle, so that he may apply the brakes on the engine.

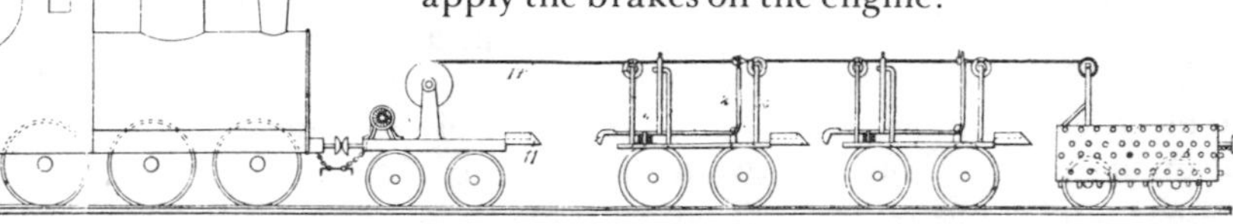

25865/1905
Elpidio Paparella *Engineer*
95 Via Principe Umberto, Rome, Italy

(Boult, Wade & Kilburn)

Robeson's Improved Motoring, Driving or Walking Coat overcomes the problems experienced by wearers of ordinary overcoats which, when exposed to a head wind, separate at the middle, thereby exposing the legs of the wearer and, what is more serious, the crutch or abdomen. The wearer thus runs considerable risk, many cases of inflammation having been traced to this cause. This improved garment has the appearance of an ordinary coat when worn for walking, since it is provided with flaps or bags which can be folded back but, when motoring, are drawn out so as to fit neatly over the knees and afford complete protection.

26290/1905
John Granville Robeson *Tailor and Robe Maker*
196 Aldersgate Street, London EC, England

(Marks & Clerk)

Sayer's Improved Cycle or other Vehicle is efficient, convenient and portable. This is achieved by constructing the vehicle with a variable, telescopic wheelbase, wheels which may be readily adapted to run on rails, and brakes which may be automatically applied at a predetermined inclination of the road.

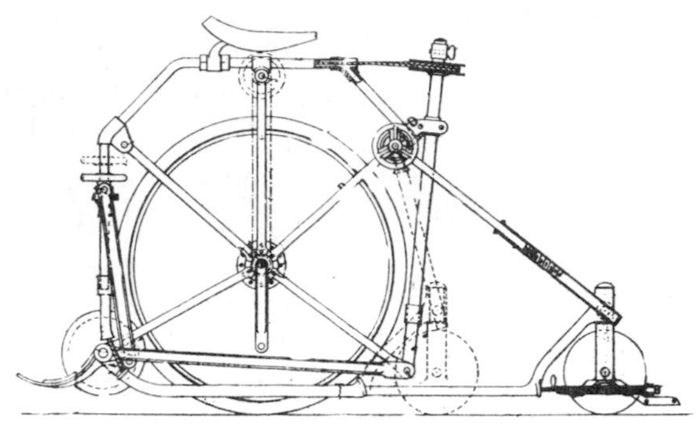

18699/1905
Robert Cooke Sayer *Civil Engineer*
11 Clyde Road, Redland, Bristol, England

Taylor's Improved Apparatus for Generating Electricity on Motor Cars or Motor Boats uses a wind motor and dynamo to enable the air, or wind resistance, or other pressure to which the vehicle concerned is subjected — especially when travelling at high speeds — to be utilised to generate electricity for charging batteries, lighting, or providing auxiliary driving power. It is obvious that instead of employing just one tube to collect the air, several tubes may be employed and the air received by them conveyed to a large number of wind motors or turbines, which may be arranged in tandem fashion if desired.

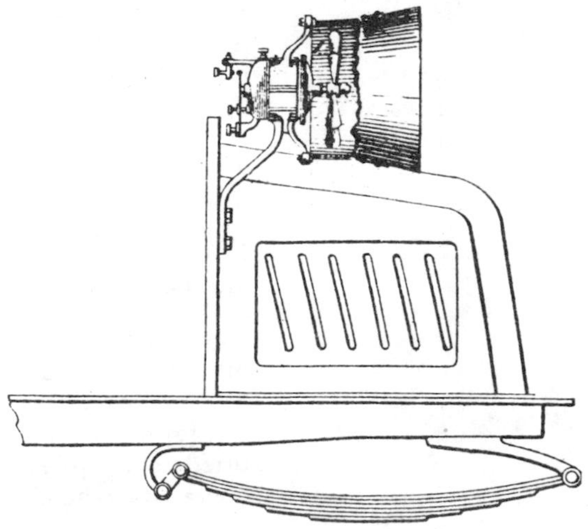

25353/1905
Joseph Taylor *Engineer*
18 Ingleby Road, Ilford, Essex, England

(Brewer & Son)

1905

Lady Lampson's Improved Carving Knife or Slicing Device has a number of blades secured by a pivot pin and a locking pin. It is of use in slicing meat, fish, game, bread or vegetables. An adjustable guide may be secured to the outside blade. The blades may fold down, wholly or partly, into the handle.

22006/1905
Sophia, Lady Lampson *Widow*
70 Knightsbridge, London, England

(Jensen & Son)

Serné and Parker's Apparatus for Forcing Air employs one of the explosion cylinders of a self-propelled vehicle for inflating the tyres when necessary. A throttle valve disconnects the explosive gas supply, and another valve puts the cylinder into communication with the tyre.

22105/1905
Leon Serné *Manufacturer*
88 Seymour Street, Euston Road, London, England

Mark Parker *Merchant*
31 Tavistock Square, London, England

(Haseltine, Lake & Co)

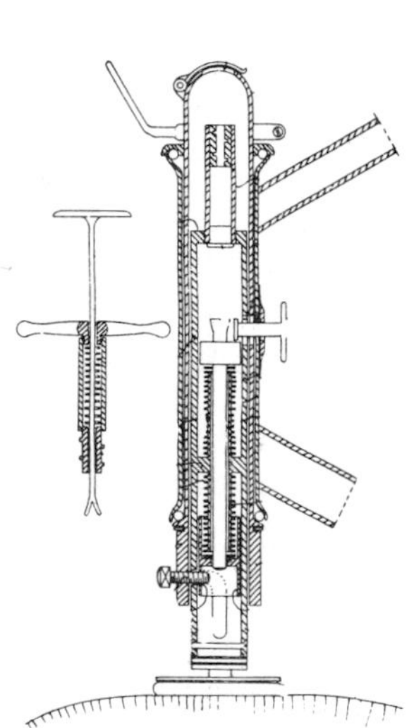

Donndorf's Improved Alarm for Cycles and the like is designed to protect them from theft by giving an audible signal upon an unauthorised person trying to use the wheel, and comprises an explosive alarm. The wheel is locked by a removable key, and the thief, in endeavouring to free the steering-head, detonates a cap. When the alarm is set, a plate is pressed on to the tyre. A thief, thinking that the locking pin is responsible for holding the plate down, turns it; this releases the firing pin and surprises the malfeasant.

23576/1905
Albert Donndorf *Blacksmith*
10 Bahnhofstrasse, Dobrilugk, Germany

(Paul E Schilling)

McGill's Improved Eraser is provided with pneumatic means for clearing away the erasings; it is formed with one or more air passages and is fitted to a squeeze ball or collapsing member operated by the hand.

24547/1905
Charles Edward McGill
Owensboro, Kentucky, USA

(W P Thompson & Co)

Prince Hozoor Meerza's Improved Suspended Rope Railway provides a novel device for guiding the suspended rail or tramcarriages past their supports. The carriage bodies may be of any suitable construction for transporting a load or passengers and may be hauled or self-propelled.

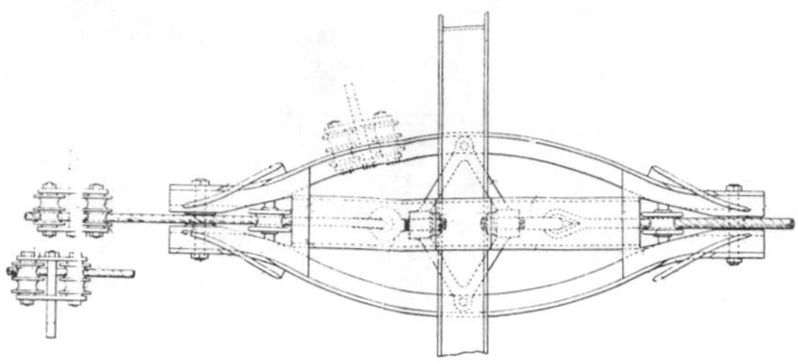

18064/1905
Prince Hozoor Meerza *Gentleman*
The Palace, Murihidabad, Bengal, India

(Abel & Imray)

Burn's Improved Lavatory Appurtenance is a little device consisting of a soap tray with fluted inclined sides and a removable hollow central part with a hole to receive toothbrush handles. The soap and nail-brush rest in the annular groove. Its object is to hold the soap, nail-brush etc. so that perfect drainage shall ensue, enabling the soap tablet after use to become almost immediately dry and hard and thereby avoid the wastage and uncleanliness customary in soap dishes where the tablet is allowed to remain soaking in water on a level surface.

18131/1905
John Sowerby Burn *Merchant*
Carlton Hotel, Tunbridge Wells, Kent, England

(W P Thompson & Co)

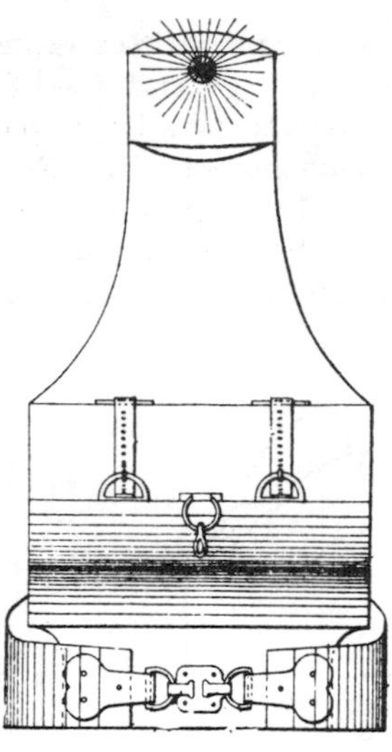

Focketyn's Improved Life Belt Apparatus has front and back floating members, and a lamp and hood affixed to the back member. Its light and simple construction means that people at sea in times of unusual peril can wear it whilst sleeping, without being caused any great amount of inconvenience thereby.

26487/1905
Jack Focketyn *Architect*
30 Boulevard van Iseghem, Ostend, Belgium

(Marks & Clerk)

Wynne-Ffoulkes's Improved Ladies' Hat solves what may be termed the "matinée hat" problem, that is to say to provide means whereby ladies wearing broad-brimmed or so-called "picture" hats may be able, at a moment's notice, and without disarranging the coiffure, to divest themselves of that portion of the hat which blocks the view of persons sitting behind them, and appear attired in a head-dress which is complete in itself and of a becoming character. To this end, ladies' "picture" hats for use in theatres etc are made with detachable brims. The brim has a stiff flange adapted to fit friction-tight to the separate crown.

The crown is of approximately hemispherical form, adapted to fit closely to the head. It is made of stiff material, such as buckram and suitably trimmed; it may have an opening at the top to be fastened up when required, or may be of any other suitable design, such as a "Juliet" cap which, when the brim is removed, would constitute a complete head-dress such as a lady might wear indoors under ordinary circumstances.

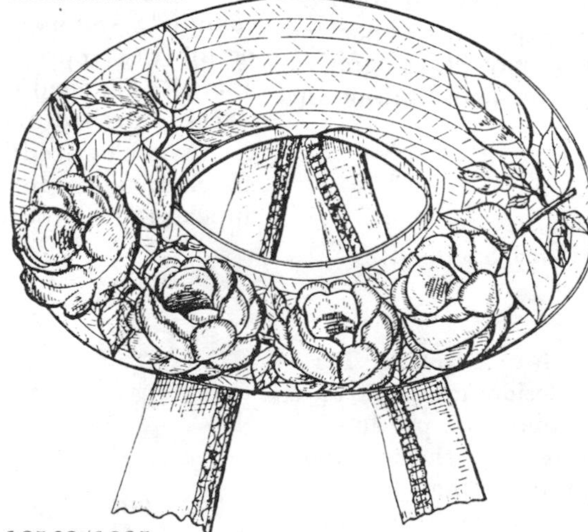

18568/1905
Louisa Florence Wynne-Ffoulkes
117 Beaufort Mansions, Chelsea, London SW, England

(A M & WM Clark)

Truman's Improved Holders for Trays are used for receiving the trays used by barmen and others in serving beer in public houses. They provide a device which will enable a dry tray always to be at hand, will collect any liquid spilt during serving customers and, if desired, make it possible for the beer which is slopped on to the trays to be collected and sent back to the tank.

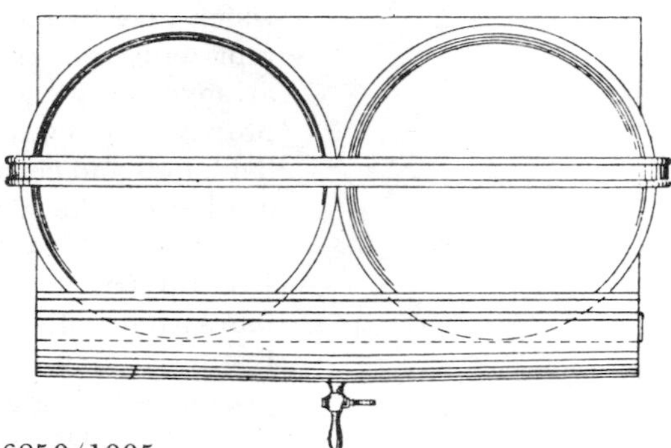

16850/1905
Joseph Truman *Licensed Victualler*
Butchers Arms Hotel, 17 Monk Street, Birkenhead, Cheshire, England

(W P Thompson & Co)

L'Estrange Burges's Improved Game Imitating the Strokes of Golf and the nomenclature and general surroundings of Cricket.

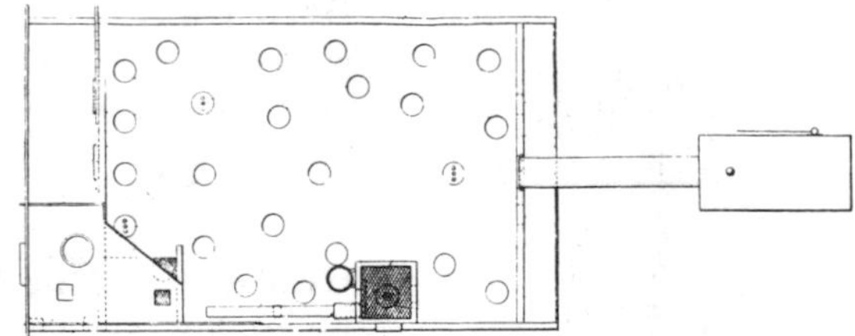

19326/1905
John Charles Walter L'Estrange Burges *Clerk in Holy Orders*
26 Cumberland Terrace, Regent's Park, London NW, England

Hahn's Improved Tooth Brush has a slot in its head so that cleansing the bristles is facilitated, thus obviating the defect of all tooth-brushes ordinarily employed at present, which are either difficult or impossible to cleanse, so that particles of food left in the teeth become rubbed into the brush, remain there and decompose.

19886/1905
Dr Gustav Hahn *Dentist*
74 Kurfurstenstrasse, Berlin, Germany

(Haseltine, Lake & Co)

1905

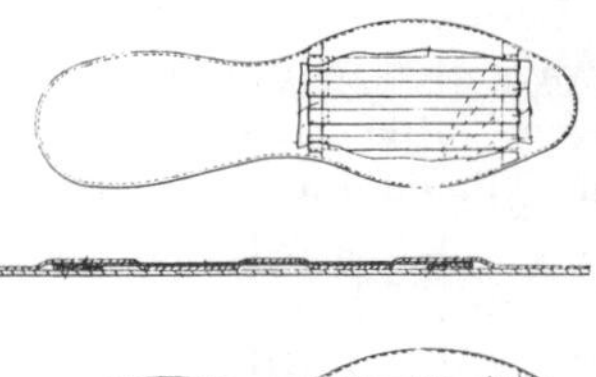

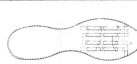

Romain and Anidjah's Improved Electropathic Socks for Boots and Shoes are made with copper and zinc strips spaced apart and interlaced with the felt of the sole, and mounted on a backing of cork. The strips may be connected in series or parallel.

The object of the invention is to provide a means for generating a slight electric current which may be passed through the body, and it represents an improvement on the various arrangements which have hitherto been proposed for obtaining such a result, for instance hat bands and corsets with copper and zinc strips or wires interlaced with the fabric employed.

19125/1905
Jonas Anidjar Romain *Manufacturer*
Brunswick House, Brunswick Place, City Road, London EC, England

Stella Sarah Anidjah *Spinster*
355 Essex Road, Canonbury, London N, England

(Marks & Clerk)

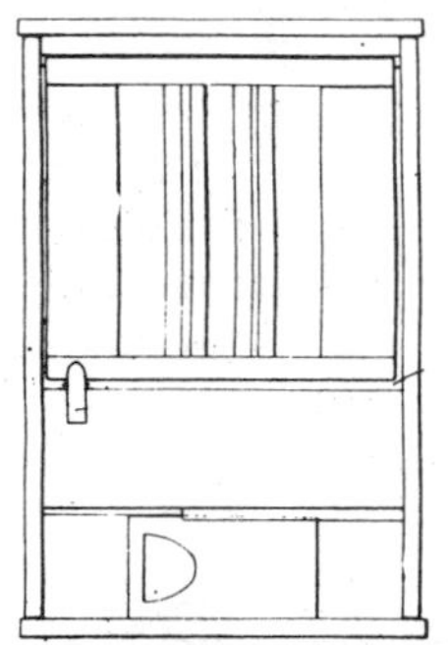

Stewart's Improved Poultry Trap Nest is so arranged that the door closes if the occupant should lay an egg, thus indicating that the event has taken place and detaining the hen until the egg has been removed.

20177/1905
Frederick Stewart *Gentleman*
26 East Dulwich Road, London, England

(Hughes & Young)

Richardson's Ring Door Knocker to Wear on the Finger has a knob or projection for knocking on doors; the projection may be integral or detachable and the whole made of brass or any other suitable material.

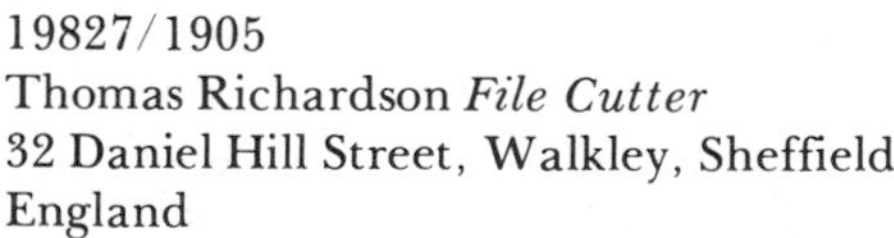

19827/1905
Thomas Richardson *File Cutter*
32 Daniel Hill Street, Walkley, Sheffield, England

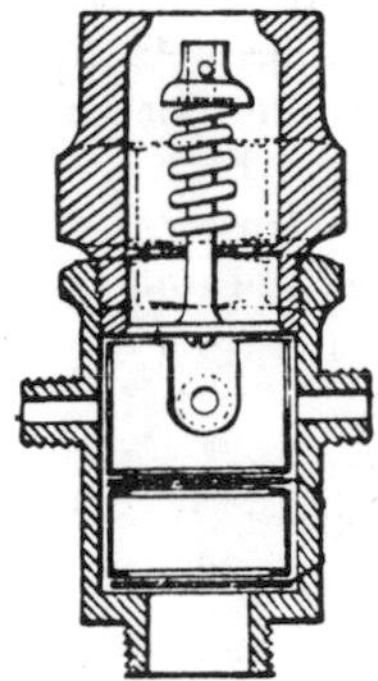

Macquaire's Apparatus for Inflating Tyres converts one of the explosion cylinders of a motor-car engine into a pump by providing a spring weaker than those of the ordinary admission valves, and a filter so that dirt, soot, or oil is intercepted and not passed to the tyres.

22000/1905
Léon Macquaire
21 Rue de Malte, Paris, France

(Boult, Wade & Kilburn)

Mahs and Kochmann's Improved Means for Weighing the Contents of Bags and Baskets allows of the testing, unobserved, in a basket the weight of goods bought, such as meat, vegetables or the like. The basket or string bag is combined with an ordinary spring balance whose readings may be surreptitiously observed through an opening in the bag.

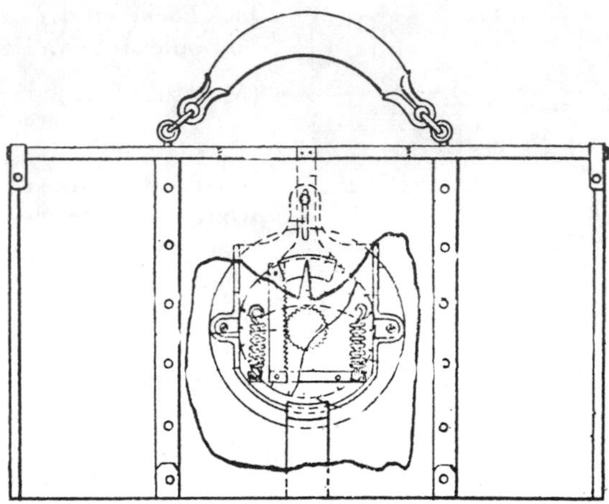

17557/1905
David Mahs *Saddler*
Biala, Galicia

Berthold Kochmann *Merchant*
6 Friedrichstrasse, Kattowitz, Upper Silesia

(Herbert Haddan & Co)

Stein and Bucher's Improved Toy is simple, inexpensive, and affords considerable harmless amusement. It comprises a wheel in multi-coloured sectors, which carry near their peripheries representations of mice or other animals. Fastened to the handle with which the wheel is pushed along is a representation of a cat. As the device is moved over the ground, the appearance of running is obtained and the cat appears to reciprocate as if chasing the mice.

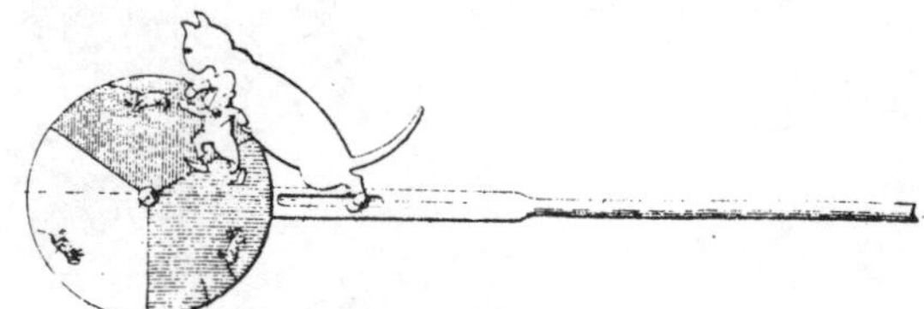

17692/1905
Anton Stein *Gentleman*
Kennett, California, USA

Joseph Bucher *Gentleman*
Redding, California, USA

(Victor J Evans)

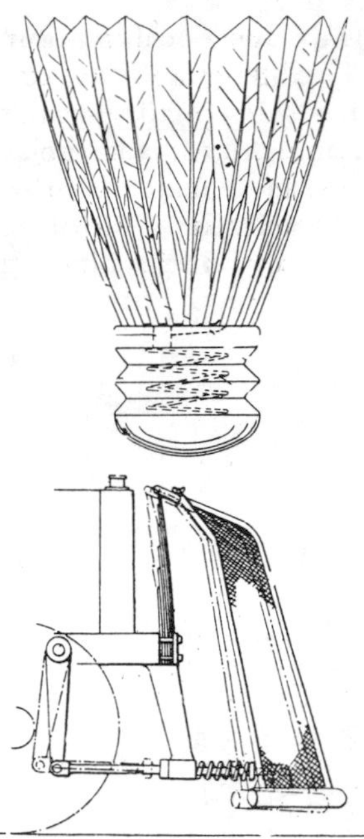

Sibbring's Improved Shuttlecock emits a musical note during its flight, since it has a bellows in the base compressed when it is struck, and a reed actuated by the expanding bellows as it flies.

25803/1905
Mary Sibbring *Married Woman*
17 Chester Road, Tuebrook, Liverpool, England

(J A Coubrough)

Ansell's Improved Protective Device for use with Self Propelled and other Vehicles consists in a light framework covered with canvas, sheet metal, woven wire or other flexible material, and inclined backwards and towards each side, the object being to minimise the effects resulting from an accident incidental to the vehicle colliding with obstructions such as human beings or animals.

The lower edge of the guard is fitted with a cushion of cork or india-rubber, and it may be arranged to open so as to give access to the starting handle or other parts, and provided with glass panels in front of the lamps. The whole device is arranged at such a distance from the ground level that, in the ordinary course, the impact with the obstruction, say in the case of a human being, would take place below the knees, causing the individual to be caught and thrown upon the device and, owing to the character thereof, to be deflected and left upon the side of the vehicle and, as regards the latter, out of harm's way.

21536/1905
John Evelyn Ansell *Barrister*
4 Glenloch Road, Hampstead, London, England

(Haseltine, Lake & Co)

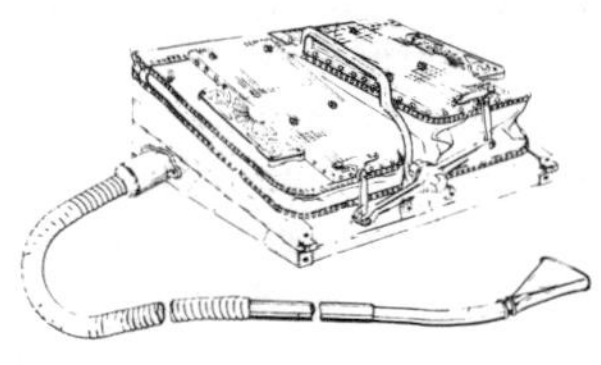

Griffiths's Improved Vacuum Apparatus for Removing Dust from Carpets is a portable appliance which can be operated, and the power required readily obtained by, any one person (such as the ordinary domestic servant) and in such a manner that the weight of the operator is utilised for assisting in obtaining the requisite power. At the same time, the whole of the parts are constructed so as to be easily removable for cleansing and can also be attended to by any ordinary person. The apparatus has a flexible pipe, to which may be attached a variety of shaped nozzles, connected to a pair of exhausting bellows operated by the feet of the maid's assistant.

While being very effective in operation, such apparatus can be manufactured to be put on the market at a price within reach of most householders and when not in use occupies but very little space.

21304/1905
Walter Griffiths *Manufacturer*
72 Conybere Street, Highgate, Birmingham, England

(Lewis William Gould)

The Zinns' Improved Safety Razor relates to those of the type where the blade is narrow, thin and disposable as in the kind described in the Specification of Letters Patent No. 28763, AD1902, granted to King Camp Gillette. It provides a means of easily inserting and withdrawing such blades from their holder and retaining them accurately in place to give an ordinary, medium close, or close shave as desired. Accordingly, the blade is clamped in position by a spring, and the frame is curved to form a receptacle for the lather.

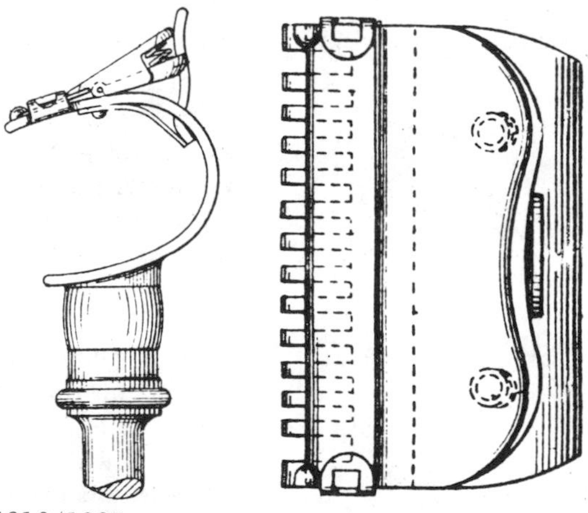

24010/1905
Mary Zinn
54 West 96th Street, New York, USA

Martin Zinn
121 St Nicholas Avenue, New York, USA

Arthur Simon Zinn
218 West 139th Street, New York, USA
Manufacturers

(A M & WM Clark)

1905

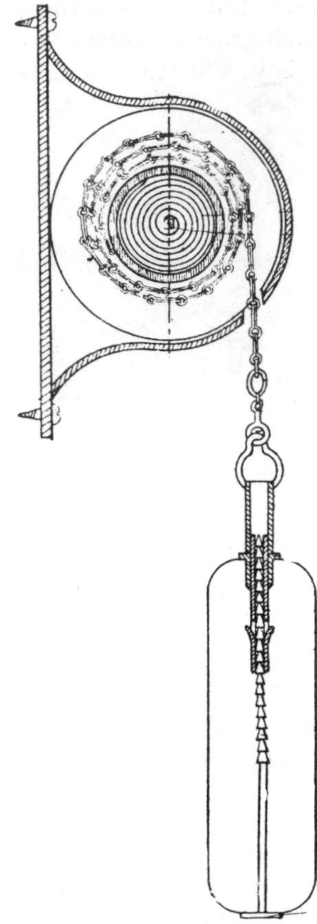

Baron's Improved Suspension and Securing Device for Toilet Soap has a chain wound on a drum, by which the soap is suspended, and a spiral spring, to prevent the soap from being withdrawn during use.

26051/1905
René Joseph François Samuel Baron
Manufacturer
65 Rue Ste Anne, Paris, France

(Boult, Wade & Kilburn)

Blättner's Selfacting Egg Lifter raises boiled eggs from the water after a predetermined time, thus avoiding the need for a person to have to stand and watch an hour-glass or listen for a bell if the eggs are not to boil for too long. The eggs are put in a pan, which is counterbalanced by means of a sliding weight on the beam which carries it. The beam is tilted so that the eggs are immersed in the boiling water, and a graduated slide adjusted to the boiling-time desired. The slide is gradually withdrawn, either by the fall of sand in an hour-glass, or by clockwork. When the preset time has elapsed the beam is released and the eggs are lifted from the pan.

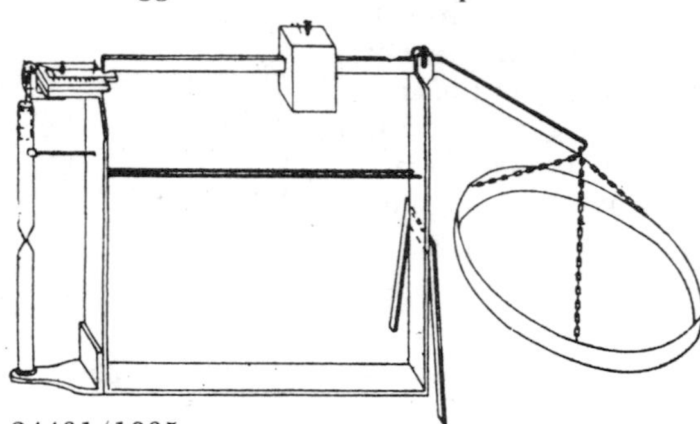

24401/1905
Louis Blättner *Manufacturer*
5 Jägerstrasse, Cassel, Germany

(Otto Klauser)

Cribb's Lid Connector for Tea-pots makes it possible for the lid to be easily opened and easily detached but may not be liable to fall off, as is the case with pots ordinarily constructed, and whereby many lids of china and porcelain are broken. The device comprises a forked metal arm to grip the knob of the lid, hinged to a loop which encircles the body of the pot, and tightened with a screw.

25514/1905
James Preston Cribb *Photographer*
127 Chichester Road, North End, Portsmouth, England

(Browne & Co)

Marshall's Improved Toy for Soothing and Amusing Infants is a flexible arch of wire or cane, sprung between the sides of the infant's chair or cot, and carrying dolls, balls, bells, pompom rosettes, loops, festoons, movable figures or other amusing objects, the whole being free to swing and vibrate and cause a pleasing and musical effect to a child of tender age.

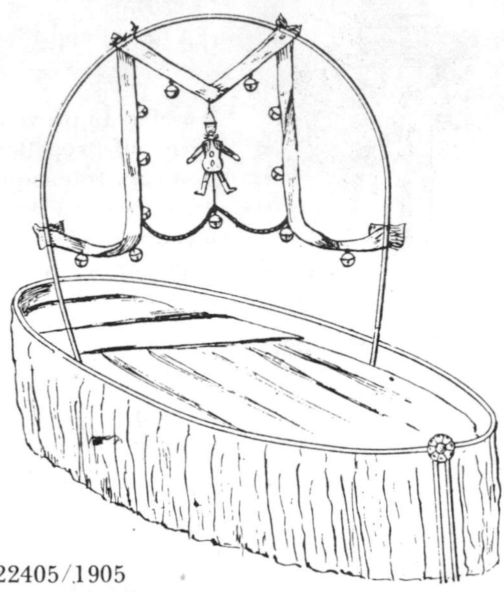

22405/1905
Dale Marshall *Engineer*
30 Winchcombe Street, Cheltenham, Gloucestershire, England

Edwick's Improved Ladies' Hat Fastener does away with the long pins which must normally be used to secure the hat, which cause the hat or bonnet to become unsightly and shabby after but a short time because of the numerous holes produced by said pins, which are also a source of great danger to others who may happen to come in contact with their projecting ends by accident. This invention does away with these disadvantages and consists in a single pin, preferably curved, provided with notches which engage with a clasp. The pin is attached to eyes sewn to the interior of the hat, one on each side. The device may be made of bone, ivory, horn, metal or other suitable material and, if required, parts may be jewelled.

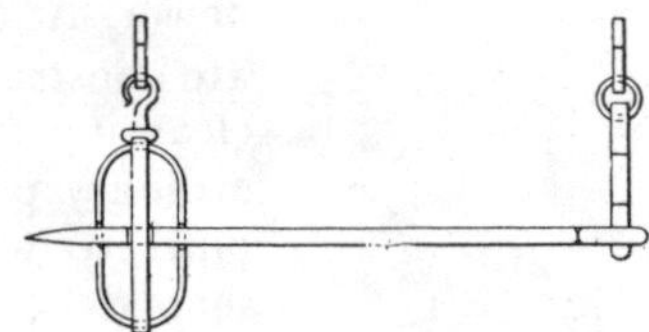

12918/1905
Ellen Elizabeth Edwick *Gentlewoman*
133 Mile End Road, London, England

(Crozier & Co)

Stawartz's Improved Hat Fastener consists in a pair of spring-loaded combs attached to a yoke in the crown of the hat. The combs are sprung so that they normally grip the hair; finger-grips enable them to be retracted during the positioning of the hat on the head. The pins thus ordinarily employed are entirely dispensed with and the securing means are also hidden from view.

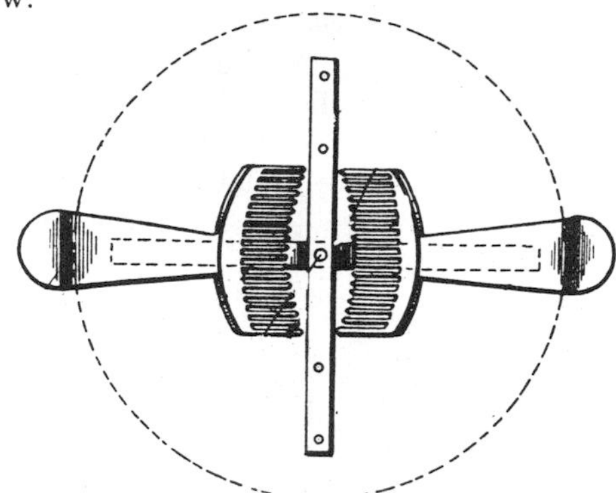

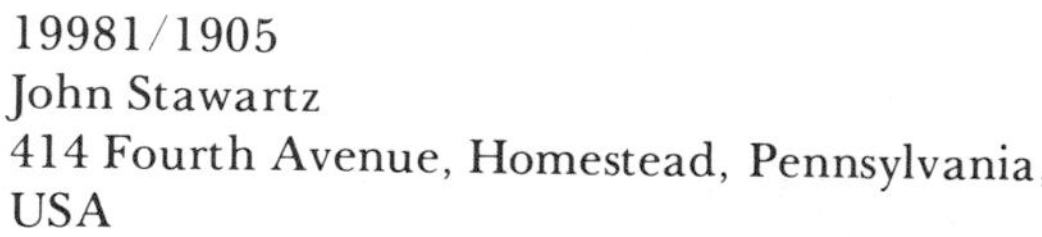

19981/1905
John Stawartz
414 Fourth Avenue, Homestead, Pennsylvania, USA

(Herbert Haddan & Co)

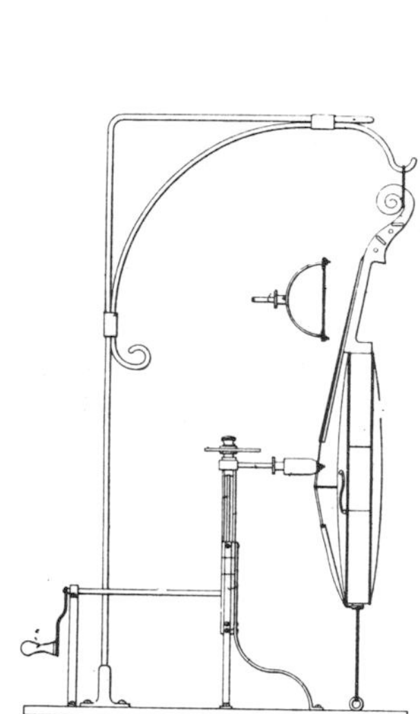

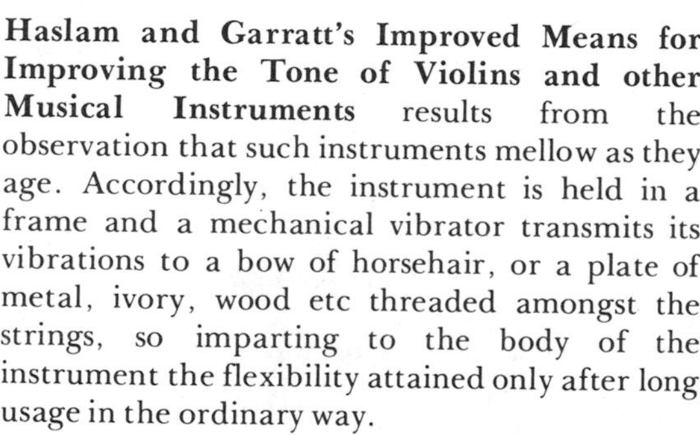

Haslam and Garratt's Improved Means for Improving the Tone of Violins and other Musical Instruments results from the observation that such instruments mellow as they age. Accordingly, the instrument is held in a frame and a mechanical vibrator transmits its vibrations to a bow of horsehair, or a plate of metal, ivory, wood etc threaded amongst the strings, so imparting to the body of the instrument the flexibility attained only after long usage in the ordinary way.

25683/1905
William Doidge Haslam *Medical Practitioner*
87 Park Lane, Croydon, Surrey, England

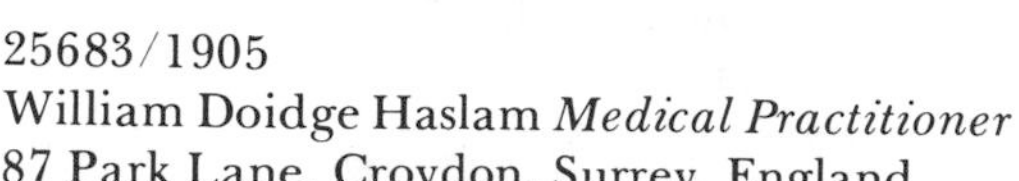

John Edwin Garratt *Manufacturer*
124 Southwark Street, London, England

(Marks & Clerk)

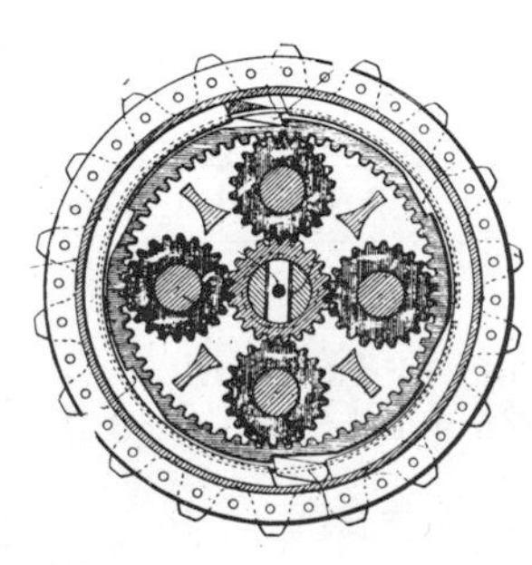

Archer's Improved Variable Gear and Brake Mechanism for Velocipedes incorporates a back-pedal brake into Sturmey's invention.

25799/1905
James Archer *Engineer*
35 Mayors Road, Peterborough, England

The Three Speed Gear Syndicate Ltd *Engineers*
Faraday Road, Kenton, Nottingham, England

(John G Wilson & Co)

Howgrave-Graham and Gavin's Improved Foot-rest for Indoor Use is suspended by rods or chains from the mantel-shelf. The rods or chains may be attached to the shelf either by specially-shaped hooks, or by clamps.

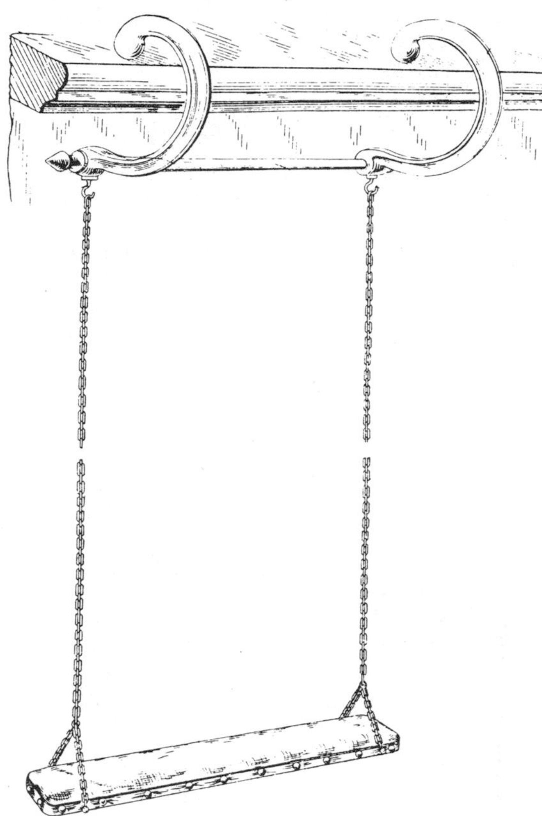

22207/1905
Hamilton Maurice Howgrave-Graham *Civil Servant*
12 Willow Road, Hampstead, London NW, England

William Gavin *Clerk*
65 South Hill Park, Hampstead, London NW, England

Harris's Improved Device to Prevent Side-slip of Road Vehicles — particularly motor road vehicles — has a pair of wheels whose axles are arranged in the same plane as that of the rear axle of the vehicle, but at 45°C to the road surface. When there is a danger of side-slip, these bevelled wheels may be let down and held against the road surface by means of a spring. They may be lifted from the ground by any convenient means when not in use.

26702/1905
Samuel Harris *Gentleman*
29 Lincoln's Inn Fields, Holborn, London, England

(Phillipss)

1905

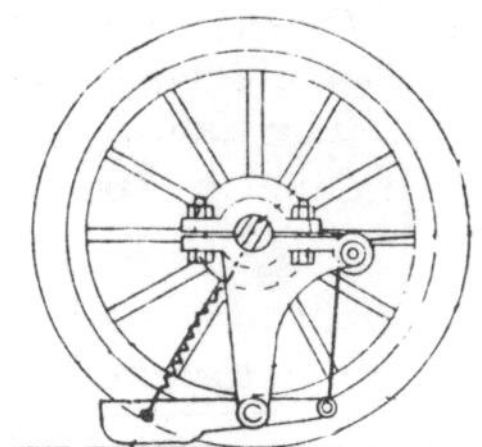

Buckingham's Improved Means for Preventing Vehicles from Slipping Sideways or Skidding when turning corners, or swerving to avoid other vehicles, is a brake consisting in blades carried by a lever, and normally kept out of contact with the road surface. The blades are splayed outwards and sharpened, and arranged on each side of a rubber, vulcanized-fibre or other pad. The blades and pad are curved longitudinally to prevent too violent action of the anti-skidder, and to permit the backing of the vehicle.

22993/1905
Edward John Buckingham *Engineer*
193 Upland Road, East Dulwich, London, England

(T E Halford)

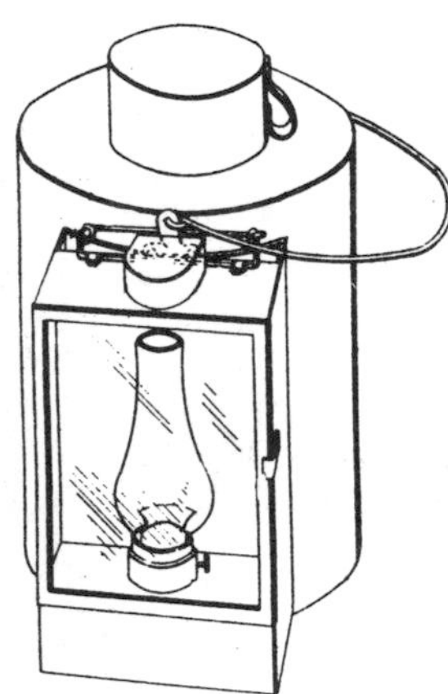

Malmo's Improved Lantern and Dinner Pail is a combination of the two articles whereby the heat from the former may be transmitted to the contents of the latter. It not only keeps warm the victuals inside but also provides an illuminating means whereby miners or others who work among similar surroundings may find their way at all times and avoid obstructions and dangerous places.

23971/1905
Christina Malmo *Gentlewoman*
Walkerville, Montana, USA

(Victor J Evans)

The Maschins' Improved Trick Device or Lungtester causes a spray of some pulverulent material, such as lamp-black, flour, water, etc, to be blown in the face of an uninitiated operator. When a person not knowing the construction of the device blows into the mouthpiece, the material is carried up and distributed by means of the windmill he is endeavouring to rotate. The initiated operator, however, closes the obvious aperture with the tip of his tongue as he blows, thereby causing the windmill to revolve by air entering a hidden aperture, but not disturbing the trick material in the canister. The device is both tempting and deceptive in appearance and so simple in construction as to be cheaply manufactured and placed upon the market.

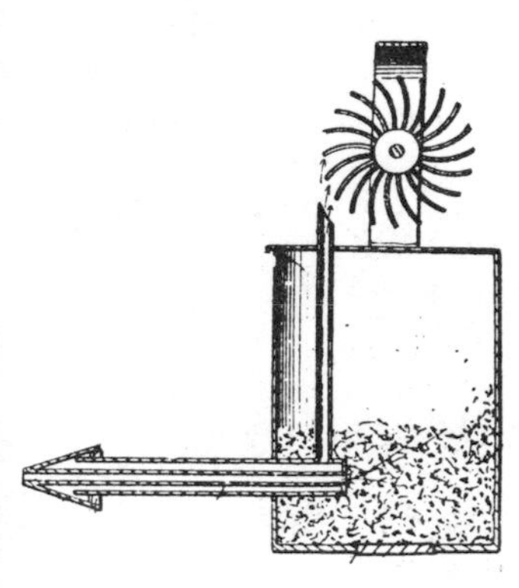

24820/1905
George Leslie Maschin
Emil Tony Maschin
24 Dubois Street, Westfield, Massachusetts, USA

(Boult, Wade & Kilburn)

Tuck and Boehme's Improved Mask is designed to be worn at parties and other gatherings and will produce curious effects in the wearer thereof and render him or her the cause of much amusement and delight to the beholders.

The mask is designed to cover part of the face only and to bear a representation which, together with the part of the visible face of the wearer, will produce a startling and amusing effect: if, for example, a mask representing a woman's face and bonnet is worn by a gentleman having a beard or moustache or both, the resulting incongruity will be ludicrous.

Other representations may also be employed, such as a clown, or a North American Indian in full war-paint and feathers, or a demon surmounted by horns.

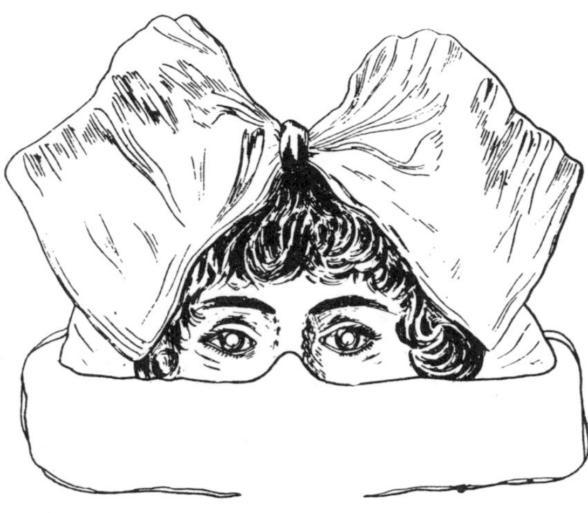

20054/1905
Adolph Tuck of Raphael Tuck *Fine Art Publishers*
Raphael House, Moorfields, London, England

Alfred Boehme *Colour Printer*
158 Reichenberger Strasse, Berlin, Germany

(Newnham Browne & Co)

Pryce-Jones's Improved Hair-pin has cranked ends to minimise the risk of its working loose, or its being accidentally driven into the head of the wearer.

25873/1905
Rosina Ida Pryce-Jones *Married Woman*
St Davids, Newtown, Montgomery, North Wales

(G G M Hardingham)

McGahan's Improved Method of Treating Leaves renders them suitable for souvenirs, presents, and mementoes and as a basis for seasonal greetings, mottoes, figures or such like.

Leaves are dried, and perhaps heated or boiled, so that when they are struck repeatedly with a brush of bristle or wire, the pulpy portion so produced may be removed from the skeleton of the leaf, which may then be arranged with a suitable plain or decorated backing or cover, gilded, painted, lettered or ornamented with a photograph. Shaped brushes or stencils may be used to form the design, and tobacco, coffee, or any other suitable leaves may be treated to form advertisements for shop windows and many other purposes.

19233/1905
George Ronald McGahan *Engineer*
11 Gladstone Avenue, Barrhead, Renfrewshire, Scotland

(W R M Thomson & Co)

Arlt and Fricke's Improved Hair Drying Apparatus uses air heated by the flame of a spirit lamp, the products of whose combustion are blown on to the hair by a fan, together with a current of fresh air drawn in at the same time. The fan is operated by gearing from a pawl and ratchet mechanism operated by the thumb or a finger of the hand holding the apparatus, and is so constructed that little exertion is required to produce a strong current of air. Furthermore, the combination of fresh and cool air with heated air from the lamp obviates the inconvenience of the air blown on to the hair becoming too hot, which would produce headaches in sensitive persons.

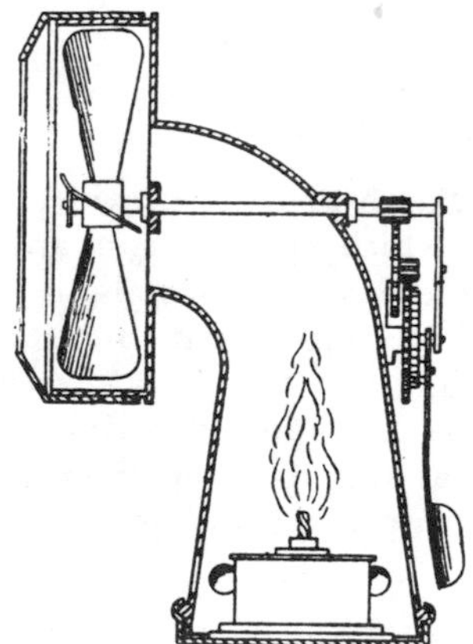

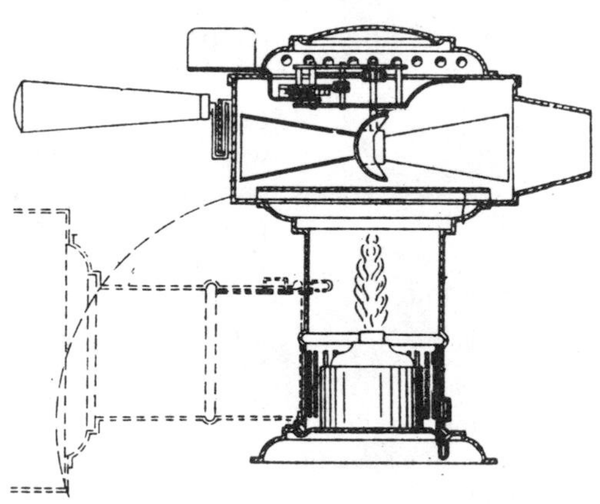

24170/1905
Arlt and Fricke
12 Schinkestrasse, Berlin, Germany

(C Bollé)

Lovell's Fastener for Bed Clothes has been devised because it is well known that much inconvenience arises from bedclothes slipping off a person when asleep. The object of this invention is to construct a fastener which will retain the bed clothes in a proper position on a sleeper. The device can be readily fastened after a person has got into bed, and may be readily unfastened to enable a person to get out of bed with ease. It consists in a length of chain, or a loop of tape, or rubber, with a bar or other fastening means at one end and a safety pin or the like at the other. One end is passed through tapes, loops etc on the bedclothes; the other is pinned to the lower bedclothes, mattress etc, or the bed.

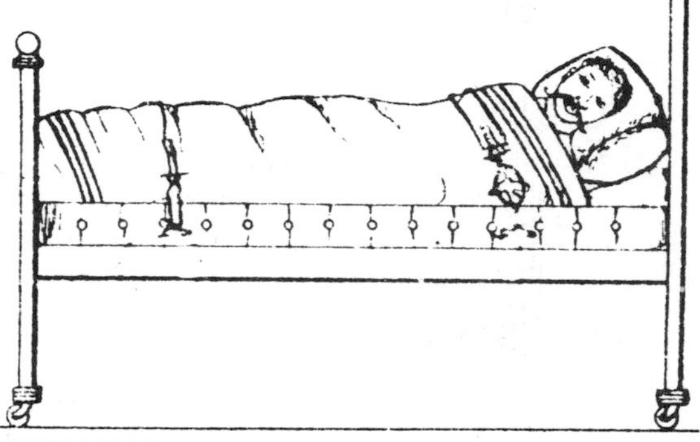

20300/1905
Edward James Lovell *Manager of Works*
Rose Cottage, Mistley, Essex, England

(Harris & Mills)

Douglas's Improved Toy is made in the form of a pump, or animal, which delivers sweets. The handle of the pump, or the tail of the animal, is operated, whereupon a sweet is delivered from the spout, or mouth, as appropriate.

21155/1905
William Percy Douglas *Commercial Traveller*
139 Brooke Road, Stoke Newington, London, England

(Browne & Co)

1905

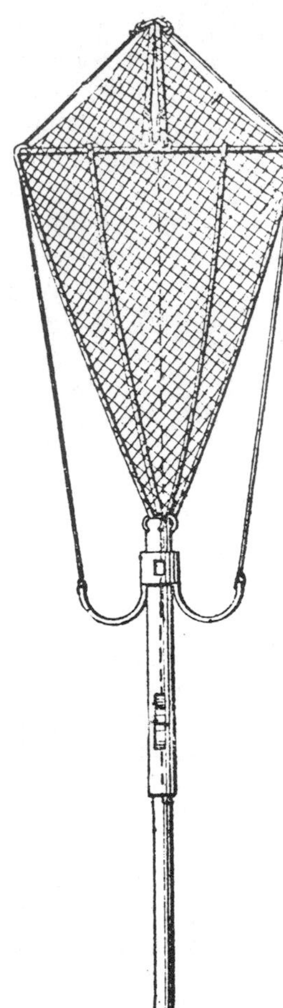

Terletzky's Improved Insect Catching Device employs arrows fitted with baskets or nets, provided with doors adapted to be closed automatically when the arrows have reached the end of their flight.

The arrows are fired at the insects.

18008/1905
Max Terletzky *Gentleman*
Goble, Oregon, USA

(Victor J Evans)

Johnson's Improved Motor Cycle has its parts arranged so as to conduce to special economy in construction and safety in use. The rider stands astride the rear wheel on two platforms supported slightly above the ground. The rear wheel preferably has two treads, with its driving belt or chain acting between them, a combination which, with the standing attitude of the rider, both lowers the centre of gravity of the machine and allows it to be easily shifted laterally, so that the cycle will stand erect when at rest and its normal liability to skid is lessened.

24229/1905
Walter Claude Johnson *Engineer*
Broadstone, Forest Row, Sussex, England

(Phillips & Leigh)

Rush's Improved Vehicle Tyre has its outer circumference protected with cork, and a strip of metal between this and the pneumatic tube.

5721/1905
Martin van Buren Rush *Artisan*
Anderson, Indiana, USA

(Jensen & Son)

Elliott's Improved Apparatus for Producing Spectacular and other Effects in Theatres uses three separate organs, or like keyboard instruments, supplied with air from one bellows, and provided with contact makers and circuits containing coloured electric lamps adapted to produce harmonious, scenic or decorative effects. The lamps may vary according to the music, or be grouped to form stars or designs. Stages, stairs, and an arched opening are provided, so that the performers can group themselves in fancy positions, form living pictures, or gain access to the keyboards.

24653/1905
R Elliott
Middlesborough Boiler Works, Middlesborough, Yorkshire, England

J Elliot
12 Park Street, Middlesborough, Yorkshire, England *Executors of*
James Bedford Elliott *Public Entertainer*

(W P Thompson & Co)

Cochrane's Improved Tooth-brush is provided with a resilient shield which springs into an annular groove in its handle. A case may also be provided to spring over the shield and protect the head of the tooth-brush.

This means that in use saliva and dentifrices are prevented from gravitating from the mouth to the handle or hand and being disagreeable and insanitary to the user.

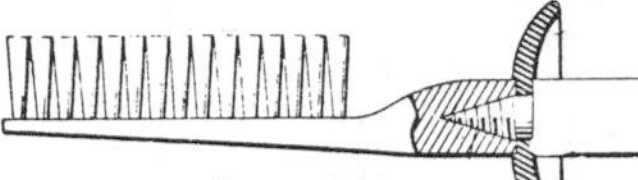

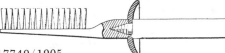

17749/1905
James Anson Cochrane *Inventor*
Cincinnati, Ohio, USA

(Wheatley & Mackenzie)

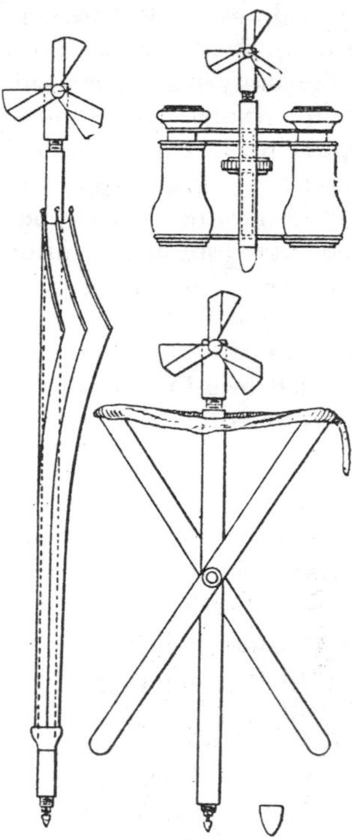

Kaiser's Improved Portable Fan or Ventilator is designed to be fastened to a walking stick, umbrella, opera glass, tripod or similar object, and is driven by means of a spring.

13194/1905
Gabriel Kaiser *Manufacturer*
24 Rue des Petites Ecuries, Paris, France

(Jean Bonnicart)

Margolinsky's Improved Protector against Gout and Rheumatism can be used as a vest, or as a knee, leg, shoulder, back or chest protector, as well as in the form of gloves or other ways. It consists in layers of impregnated camel's hair and preferably also of paper, arranged between silk or woven material and sewed together in squares.

The protector is impregnated first with strong tar vapours, then with extract of henbane and camphor dissolved in rapeseed oil, and pure olive oil, next with dried camphor and spirit of camphor, and finally with a decoction of spruce twigs.

10418/1905
Wulff Salomon Margolinsky *Merchant*
11 Westend, Copenhagen, Denmark

(Wheatley & Mackenzie)

Simms's Improved Flying Machine is balanced by means of a gyroscope driven by a petrol or other motor, both in the air, when the aëronaut can vary the inclination of the gyroscope rotor by pulling cords, and whilst on the ground. Any of the ordinary ways, as by means of aeroplanes for example, can be used to lift the machine once it has attained the desired speed, and the gyroscope will then maintain its equilibrium.

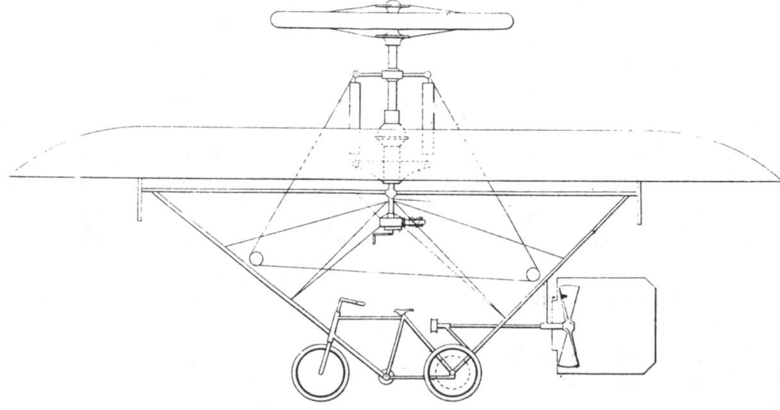

25395/1905
Frederick Richard Simms *Consulting Engineer*
Welbeck Works, Kimberley Road, Willesden Lane, Kilburn, Middlesex, England

(G F Redfern & Co)

Kursheedt's Finger-exercising Apparatus for use by Musicians consists in a base plate on which are adjustably mounted a hand-rest and a number of plates adapted to keep the fingers apart; these may be fixed into a multiplicity of holes, thus adapting the device to different users. The device exercises and spreads the fingers for the purpose of making them nimble and extending their lateral reach; by the use of this hand-expander the fingers will attain the same extension within a few weeks which formerly required several years.

18049/1905
Edmund Barrow Kursheedt *Gentleman*
16 Brighton Avenue, East Orange, New Jersey, USA

(W P Thompson & Co)

Bhisey's Improved Bust-Improver provides a hygienic means for imparting a graceful and full appearance to the bust whilst not interfering with the comfort and health of the wearer. It is intended to be worn below the natural bust, and secured in place by means of tapes, straps or strings, or by attachment to the corsets.

It consists in a non-elastic concave plate to which is attached a top piece of great elasticity and capable of the desired extension when inflated. The edge of the device which comes into contact with the skin is corrugated to provide air passages, and perfect ventilation is assured by the presence of a central ventilating tube, crowned with a perforated nipple and reinforced with a spiral spring to prevent lateral compression and consequent closing of said air passages. Two of the devices are provided, connected by a flexible tube, with a suitable valve enabling the necessary inflation to be performed.

The devices may be given a circular, oval or other suitable shape, and will be hidden below the natural bust when a low-cut bodice is worn for evening wear.

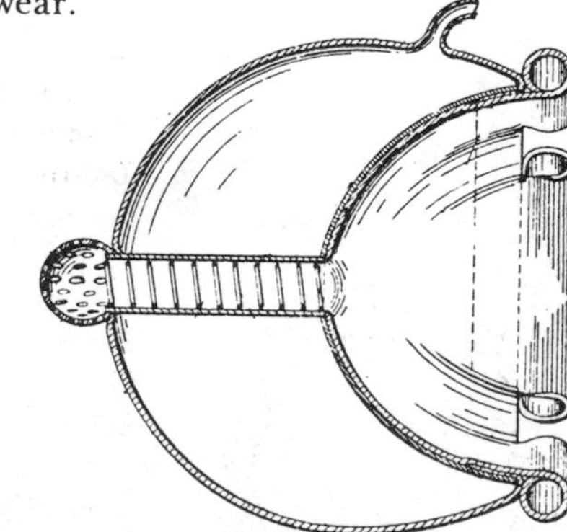

12358/1905
Shanker Abaji Bhisey *Engineer*
323 Essex Road, London N, England

(Hughes, Son & Thompson)

1905

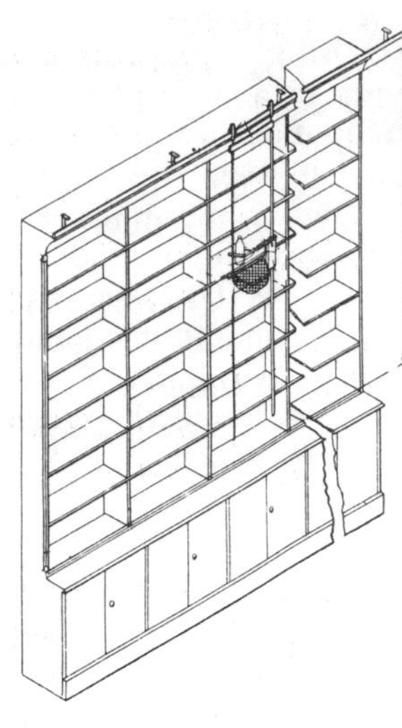

Ward's Appliance for Removing Articles from Elevated Shelves obviates the necessity for a ladder, which is always more or less in the way, and saves valuable store space so that the shelving may be carried right up to the ceiling.

It comprises a carriage which runs along the tops of the shelves, from which depend jaws with a safety net beneath, adjustable in height.

It is thus possible to position the jaws anywhere before the shelves, withdraw the desired article, and bring it to a height where it may be conveniently retrieved from the net.

The device can also be used for placing goods on shelves, its mode of use in this circumstance being so obvious as not to need explanation, but such use would be profitable only in isolated cases.

22587/1905
George Ward *Storekeeper*
Wyndham Street, Roma, Queensland, Australia

(Boult, Wade & Kilburn)

Shuttleworth-Brown's Improved Bed-sheet and Pillow-slip are designed to provide means of protection against contagion, whether arising from the skin, or the mucous membrane, from the lungs, or the breath or from the eyes, and also as a means of assurance that those parts of the sheets which have been at the foot, shall not be placed at the head of the bed.

The danger of contagion is due to the fact that, in (conscientiously) re-making the bed, the sheets are liable to be reversed, and then the sleeper lies in contact with those sides of the sheets which have been in contact with the blankets, or with a bed that may have been slept upon by persons affected by every form of contagious disease.

The means of preventing these evils are extremely simple: the top sheet bears a device (a peculiar form of hem, for example) which may make it recognisable, equally, to the chambermaid and the would-be sleeper; the device indicates that it is the top sheet, that it is the top side up, and that its top end is at the head of the bed. The bottom sheet must also bear a similar device distinguishing it as the bottom sheet.

For convenience, the device may be located at the point where monograms or crests are embroidered on luxurious sheets — the position of such a device on the bottom sheet would be just from under the pillow — so that, the moment the clothes are turned down a little, it is visible, or can be felt, and the would-be sleeper is assured. The pillow-slip prevents the evils of conventional pillow-slips, which leave the sleeper's head unprotected, and, if he should seek protection against cold or draught, he may draw the sheet and other things over his head and injure himself, and possibly infect them and the pillow, by confining his breath.

To preclude this possibility, the improved pillow-slip is provided with a flap as long, and nearly as wide, as the pillow, which the sleeper may draw over his head; but respiration will be free.

10841/1905 & 10841A/1905
David Harry Shuttleworth-Brown *Engineer*
Belle Vue Villa, South Wimbledon, Surrey, England

Stock's Improved Musical Toy for Christmas Trees consists in a member bearing figures of angels which is caused to rotate by a fan acted upon by hot air rising from candles placed beneath. The angels in their rotation strike the bells on the standard and so produce a lovely music.

If the toy is placed on a table, the foot may take the form of a house containing the Holy family.

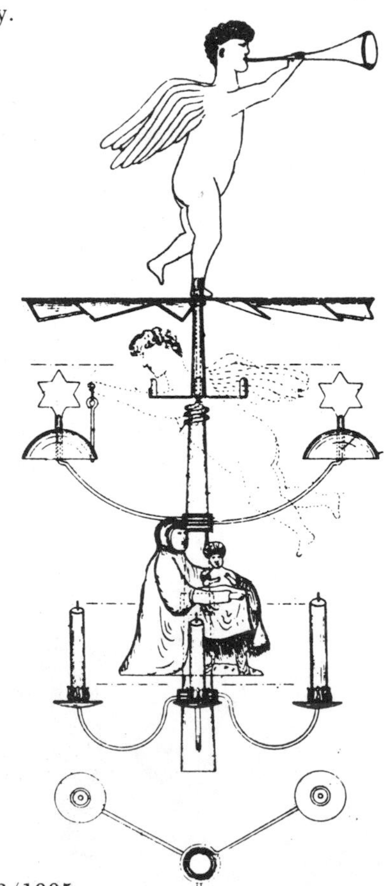

27033/1905
Walter Stock *Manufacturer*
38 Scheidterbergerstrasse, Solingen, Rhenish Prussia

(A Daumas)

Bibliography

This is necessarily a selective list, but it gives the main sources on which we have based our work, apart from the Patent Specifications themselves.

1 ANDREWS, Allen
The Follies of King Edward VII London 1975

2 **The Annual Register**

3 **Annual Reports of The Comptroller-General of the Patent Office**

4 BAKER, R
New and Improved London 1976

5 BEETON, Isabella Mary
Book of Household Management London 1888

6 BETJEMAN, John
A Pictorial History of English Architecture London 1972

7 BISHOP, James
The Illustrated London News Social History of Edwardian Britain London1977

8 BOHIM, Klaus, in collaboration with Audrey Silberston
The British Patent System Cambridge 1967

9 BOLTON, Gambier
A Book of Beasts and Birds London 1903

10 CAMPLIN, Jamie
The Rise of the Plutocrats London 1978

11 CHANDOS, Lord
From Peace to War London 1968

12 CROW, Duncan
The Edwardian Woman London 1978

13 **Encyclopaedia Britannica** (various editions)

14 ENSOR, R C K
England 1870-1914 Oxford 1936

15 GARRISON, Fielding H
An Introduction to the History of Medicine London 1917

16 GIBBS, Philip
The Pageant of the Years London 1946

17 GLASSTONE, Victor
Victorian and Edwardian Theatres London 1975

18 GLYNN, Prudence, with Madeleine Ginsburg
In Fashion London 1978

19 GOMME, A A
Patents of Invention London 1946

20 GORDON, Colin
A Richer Dust London 1978

21 GRACE, H W
A Handbook on Patents London 1971

22 GROSSMITH, George & Weedon
The Diary of a Nobody Bristol 1892

23 HARDING, H
Patent Office Centenary London 1952

24 HARE, Hobart Amory
A Text-book of the Practice of Medicine London 1905

25 HIBBERT, Christopher
Edward VII London 1976

26 HOBHOUSE, Hermione
A History of Regent Street London 1975

27 HOLMES, Oliver Wendell
Medical Essays Boston 1883

28 HULME, E Wyndham
The Early History of the English Patent System Boston 1909

29 **The Illustrated Official Journal of Patents**

30 JACKSON, Alan A
Semi-detached London London 1973

31 JULLIAN, Philippe (*trans* Peter Dawnay)
Edward and the Edwardians London 1967

32 JULLIAN, Philippe (*trans* Stephen Hardman)
The Triumph of Art Nouveau London 1974

33 LEES, Clifford
Patent Protection London 1965

34 MACQUEEN-POPE, W
Back Numbers London 1954

35 **Twenty Shillings in the Pound** London 1948

36 MANCHESTER, The Duke of
My Candid Recollections London 1932

37 MIDDLEMAS, Keith
The Life and Times of Edward VII London 1972

38 ORD-HUME, Arthur W J G
Perpetual Motion London 1977

39 PEARSALL, Ronald
Edwardian Life and Leisure Newton Abbot 1973

40 PEEL, John & Malcolm Potts
Contraceptive Practice Cambridge 1969

41 PRIESTLEY, J B
The Edwardians London 1970

42 READ, Donald
Edwardian England London 1972

43 ROBERTSON, Patrick
The Shell Book of Firsts London 1974

44 ROUTLEDGE, George
Discoveries and Inventions of the Nineteenth Century London 1903

45 SACKVILLE-WEST, Vita
The Edwardians London 1930

46 SCOTT-MONCRIEFF, David
Veteran and Edwardian Motor Cars London 1955

47 SERVICE, Alastair
Edwardian Architecture London 1977

48 **Edwardian Architecture and its Origins** London 1975

49 THOMPSON, PAUL
The Edwardians London 1977

League table of patent agents

PATENT AGENTS— League table for our selected inventions 1901-1905

The following list gives the number of times the agents named appear in connection with our inventors. The number of different firms of agents is 151 — it will be seen that there is some movement of individuals from firm to firm. Out of a total of some 650 patents in our book, over 560 were filed with the assistance of an agent.

37 W P Thompson

35 Haseltine, Lake & Co

33 Boult, Wade & Kilburn

32 Hughes & Young

26 Herbert Haddan & Co

25 Wheatley & Mackenzie

20 Marks & Clerk

12 A M & W M Clark

11 Abel & Imray

10 Jensen & Son

9 Ferdinand Nusch
t/a F G Harrington & Co

8 G F Redfern & Co
P R J Willis

7 S S Bromhead
Browne & Co
Edward Evans & Co
Johnsons

6 Cheeseborough & Royston
Cruikshank & Fairweather
Lewis Wm Gould

5 Charles Bauer, Imrie & Co
H D Fitzpatrick
F Wise Howorth
J E Evans Jackson & Co

4 Allison Bros
George Barker
W H Beck
Cassell & Co
J A Coubrough
George Davies & Son
Ernest de Pass
E Eaton
J G Lorrain
Mewburn, Ellis & Pryor
A F Spooner
W R M Thomson & Co
John G Wilson & Co

3 J P Bayly
Chatwin Herschell & Co
Chas Coventry
Victor J Evans
F W Golby
Harris & Mills
Hughes Son & Co
Geo Thos Hyde
Benj T King
Newton & Son
J Owden O'Brien
Phillipss
Rayner & Co
B Reichhold
Paul E Schilling

2 E P Alexander & Son
Reginald W Barker
Carl Bollé
F Bosshardt & Co
Bottomley & Liddle
H F Boughton
Brewer & Son
Wm Brookes & Son
T B Browne Ltd
Edward Buttner
Carpmael & Co
G J Clarkson
Day, Davies & Hunt
Dewitz-Krebs & Co
George Downing & Son
Albert E Ellen
Hy Fairbrother
J B Fleuret
R Core Gardner
G G M Hardingham
W E Heys & Son
Philip M Justice
Charles Bosworth Ketley
J W Mackenzie
J A Nees
John P O'Donnell
Page & Rowlinson
Phillips & Leigh
Frederic Prince
Geo H Raynor
Walter Reichau
W D Rowlingson
Stanley, Popplewell & Co
W Swindell
Tongue & Birkbeck
Vaughan & Son
W Lloyd Wise
D Young & Co

1 Oskar Arendt
H B Barlow & Gillett
Witold Baronowski
Jean Bonnicart
Rowland Brittain
B Brockhues
William Brookes & Son
C Barnard Burdon
C H Burgess
Edward Buttner
Frederick J Cheesebrough
Thos S Crane
Crozier & Co
A Daumas
H W Denton Ingham
Dewitz, Morris & Co
Dracup & Nowell
Bernhard Dukes
Edwards & Co
Elt & Co
Fairfax & Wetter
Fell & James
H D Fitzpatrick
P Follin
H C Fowler
George T Fuery
William Gadd
Gedge & Feeny
Gerson & Sachse
T E Halford
F G Harrington
Herschell & Co
W E Heys
Imrie & Co
J C Jackson
R W James
J Clark Jefferson
Johnsons & Willcox
Louis E Kippax
Otto Klauser
Ferdinand Klostermann
C Kluger
Douglas Leechman
F W le Tall
W Bestwick Maxfield
Max Menzel
George T Millard
W H Potter
Charles T Powell
Jno H Raynor
A Archd Sharp
Castle Smith
James G Stokes
Tasker & Crossley
Alfred William Turner
R J Urquhart
John Hindley Walker
Stephen Watkins & Groves
John Waugh
White & Woodrington
Warwick Henry Williams

Index